Exploring the Thalamus

Exploring the Thalamus

S. Murray Sherman
Department of Neurobiology
State University of New York
Stony Brook, New York

R. W. Guillery
Department of Anatomy
University of Wisconsin School of Medicine
Madison, Wisconsin

ACADEMIC PRESS
An Imprint of Elsevier
San Diego San Francisco New York Boston London Sydney Tokyo

The following items were taken with permission from previously published sources:

Chapter II: Fig. 1 from Kaas et al., © 1972 Karger, Basel. Fig. 3 from Rodieck and Brening, © 1983 Karger, Basel, and based on data of Boycott and Wässle (1974). Fig. 6A,B from Bartlett and Smith (1999). Fig. 6C,D from Winer and Morest, *J. Comp. Neurol.* **221,** 1–30, © 1983 John Wiley & Sons. Reprinted by permission of Wiley-Liss, Inc., a subsidiary of John Wiley & Sons, Inc. Fig. 6E,F from Guillery, *J. Comp. Neurol.* **128,** 21–50, © 1966 John Wiley & Sons. Reprinted by permission of Wiley-Liss, Inc., a subsidiary of John Wiley & Sons, Inc. **Chapter III:** Fig. 1 from Lachica and Casagrande (1988). Reprinted with the permission of Cambridge University Press. Fig. 2 from Pallas *et al., J. Comp. Neur.* **349,** 343–362, © 1994 John Wiley & Sons. Reprinted by permission of Wiley-Liss, Inc., a subsidiary of John Wiley & Sons, Inc. Fig. 3 from Sur *et al.* (1987). Fig. 4 from Guillery, *J. Comp. Neurol.* **128,** 21–50, © 1966 John Wiley & Sons. Reprinted by permission of Wiley-Liss, Inc., a subsidiary of John Wiley & Sons, Inc. Fig. 10 from Cox *et al., J. Comp. Neurol.* **366,** 416–430, © 1996 John Wiley & Sons. Reprinted by permission of Wiley-Liss, Inc., a subsidiary of John Wiley & Sons, Inc. Fig. 11 from Uhlrich *et al.* (1991). **Chapter VII:** Fig. 1A,B based on Cajal (1995), "Histology of the Nervous System of Man and Vertebrates" (translated by N. and L. W. Swanson), Oxford University Press, Footnote 2 from Walls (1953), University of California Publications in Physiology Vol. 9. **Chapter IX:** Fig. 1A from Fig. 3A of Mastronarde (1997). Fig. 1B redrawn from Fig. 2A of Tsumoto *et al.* (1978). *Exp. Brain Res.* **32,** 345–364, © 1978 Springer-Verlag. Data in Figs. 1C,D kindly provided from the authors for replotting: Fig. 1C redrawn from Usrey *et al.* (1998). *Nature* **395,** 384–386. Fig. 1D redrawn from Reid and Alonso (1996). *Curr. Opin. Neurobiol.* **6,** 475–480, © 1996, with permission from Elsevier Science.

This book is printed on acid-free paper. ∞

Copyright © 2001 by S. Murray Sherman and R. W. Guillery

All Rights Reserved.
No part of this publication may be reproduced or transmitted in any form or by any means, electronic or mechanical, including photocopy, recording, or any information storage and retrieval system, without permission in writing from the publisher.

Transferred to Digital Printing 2009

Permissions may be sought directly from Elsevier's Science and Technology Rights Department in Oxford, UK. Phone: (44) 1865 843830, Fax: (44) 1865 853333, e-mail: permissions@elsevier.co.uk. You may also complete your request on-line via the Elsevier homepage: http://www.elsevier.com by selecting "Customer Support" and then "Obtaining Permissions".

Academic Press
An Imprint of Elsevier
525 B Street, Suite 1900, San Diego, California 92101-4495, USA
http://www.academicpress.com

Academic Press
A Harcourt Science and Technology Company
http://www.academicpress.com

Library of Congress Catalog Card Number: 00-104281

ISBN-13: 978-0-12-305460-9

Brief Contents

CHAPTER I   *Introduction*   1

CHAPTER II   *The Nerve Cells of the Thalamus*   19

CHAPTER III   *The Afferent Axons to the Thalamus*   59

CHAPTER IV   *Intrinsic Cell Properties*   109

CHAPTER V   *Synaptic Properties*   143

CHAPTER VI   *Function of Burst and Tonic Response Modes in the Thalamocortical Relay*   169

CHAPTER VII   *Maps in the Brain*   197

CHAPTER VIII   *Two Types of Thalamic Relay*   229

CHAPTER IX   *Drivers and Modulators*   249

CHAPTER X   *Overview*   267

Contents

Preface xiii

Abbreviations Used xvii

CHAPTER I *Introduction*

A. Thalamic Functions: What Is the Thalamus and What Is It For? 1
B. The "Thalamus" as a Part of the Diencephalon: The Dorsal Thalamus and the Ventral Thalamus 8
  1. The Dorsal Thalamus 10
  2. The Ventral Thalamus 15
C. The Overall Plan of the Next Nine Chapters 16

CHAPTER II *The Nerve Cells of the Thalamus*

A. **On Classifying Relay Cells** 19
  1. Early Methods of Identifying and Classifying Thalamic Relay Cells 19
  2. On General Problems of Cell Classification 24
  3. On the Possible Functional Significance of Cell Classifications in the Thalamus 26
  4. Classifications of Relay Cells Based on Dendritic Arbors and Perikaryal Sizes 29
  5. Laminar Segregations of Distinct Classes of Geniculocortical Relay Cells 35

6. The Cortical Distribution of Synaptic Terminals from Relay Cell Axons   38
7. Perikaryal Size and Calcium-Binding Proteins   45

B. Interneurons   47
　1. Interneuronal Cell Bodies and Dendrites   47
　2. On Distinguishing Interneuronal Axons and Dendrites   49
　3. The Axons of the Interneurons   51
　4. Classifications of Interneurons   52

C. The Cells of the Thalamic Reticular Nucleus   54
D. Summary   57
E. Some Unresolved Questions   57

CHAPTER III *The Afferent Axons to the Thalamus*

A. A Functional Classification of Afferents to the Thalamic Nuclei   60
B. Afferent Axon Types as Seen Light Microscopically   62
　1. Light Microscopy of Ascending Afferents and of Cortical Afferents in First Order Nuclei   62
　2. Light Microscopy of Cortical Afferents in Higher Order Nuclei   68
　3. Topographical Organization of Corticothalamic Afferents   72
　4. Other Afferents to Thalamic Nuclei   75
　5. Afferents from the Thalamic Reticular Nucleus to First and Higher Order Nuclei   77
　6. Connections from Interneurons to Relay Cells   82
　7. The Heterogeneity of the Several Afferent Systems   82

C. Electron Microscopic Appearance of the Afferent Axon Terminals and Their Synaptic Relationships   82
　1. General Description   82
　2. The RL Terminals in First Order Nuclei   87
　3. The RL Terminals in Higher Order Nuclei   90
　4. The RS Terminals   91
　5. The F Terminals: F1 and F2   93
　6. Quantitative Relationships   94
　7. The Glomeruli and Triads   97

D. Afferents from Interneurons and Reticular Cells   99
E. GABA Immunoreactive Afferents   100

Contents

  F. Afferents to the Thalamic Reticular Nucleus   101
  G. Some Problems of Synaptic Connectivity Patterns   102
      1. The Relationship of Two Driver Inputs to a Single Thalamic Nucleus: Does the Thalamus Have an Integrative Function?   102
      2. The Complexity of Multiple Layer 6 Afferents to the Thalamus and Some Related Problems   103
  H. Summary   106
  I. Some Unresolved Questions   107

CHAPTER IV *Intrinsic Cell Properties*

  A. Cable Properties   109
      1. Cable Properties of Relay Cells   111
      2. Cable Properties of Interneurons and Reticular Cells   114
      3. Implications of Cable Properties for the Function of Relay Cells and Interneurons   115
  B. Membrane Conductances   117
      1. Voltage-Independent Membrane Conductances in Relay Cells   118
      2. Voltage-Dependent Membrane Conductances in Relay Cells   118
      3. Interneurons   136
      4. Reticular Cells   137
  C. Summary and Conclusions   140
  D. Some Unresolved Questions   141

CHAPTER V *Synaptic Properties*

  A. Ionotropic and Metabotropic Receptors   144
      1. Different Types of Metabotropic and Ionotropic Receptors   144
      2. Functional Differences between Ionotropic and Metabotropic Receptors   145
  B. Synaptic Inputs to Relay Cells   147
      1. Driving Inputs to Relay Cells   147
      2. Inputs to Relay Cells from Interneurons and Cells of the Thalamic Reticular Nucleus   151
      3. Inputs from Cortical Layer 6 Axons to Relay Cells   151

   4. *Brain Stem Modulatory Inputs to Relay Cells*   156
   5. *Other Inputs*   157
 C. **Inputs to Interneurons and Reticular Cells**   158
   1. *Glutamatergic Inputs*   160
   2. *Cholinergic Inputs*   162
   3. *GABAergic Inputs*   164
   4. *Noradrenergic Inputs*   165
   5. *Sertonergic Inputs*   166
   6. *Histaminergic Inputs*   166
 D. **Summary**   166
 E. **Some Unresolved Questions**   167

CHAPTER VI   *Function of Burst and Tonic Response Modes in the Thalamocortical Relay*

 A. **Rhythmic Bursting**   169
   1. *Visual Responses of Geniculate Relay Cells*   171
 B. **Effect of Response Mode on Transmission from Relay Cells to Cortical Cells**   179
   1. *Paired-Pulse Facilitation and Depression*   179
   2. *Possible Relationship of Firing Mode to Thalamocortical Terminal Patterns*   185
 C. **Control of Response Mode**   188
   1. *Brain Stem Control*   188
   2. *Cortical Control*   190
 D. **Summary**   192
 E. **Some Unresolved Questions**   194

CHAPTER VII   *Maps in the Brain*

 A. **Introduction**   197
   1. *The Functional Significance of Mapped Projections to Thalamic Relays*   197
   2. *The Nature of Thalamic and Cortical Maps*   199
 B. **Early Arguments for Maps**   200
 C. **Clinical and Experimental Evidence for Maps in the Geniculocortical Pathway**   203
   1. *Establishing That There Are Maps*   203
   2. *The Alignment of Maps with Each Other*   206

Contents

    D. Multiple Maps in the Thalamocortical Pathways   209
        1. The Demonstration of Multiple Maps   209
        2. Mirror Reversals of Maps and Pathways   210
    E. Abnormal Maps in the Visual Pathways   213
        1. Abnormal Pathways in Albinos   213
        2. Experimental Modifications of the Thalamocortical Pathway   220
    F. Maps in the Thalamic Reticular Nucleus   220
    G. Summary   226
    H. Some Unresolved Questions   227

CHAPTER VIII   *Two Types of Thalamic Relay*

    A. The Basic Categorization of Relays   229
    B. The Evidence That There Are Two Distinct Types of Afferent That Go from Neocortex to Some Thalamic Nuclei   234
        1. The Structure and Laminar Origin of the Corticothalamic Axons   235
        2. The Functional Evidence for Two Distinct Types of Corticothalamic Afferent   237
        3. What Is the Functional Significance of a Higher Order Thalamic Input?   240
    C. The Relationship of First and Higher Order Relays to the Thalamic Reticular Nucleus   245
    D. Summary   246
    E. Some Unresolved Questions   247

CHAPTER IX   *Drivers and Modulators*

    A. Drivers and Modulators in the Lateral Geniculate Nucleus   250
        1. General Differences between Drivers and Modulators   251
        2. Receptive Fields and Cross-Correlograms   253
        3. Why Are Driver Inputs So Small in Number?   257
    B. The Geniculate Input to Cortex as a Driver   259
    C. Tonic and Burst Modes in Thalamic Relay Cells   260
    D. Key Differences between Drivers and Modulators   260
    E. The Sleeping Thalamus   261
        1. Slow Wave Sleep   261
        2. REM Sleep   263

F. Can Extradiencephalic GABAergic Inputs to Thalamus
   Be Drivers?   263
G. Summary   264
H. Some Unresolved Questions   266

CHAPTER X *Overview*

Bibliography   273
Index   303

Preface

The title of this book, *Exploring the Thalamus,* is intended to convey the sense that there is a great deal still unknown about the thalamus and that our aim in writing has been to show some of the major and minor roads, the footpaths, and the frank wildernesses that still need to be explored. The thalamus, in terms of its detailed connectivity patterns, the functioning of its circuitry, and, perhaps most interestingly, its functional relationship to the cerebral cortex, is largely *terra incognita*. The cortex depends critically on the messages it receives from the thalamus. It receives very little else. Understanding cortical functioning will depend on understanding the thalamic inputs that are the necessary first step in cortical processing. There is a serious sense in which one can regard the thalamus as the deepest layer of the cortex. That is, cortex and thalamus depend very closely on each other; neither would amount to much without the other, and if we are to understand the workings of either, we must, as we try to show in what follows, be able to understand the messages that each is sending to the other. Our main focus on the thalamus in this book will be on its relation to cortical function. Perhaps this focus on thalamocortical interrelationships should have been a part of our title, but we wanted a brief title and we had in mind a book on the thalamus, not one on the cortex. We are looking at ways to explore how thalamic organization should influence our views of cortical function; we are not looking to provide a general account of the thalamus as an entity in itself.

In a preface, authors are expected to say for whom the book has been written. We hope that this book will serve to introduce graduate students, postdoctoral fellows, and investigators who need to learn about the thalamus to some of the interesting aspects of the subject. Also, since many view the thalamus as an uninteresting, mechanical relay of peripheral messages to cortex that is already well understood, we have tried to explain that this view is far too simplistic and that there are many important problems about the function and the structure of thalamus that remain unrecognized and unresolved. One of our aims has been to persuade colleagues that these problems are of interest and worth significant research investment. We have tried to make each chapter more or less independent, so that perhaps one or another can be assigned as course reading for graduate students. This entails some repetition from one chapter to another; we trust that it is not excessive for those who are motivated to read the whole thing.

Our thoughts about who would read the book were not our main focus when we first started discussions and rough drafts almost 10 years ago. Rather, we undertook the task initially because we found that the thoughts, the discussions, and the arguments that have accompanied the job of writing were sufficient stimulus in themselves. As we wrote, exchanged drafts, sent each chapter back and forth many times to be annotated, corrected, reannotated, and recorrected, we gradually learned a great deal about our subject (and about each other). The real truth is that we wrote the book for ourselves, and once we had that aspect of the writing fairly in hand, we worked hard to make it accessible to others.

We have not attempted to present a complete and coherent view of everything that is known about the thalamus. We have instead followed arguments and lines of inquiry that can lead to new questions, interesting thoughts, or new experimental approaches. Knowledge of the thalamus is extraordinarily patchy. The thalamus is divided into many different "nuclei." There are some thalamic nuclei that have been studied in considerable detail, and others about which we know almost nothing. Our plan in writing the book has been to assume that there is a basic ground plan for the thalamus. Although there are often important differences between one thalamic nucleus and another, in one species or between species, there is yet a common pattern of organization seen over and again in essentially all thalamic nuclei. We have tried to explore the nature of this common pattern and to ask questions about its functional significance.

We have both spent the greater part of our careers studying the visual pathways, and it won't take a very subtle reading of the book to recognize this. We turn to the visual relay in the thalamus repeatedly not only because this is the part we know best but also because in our readings and in our discussions with colleagues we find that the visual relay has, time and again, received more detailed experimental study than other thalamic relays. The visual relay may well have some special characteristics that distinguish it from other relays. In some instances this is clear, and we recognize it. However, in many instances it is reasonable to treat the visual relay as an exemplar of thalamic relays in general, and in many parts of the book, that is how we have approached the analysis of thalamic functions. This approach raises important questions about nonvisual parts of the thalamus, and our expectation is that the comparisons will stimulate further study of these questions.

We have stressed that this book is not a complete inventory of all that is known about the thalamus. There are many important references we have not cited, and there are several lines of inquiry that we have not included. We say virtually nothing about the development or the comparative anatomy of the thalamus even though each is an extremely interesting subject in its own right. They should perhaps form the nucleus of another book. Nor do we cover the clinical aspects of thalamic dysfunction, another potentially interesting area, although it seems likely to us that this will become of greater interest once we know more about some of the basic ground rules of thalamic function and connectivity that are still missing from our current knowledge. For instance, the thalamus has long been implicated in epilepsy and certain sleep disorders, it is related to the production of pathological pain, and there is new interest in the thalamus as a particularly interesting site of pathology in schizophrenia and other cognitive problems. The complexity of the two-way links between thalamus and cortex and the limited nature of our knowledge about these links, especially in the human brain, make interpretations of clinical conditions extremely difficult and often rather tenuous, and we have not addressed them in this book.

We have tried to achieve two major aims in the book. The first is to look at many of the outstanding puzzles and unanswered questions that arise as one studies the structural and functional organization of the thalamus. The second aim, growing out of a small proportion of these questions, is to move toward an understanding of the possible role(s) of the thalamus in cortical functions, so that some coherent suggestions about this role could be presented as the book proceeds. The first aim is summarized to a limited extent by a short list of "Some Unresolved Questions" that appears at the end of each chapter. These are not questions to which a student can find answers in the text. They are, rather, designed to focus on some of the issues that need to be resolved if we are to advance our understanding of the thalamus. They do not represent an exhaustive list, and the interested reader is likely to find a number of other questions that are currently unanswered and often unasked. The listed questions should be seen as representing a state of mind, and they are an important part of the book as a whole. They should lead to more questions, and they should point to paths that have perhaps never been explored or along which our predecessors have been lost in the past. We hope that by stimulating a questioning attitude to thalamic organization we will encourage a view of the thalamus as far more mysterious than is commonly taught. This clearly implies that our second aim, to understand the role of the thalamus, which we present in detail in the later chapters, can at best be only partially achieved. We present a view of the thalamus that is based on the classical view of it as a relay of ascending messages to cortex. However, we see it as a continually active relay, serving sometimes as a "lookout" for significant new inputs and at other times as an accurate relay that allows detailed analysis of input content in the cortex. This is based on the recognition of two distinct types of input to the thalamus, the "drivers" that carry the message and the "modulators" that determine how the message is transmitted to cortex. The former can carry ascending messages from the periphery as well as descending messages from cortex itself. These messages are generally mapped, giving them a definite locus in the environment or in some other part of the brain. In contrast, the latter either can be mapped and thus act locally like the drivers or can lack a mapped organization and then act globally. Recognizing that drivers can take origin in the cortex leads to an interesting new view of corticocortical communication because it stresses that messages that pass from one cortical area to another may be under the same set of modulatory controls in the thalamus as are the inputs that are passed to the cortex from the peripheral senses.

It is probable that many of the ideas we present in this book will prove wrong. Whether they are right or wrong, we have tried to make them stimulating. To quote Kuhn (1963) quoting Francis Bacon, "Truth emerges more readily from error than from confusion," which provides our best justification for writing this book about a subject that in terms of the currently available literature is often extremely confusing.

Finally, both authors owe thanks and the book itself owes its existence to many people and organizations. A number of colleagues read an early draft of the book, and the comments and critical points that they raised have helped us to reorganize and correct a great deal of the book in terms of style, order of presentation, and content. We thank Paul Adams, Joe Fetcho, Sherry Feig, Lew Haberly, Carsten Hohnke, Jon Levitt, John Mitrofanis, and Phil Smith for their helpful comments. We recognize the amount of time and effort that they have contributed; we are most grateful for it and for the significant improvements that their careful readings have produced. The final version of this book is, of course, entirely our responsibility. All of our colleagues will likely find many places where they can write further instructive comments in the margins, and perhaps some of these will lead to useful explorations of the thalamus in the future. Majorie Sherman helped with the proof reading. Sherry Feig helped one of us (R.W.G.) learn how to draw on a computer. Both authors received support from the NIH while this book was being written (Grants EY03038, EY11409, and EY11494), and at the early stages R.W.G., while in the Department of Human Anatomy at Oxford, was supported by the Wellcome Trust. The initial stimulus for planning the book came from the year S.M.S. spent as a Newton-Abraham Visiting Professor at Oxford in 1985–1986.

S. Murray Sherman
R. W. Guillery

Abbreviations

We have, as far as possible, avoided the use of abbreviations, except for a few that are commonly used and widely recognized for complex names. They are the following: AMPA, (R,S)-α-amino-3-hydroxy-5-methyl-4-isoxazolepropionic acid; EPSP, excitatory postsynaptic potential; GABA, γ-aminobutyric acid; GAD, glutamic acid decarboxylase; IPSP, inhibitory postsynaptic potential; NMDA, N-methyl-D-aspartate.

CHAPTER I

Introduction

A. Thalamic Functions: What Is the Thalamus and What Is It For?

The thalamus is the major relay to the cerebral cortex. It has been described as the "gateway" to the cortex. Almost everything that we can know about the outside world, or about ourselves, is based on messages that have to pass through the thalamus. It forms a relatively small cell group on each side of the third ventricle and can be seen most readily in a midsagittal section of the brain (Figure 1). It can be divided into several distinct nuclei or nuclear groups, each concerned with transmitting a characteristic type of afferent signal (visual, auditory, somatosensory, cerebellar, etc.) to a cytoarchitectonically distinct and functionally corresponding area or group of areas of the cerebral cortex (Figures 2 and 3) on the same side of the brain. This is the view of the thalamus that was developed during the 70 (plus) years up to circa 1950. It has served us well and is still the view presented in most textbooks. It was based on clinical observations related to postmortem study of the brain and on relatively crude experimental neuroanatomical methods: the Marchi method for staining degenerating myelin in pathways that had been cut or injured, and the Nissl method, which shows nerve cells grouped into functional nuclei or undergoing retrograde degeneration after their axons have been cut. These methods give results in terms of large populations of cells or axons and large areas of thalamus or cortex. Perhaps it was fortunate that modern methods for studying the detailed connectivity patterns of single cells or small groups of cells were not available when the thalamic connections were first being defined. If they had, it is probable that no one would have been able to see the larger thalamic forest for the details of the connectional trees. We shall start with the forest.

The schematic view of thalamocortical relationships, summarized in Walker's great book (1938) or in Le Gros Clark's earlier review (1932), provided a powerful approach to thalamic function. Even though it was heavily dependent on relatively gross methods, it showed us how to divide up the thalamus and how to relate each of the resulting major

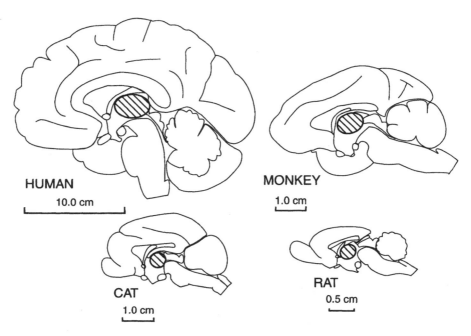

Figure 1
Midsagittal view of the cerebral hemispheres of a human, a monkey, a cat, and a rat (in size order) to show the position and relative size of the thalamus, which is indicated by diagonal hatching.

thalamic nuclei or nuclear groups to one or another part of the cerebral cortex (see Figures 2 and 3). Above all, it demonstrated the extent to which the functions of any one part of the cerebral cortex must be dependent on its thalamic inputs. It should be noted that the thalamus is connected to the so-called neocortex. Olfactory cortex and hippocampal cortex are not neocortex and do not receive comparable thalamic afferents. Olfactory afferents represent the only pathway of a sensory system that does not have to go through the thalamus before it can reach the cortex. The following account deals with the relationships between thalamus and neocortex only.

Figure 1 shows the thalamus in relation to the rest of the cerebral hemisphere and shows that the thalamus is small relative to the whole cerebral hemisphere in all mammals. There are a great many more cortical cells than there are thalamic cells even though the neocortex depends on the thalamus for its major inputs.[1] Each major neocortical area

---

[1] From the evidence available for the geniculocortical pathway to the primary visual cortex (variously called V1, area 17, or striate cortex) it appears that there are about 350 to 460 × 10^6 cortical nerve cells in V1 of each hemisphere in the monkey, and about 55 to 70 × 10^6 in the cat. The numbers of nerve cells for one lateral geniculate nucleus are about 1.6 × 10^6 and 0.45 × 10^6, respectively. Because not all geniculate cells project to V1, the projecting geniculate cells represent 0.5% or less of the total number of cortical cells in the area receiving the projection. See Rockel *et al.* (1980) for cell densities in cortex; Duffy *et al.* (1998) for area of V1; Matthews (1964) for cell numbers in the monkey lateral geniculate nucleus; and Bishop *et al.* (1953) for the cat.

depends on a well-defined thalamic cell group, that is, on a thalamic nucleus or group of nuclei. In the evolutionary history of mammals, an increase in the size of any one part of cortex generally relates to a corresponding increase in the related thalamic nuclei. The functionally best-defined cortical areas (visual, auditory, motor, etc.) depend for their functional specialization on the nature of messages that pass to that cortical area from the thalamus. The visual cortex is visual because it receives visual messages from the retina through its thalamic relay, and the same is true of the other thalamic nuclei outlined in bold and hatched in Figure 2.

Figure 2 shows some of the major thalamic nuclei in a simplified, schematic form for a generalized primate. Details differ for each species, and a number of nuclei are not included in the figure because they play no role or only a minor role in the rest of this book. However, the general relationships shown apply to all mammals. Figure 3 shows how some of these major thalamic nuclei are linked to specific, functionally or structurally defined cortical areas. Further details on individual thalamic nuclei and their connections can be found in Berman and Jones (1982) and Jones (1985). Figures 2 and 3 show that for some, but by no means for all of the thalamic nuclei, we can define the dominant or functionally "driving" afferents. That is, Figure 3 shows that the lateral geniculate nucleus is visual and the medial geniculate nucleus is auditory. The ventral posterior nucleus[2] is somatosensory, which is to say that the ascending pathways concerned with tactile stimuli and with stimuli related to the position and movements of body parts (kinesthesis) go to this nucleus as do pathways concerned with pain and temperature. For reasons detailed in Chapter III we treat the afferents from the cerebellum, related to movement control, that go to the ventrolateral and ventral anterior cells groups as drivers and also the mamillothalamic tract that sends information to the anterior thalamic nuclei about ongoing activity in the hippocampal formation and in the midbrain. These represent the major, known ascending driver inputs to the thalamus. We shall treat these ascending afferents as the "drivers" of the thalamic relay cells that they innervate, meaning by this that they are the afferents that determine the receptive field properties of the relay cells. Other afferents, which we shall treat as modulators, can modify the way that the message is transmitted, but they are not responsible for the main qualitative nature of the message that is conveyed to the cortex by the thalamic relay cell. We shall argue that each thalamic nucleus has drivers and modulators and that identifying the drivers for many thalamic nuclei is likely to be a key to understanding their functions, which are currently poorly defined.

The afferents to the other main thalamic nuclei, shown with lighter

---

[2] The lateral and medial parts of the ventral posterior nucleus (VPM and VPL) are often referred to as part of the "ventrobasal" complex, and a distinction is made between a nuclear complex, or group of nuclei, and a thalamic nucleus that has no further subdivisions. The use of "complex" has been rather inconsistently applied in the past and is difficult to apply rigorously; the same is true when the term "nucleus" is used to apply to a cell grouping and to its subdivisions. For the purposes of this book, these problems are not important, and we will stay with the term "nucleus" throughout.

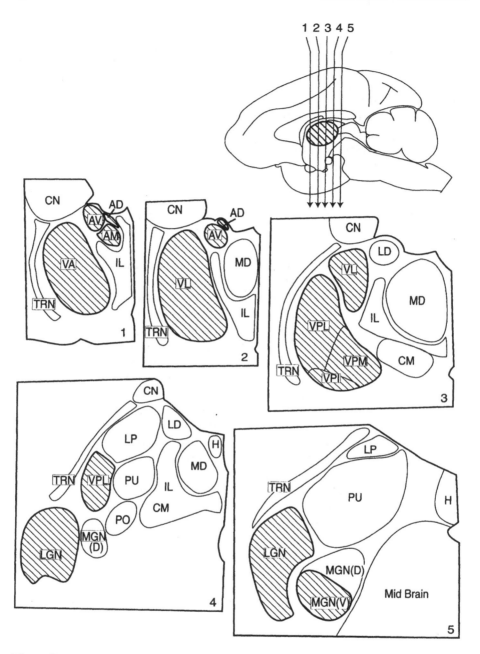

Figure 2
Schematic view of five sections through the thalamus of a monkey. The sections are numbered 1 through 5 and were cut in the coronal planes indicated by the arrows in the upper right midsagittal view of the monkey brain from Figure 1. The major thalamic nuclei in one hemisphere for a generalized primate are shown. The nuclei that are outlined by a heavier line and filled by diagonal hatching are described as first order nuclei (see text), and the major functional connections of these, in terms of their afferent (input) and efferent (output) pathways to cortex, are indicated in Figure 3. AD, anterodorsal nucleus; AM, anteromedial nucleus; AV, anteroventral nucleus; CM, center median nucleus; CN, caudate

A. Thalamic Functions: What Is the Thalamus and What Is It For?

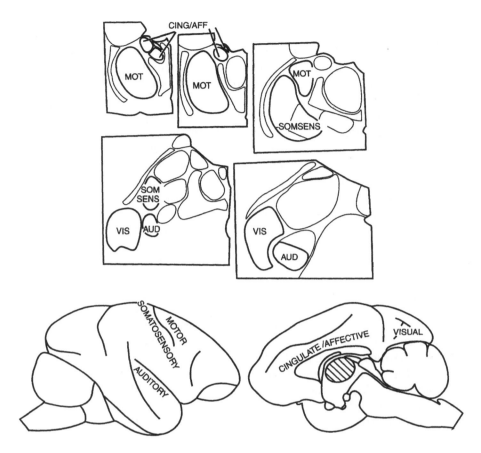

Figure 3
The upper part shows the nuclei illustrated in Figure 2, and the lower part shows a lateral (left) and a medial view of the hemisphere in a monkey to indicate the functional connections of the major first order thalamic nuclei. AUD, auditory; CING/AFF, cingulate/affective; MOT, motor; SOMSENS, somatosensory; VIS, visual.

outlines and no shading in Figure 2 and unlabeled in Figure 3, are less straightforward and will be considered later (Chapters III and VIII). We have presented evidence that these nuclei receive their major driving afferents from the cerebral cortex itself and thus serve as a relay on corticocortical pathways, not as relays of subcortical afferents to cortex (Guillery, 1995; Sherman and Guillery, 1996). That is, they contain "higher order" relays. We define first order relays as those concerned with sending messages to the cortex about what is happening in the subcortical

---

nucleus (not a part of the thalamus); H, habenular nucleus; IL, intralaminar (and midline) nuclei; LD, lateral dorsal nucleus; LGN, lateral geniculate nucleus; LP, lateral posterior nucleus; MGN, medial geniculate nucleus; PO, posterior nucleus; PU, pulvinar; TRN, thalamic reticular nucleus; VA, ventral anterior nucleus; VL, ventral lateral nucleus; VPI, VPL, VPM, inferior, lateral, and the medial parts of the ventral posterior nucleus or nuclear group.

parts of the brain and higher order relays as those that provide a transthalamic relay from one part of cortex to another. It is to be noted that in primates the nuclei that contain the higher order circuits form more than half of the thalamus. How these "indirect" corticocortical relays relate to direct corticocortical connections is a challenging question considered in Chapters VIII and IX.

Although for many of the thalamic nuclei we can show how they serve to connect different cortical areas to sensory surfaces of the body or to other parts of the brain, we cannot readily demonstrate what it is that the thalamus does for the messages that are passed from ascending pathways to the cerebral cortex. Why don't the ascending pathways go straight to the cortex? This question was always present, but was brought into striking focus in the 1960s when electron microscopists showed the complexities of the synaptic relationships in the thalamus (Szentágothai, 1963; Colonnier and Guillery 1964; Peters and Palay, 1966). Only about 20% of the synapses in the major relay nuclei like the lateral geniculate were then seen to come from the major ascending pathways (Guillery, 1969b), and recent figures show less than half that percentage contact the relay cells that are responsible for passing the information on to the cortex (Van Horn et al., 2000). Complex synaptic formations involving serial synapses and connections from local or distant inhibitory cells characterize all of the thalamic nuclei (e.g., Ralston and Herman, 1969; Jones and Powell, 1969; Morest, 1975), and most thalamic nuclei, in accordance with their shared developmental origin, have more or less the same general organizational plan. Exceptions are considered in later chapters.

The complexity of thalamocortical pathways was further increased by the demonstration of the connections shown in Figure 4. Not only is there a massive input from the deeper layers of the cerebral cortex back to the thalamus, but there is a specialized cell group adjacent to the thalamus, the thalamic reticular nucleus, that receives excitatory branches from the corticothalamic and thalamocortical axons and sends inhibitory axons back to the thalamus (Jones, 1985). The functional role of these reticular connections and of the complex synaptic arrangements found in the thalamus represented (and still represents) a challenging puzzle, a challenge that was greatly increased in recent years by the discovery of diverse transmitters, voltage- and ligand-gated ion channels, and receptors that contribute to the synaptic organization in the thalamic relay (see Sherman and Guillery, 1996; and Chapters V and VI). The functional control of membrane conductances depends on a highly complex interplay of afferent activity and local conditions that will be considered in Chapter IV. These conditions in turn determine the way in which a thalamic cell responds to its inputs and thus determine how messages that come into the thalamus are passed on to cortex. This, the manner in which a thalamic cell passes messages on to cortex, is not constant but depends on the attentive state of the whole animal (waking or sleeping) and probably on the local salience of a particular stimulus or group of stimuli as well: are they new, threatening, interesting, or merely a continuation of prior conditions? (See Chapter VI.)

A. *Thalamic Functions: What Is the Thalamus and What Is It For?* 7

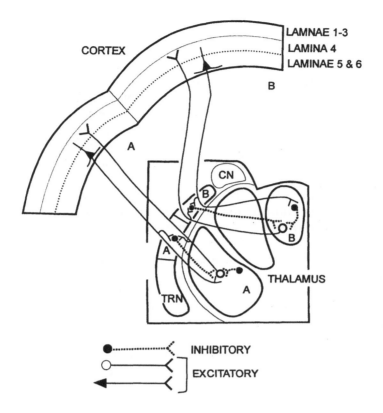

Figure 4
Schematic view of the interconnections between the first order thalamic nuclei, cerebral cortex, and the thalamic reticular nucleus (TRN) as seen on coronal sections of thalamus and cortex. The nuclei labeled A and B are connected with distinct sectors of the reticular nucleus also labeled A and B, and with distinct cortical areas. The connections are shown for components A and B only. The cortical pyramids that give off collateral branches to the reticular nucleus lie in cortical layer 6. CN, caudate nucleus.

When one considers the factors relevant to how the transfer of messages is controlled or gated in the thalamus, it is probable that more than one functionally significant mechanism will be apparent once we have a clear understanding of these aspects of thalamic organization. That is, there are likely to be several more or less distinct functional roles for the synaptic arrangements present in the thalamus. They may be active at different times, or they may have concurrent effects. Two mechanisms that have received significant attention in the recent past have roles in sleep and in the production of epileptic discharges (McCormick and Bal, 1997; Steriade *et al.*, 1993). A third that is coming into focus currently and will be addressed later in the book (Chapters VI and IX) concerns how the role of the relays may change in relation to different behavioral states and may relate to attentional mechanisms. All three involve the circuit going through the thalamic reticular nucleus that was mentioned previously (Figure 4; see Jones, 1985). We anticipate that the role of first and higher order thalamic circuits in passing messages to the cortex will follow the same basic ground rules. That is, whatever it is that the thalamus does for the major ascending pathways, it is likely to be doing something very similar for corticocortical communication. Understanding what it is that the thalamus does should not only help us to understand the functional organization of sensory pathways in relation to perception but also throw new light on perceptual and cognitive functions that have in the past been largely or entirely ascribed to corticocortical interconnections (Zeki and Shipp, 1988; Felleman and van Essen, 1991; Salin and Bullier, 1995).

There is one interesting corollary to the above. If the thalamus acts to control the way that information is relayed to the cortex, then it may be a mistake to look for it to act as an integrator of distinctive inputs as well. At present, the most detailed information available on thalamic relays shows that information from the ascending pathways is being passed to cortex without significant change in "content." That is, there are thalamic nuclei that receive afferents from more than one source, but currently there is no evidence that the two inputs in such nuclei interact to produce a significant change in the content of the input messages. The two pathways appear to run in parallel, with little or no interaction. The motor pathways that go through the ventral anterior/ventral lateral nuclei may prove an interesting exception (see Chapter III), but currently this appears to be an open question.

In this book we are concerned with exploring the way in which thalamic functions relate to cortical functions. Outputs of the thalamus that link it to other cerebral centers, particularly to the striatum and to the amygdala, represent a relatively small although important part of the thalamic relay. They play no role or only an indirect role in influencing neocortical activity, and for this reason we will not be exploring them further. We shall argue that there is likely to be a basic thalamic ground plan that represents the way in which the thalamus transmits messages from its input to its output channels. It seems probable that this ground plan will apply to all thalamic relays, and possibly, when we understand how the thalamus relates to the cortex, the nature of this thalamic relay will help us to understand the function of the currently even more mysterious pathways to other cerebral centers.

B. The "Thalamus" as a Part of the Diencephalon: The Dorsal Thalamus and the Ventral Thalamus

The term "thalamus" is commonly used to apply to the largest part of the mammalian diencephalon, the dorsal thalamus, and we will generally use it in this sense in what follows. However, it is important to recognize that there are several subdivisions of the diencephalon and to look briefly at all of them before focusing on just two subdivisions: the large dorsal thalamus and the smaller, but closely related, ventral thalamus.

B. The "Thalamus" as a Part of the Diencephalon

Figure 5 shows relationships in the diencephalon at a relatively early stage of development. The upper figure, which is a view of the lateral aspect of the brain early in development, shows that the most dorsal part

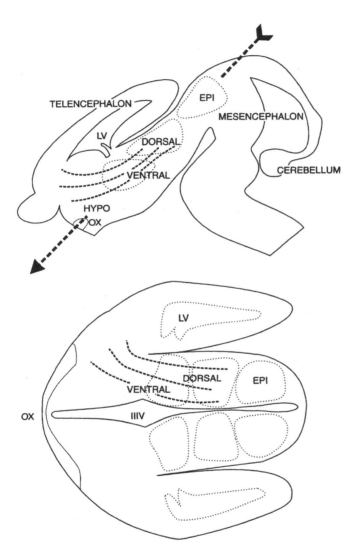

Figure 5
Schematic views of two sections through a 14-day postconception fetal mouse brain, based on photographs in Schambra *et al.* (1992). The upper figure shows a parasagittal section in which the positions of the epithalamus, the dorsal thalamus, the ventral thalamus, and the hypothalamus within the diencephalon are shown (EPI, DORSAL, VENTRAL, HYPO). The lower figure shows a section cut transversely in the oblique plane, indicated by the arrow, that includes these four diencephalic parts and the optic chiasm (OX). The subthalamus is not included in these figures. The interrupted lines show the course of the fibers that link the dorsal thalamus to the telencephalon. LV, lateral ventricle; IIIV, third ventricle.

of the diencephalon is the epithalamus, which in the adult is made up of the habenular nuclei, a few other small dorsally placed nuclei, and the related pineal body. These will not be of further concern to us, nor will two more ventral cell groups, the subthalamus, which is not shown in Figure 5 and is involved with motor pathways, and the hypothalamus, which plays a vital role in neuroendocrine and visceral functions. In this book we are concerned solely with the dorsal thalamus and with a major derivative of the ventral thalamus, the thalamic reticular nucleus.[3] These two are closely connected by the two-way links shown in Figure 4, and it is reasonable to argue that neither can function adequately without the other. Figure 5 shows that originally, during development, the ventral thalamus lies ahead of (rostral to) the dorsal thalamus. The dotted lines in the schematic views of Figure 5 stress an important relationship between the dorsal and the ventral thalamus because they show that lines of communication between the dorsal thalamus and the telencephalon, which includes the cerebral cortex, must pass through the ventral thalamus. This is a key relationship and is maintained even when the ventral thalamic derivative, the thalamic reticular nucleus, moves into its adult position lateral to the dorsal thalamus as shown in Figures 2 and 3.

1. THE DORSAL THALAMUS

In most mammalian brains, and most strikingly in the primate brain, the dorsal thalamus is by far the largest part of the diencephalon. In its size and complexity it is closely related to the development of the cerebral cortex. It can be defined as the part of the diencephalon that develops from the region between the epithalamus and the ventral thalamus. More significantly, it is the part of the diencephalon that has its major efferent connections with telencephalic structures, either striatal or neocortical. In mammals, the neocortical connections dominate, and all dorsal thalamic nuclei project to neocortex. Connections to the striatum are seen for only a few of the nuclei (primarily the intralaminar nuclei) in mammalian brains. All thalamic nuclei have relay cells, which send their axons to the telencephalon, and most have interneurons with locally ramifying axons.

a. The Afferents

We have seen that the first order nuclei of the dorsal thalamus receive a significant part of their afferent connections from ascending pathways. Some bring information about the environment to many of the major thalamic nuclei through sensory pathways such as the visual, auditory, somatosensory, or taste pathways. Others bring information about activity in lower, subthalamic centers of the brain such as the cerebellum for the anteroventral and ventrolateral nuclei or the mamillary bodies for the

---

[3] The ventral lateral geniculate nucleus is also developmentally a part of the ventral thalamus but will not play a significant role in this book. Although, like the thalamic reticular nucleus, it receives cortical afferents and does not send axons to cortex, it does not have the important connections with the dorsal thalamus that make the reticular nucleus into a key part of the thalamocortical system as a whole.

B. The "Thalamus" as a Part of the Diencephalon

anterior thalamic nuclei (Figures 2 and 3). We shall argue that these can be regarded as the driving inputs for their thalamic nuclei, determining the qualitative characteristics of the receptive fields of the thalamic cells, where these can be defined. Other inputs to first order nuclei are best regarded as modulatory or inhibitory afferents, and these affect quantitative aspects of the receptive field rather than its qualitative structure.[4] They come from the brain stem, the thalamic reticular nucleus, the hypothalamus, and from the cerebral cortex itself. The thalamic nuclei not outlined in bold in Figure 2 contain higher order circuits and appear to receive most or all of their driving afferents from the cerebral cortex itself, so that the qualitative aspects of their receptive field properties, insofar as they can be defined, depend on cortical, not on ascending, inputs. This distinction is discussed further in later chapters, particularly Chapter VIII. Here it is to be noted that the higher order thalamic relays, in addition to the driving afferents that they receive from cortex, also receive modulatory afferents from cortex and from the other structures noted earlier for the first order nuclei.

The distinction between corticothalamic axons that are drivers and those that are modulators can be made on the basis of the cortical layer from which they arise; current evidence suggests that corticothalamic afferents arising in cortical layer 5 are likely to be drivers, whereas those arising in layer 6 are likely to be modulators (Sherman and Guillery, 1996, 1998; see Chapters III and VIII). In a few instances, discussed in more detail in Chapter VIII, this distinction between drivers and modulators can be demonstrated in functional terms by recording how inactivation of the cortical afferents affects the receptive field properties of thalamic cells, but so far these instances are regrettably rare. Silencing a cortical driver produces a loss of the receptive field, whereas after a modulator is silenced, the receptive field survives. The difference between these two groups of corticothalamic afferents, the drivers and modulators, is seen not only in terms of their origin and their action on receptive field properties of dorsal thalamic cells but also in terms of the structure of the terminals that are formed in the thalamus, and probably also in terms of the nature of the synaptic transfer to thalamic cells. This will be discussed in Chapters III and V.

We shall argue that the thalamus can be regarded as a group of cells concerned, directly or indirectly, with passing information on to the cerebral cortex about essentially everything that is happening in the central or peripheral nervous system. This includes passing information about one cortical area on to another. This relay of information is subject to a variety of modulatory inputs that modify the way the information is passed to the cortex, without significantly altering the nature of that information,

---

[4] To clarify this distinction between qualitative and quantitative receptive field properties, consider the receptive field of a relay cell of the lateral geniculate nucleus. Its classical, center/surround configuration is what we would term the qualitative receptive field. Quantitative features include overall firing rate or pattern, size of the center or surround, relative strength of center or surround, etc. These quantitative features can be altered without changing the qualitative (i.e., center/surround) organization of the receptive field.

except where, as during slow wave sleep, it essentially prevents such information from reaching the cortex (see Chapter IX).

We have seen that inhibitory inputs reach thalamic relay cells from the local interneurons and from cells in the thalamic reticular nucleus. In addition, there are some other, GABA[5] immunoreactive, inhibitory afferents going to certain thalamic nuclei. The medial geniculate nucleus receives ascending GABAergic afferents from the inferior colliculus (Peruzzi et al., 1997), the lateral geniculate nucleus receives such afferents from the pretectum, and the globus pallidus and substantia nigra send GABAergic axons to the ventral anterior and the center median nucleus[6] (Balercia et al., 1996; Ilinsky and Kultas-Ilinsky, 1997). The precise role of these is not well defined and is discussed further in Chapter IX.

b. Topographical Maps

One other basic feature of the organization of the dorsal thalamus needs to be understood: Most, possibly all, thalamocortical pathways are topographically organized. This is most easily seen in the visual, auditory, or somatosensory pathways, where one can see that the sensory surfaces (retina, cochlea, body surface) are represented or mapped in an orderly way in the thalamus and in the cortex and that the pathways linking thalamus and cortex must carry these orderly maps. Even where it is not clear what it is that is being mapped, as for example, in the mamillothalamocortical pathways, which go to the cingulate cortex and are probably concerned with affective conditions or memory, or where the map appears not to be very accurate, as in many higher order circuits, we shall speak of mapped projections as having "local sign."[7] There is evidence for local sign for the whole of the pathways from the mamillary bodies through the anterior thalamus and to the cingulate cortex (Cowan and Powell, 1954), although it is not clear what function is being mapped or how accurate the map is.

Strictly speaking, a connection that shows no local sign can be regarded as a "diffuse" projection, but this term is often used rather loosely. Quite often the term is used (see Jones, 1998) to refer to a pathway that shows local sign but has significant overlap of terminal arbors or relatively large receptive fields. It is better to keep the term "diffuse" for a

---

[5] Gamma-aminobutyric acid (GABA) is the most common inhibitory neurotransmitter in the thalamus.

[6] This is often called the centre médian nucleus in accord with its French origin, or is Latinized as the nucleus centrum medianum. We have tried to stay with English names and American spelling throughout.

[7] Introduction of a new term for what appears to be an old concept needs some justification. Mapped projections that represent a sensory or cortical surface have been widely described and discussed in the past. The implication is that such "maps" are representations that can be interpreted in terms of the detailed topography of their source, and the expectation has been that such maps, to be useful, must have relatively small receptive fields. Large receptive fields have been represented as evidence for the lack of a map in a pathway. As long as receptive fields do not match the total projection, if they are arranged in a topographic order, then no matter what their size, we shall treat them as a part of a projection that has local sign. We could refer to "crude maps" and "accurate maps," but it seems that there is a reluctance to recognize a mapping in a pathway that simply allows a distinction between up and down, left and right.

pathway that demonstrably lacks local sign. This means that the relationship between the cells of origin and the terminal arbors is essentially random in topographic terms, a relationship that is not easy to demonstrate. Mostly the term has been used where experiments based on relatively large lesions or injections of tracers fail to show topography for terminals or cells of origin or where large receptive fields have been recorded and their topographic ordering has been difficult to discern. Any organization with large receptive fields and a crude local sign must be regarded as topographic rather than diffuse. It is probable that all driver afferents and many modulatory afferents have local sign. Some of the modulatory afferents coming from the brain stem will prove to be truly diffuse, but it is likely that others have local sign (Uhlrich et al., 1988).

A further distinction must be made between an afferent system that is diffuse and relatively global, terminating throughout the thalamus, and one that is diffuse but has terminals that are limited to a single thalamic nucleus or a few specific terminal zones. Those that terminate throughout the thalamus can be regarded as global from the point of view of thalamic organization in general, whereas others that are limited to a few parts of the thalamus, possibly associated with one sensory modality, are to be seen as specific, although they may prove to be diffuse in the sense of lacking local sign within their specifically localized terminal sites.

It should be clear that within a diffuse projection any one afferent fiber may be limited to a small part of the total terminal zone of that projection without this revealing anything about the nature of the projection as a whole. It could be a part of a diffuse global pathway, or it could be a part of a mapped projection to a specific terminal region. In contrast to this, a single cell that sends axonal branches to different parts of a single established map should be regarded as a part of a diffusely organized projection. As the role of the modulatory pathways in the control of thalamic functions becomes defined, so these perhaps apparently arcane distinctions are likely to prove functionally highly significant.

The mapped projections between the thalamus and the cortex are of interest not only because they show how a group of thalamic cells relates to a group of cortical cells but also because they impose important constraints on the pathways that link thalamus and cortex, and these constraints are likely to influence the connections made in the thalamic reticular nucleus as the fibers pass through it on the way to or from the cortex. If the thalamocortical and corticothalamic connections were both simple one-to-one relationships between a single thalamic nucleus and a corresponding single cortical field, then the topographic mapping of the pathways could be carried out by two simple sets of radiating connections, one coming from the thalamus and the other going to the thalamus. The connections of the reticular nucleus, lying on this pathway, would then relate to this simple radiating pattern with little interaction between adjacent sectors. However, the real-life situation is far more complex than this. Single thalamic nuclei can connect to several cortical areas and vice versa for both the driving and the modulatory connections. And many of the cortical maps are mirror reversals of each other, as are some of the thalamic maps. Figure 6 shows two adjacent cortical areas carrying mirror-reversed topographic maps (represented by 3,2,1 and 1,2,3 in the cortex) and connected to a single thalamic nucleus. In the cat, relationships in the visual pathways between areas 17 and 18 and the lateral geniculate nucleus show precisely this arrangement. Only the modulatory corticothalamic axons going from layer 6 of the cortex to the thalamus are included in the figure, which shows that the mapping between thalamus and cortex requires complex crossing of the axon pathways. It should be clear that if all of the thalamocortical and corticothalamic pathways for any one modality were included in the figure, and these, although not shown in the figures, often include several thalamic nuclei or subdivisions and several cortical areas, the result would show a complex system of crossing and interweaving axon pathways between the thalamus and the cortex. In the adult, some of this crossing occurs in the region of the thalamic reticular nucleus as shown in the figure (Adams *et al.*, 1997), and some occurs just below the cerebral cortex (Nelson and LeVay, 1985). The complex crossings are of interest because they establish a potential for connections in the reticular nucleus between the several maps present in the thalamocortical pathways of any one modality.

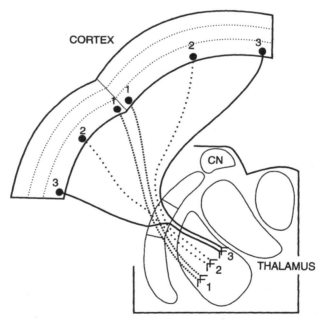

Figure 6
Schematic views of coronal sections through thalamus and cortex to show a single thalamic nucleus such as the lateral geniculate nucleus receiving corticothalamic afferents from two cortical areas. The topographic order of the projections is indicated by the numbers 1 to 3, and the two cortical representations are shown, as they often are, as mirror reversals of each other. Notice that the fibers all cross in the thalamic reticular nucleus, which is not labeled.

The key points about the dorsal thalamus are that it can be subdivided into nuclei, that each nucleus sends its outputs to neocortex, and that each nucleus receives different types of afferent, some classifiable as drivers others as modulators. Many of these connections are mapped, and the multiplicity of maps leads to complex interconnections in the thalamic reticular nucleus, which is the major part of the ventral thalamus.

2. THE VENTRAL THALAMUS

The main part of the ventral thalamus lies directly on the pathways that link the dorsal thalamus to the telencephalon, either the striatum or the neocortex. Figure 5 shows that axons do not pass in either direction between thalamus and telencephalon without going through the ventral thalamus. One important difference between the ventral and the dorsal thalamus is that the ventral thalamus sends no axons to the cortex. In mammals the major part of the ventral thalamus forms the thalamic reticular nucleus, which was briefly introduced earlier. A smaller part forms the ventral lateral geniculate nucleus, which appears to have specialized roles concerned with visually related movements. A basic organizational rule that underlies our approach to the thalamus is that a common developmental origin produces a shared organizational plan. On this basis, one should expect the reticular nucleus and the ventral lateral geniculate nucleus to share some basic structural or functional features. However, at present, it is not clear what these may be. Looking at the ventral lateral geniculate nucleus and the thalamic reticular nucleus could lead one to believe that their common developmental origin demonstrates nothing about a common organizational plan, but for the present we shall regard this either as an exception to the general rule that a shared developmental origin leads to a common structural plan or else as a reflection of our ignorance. We will concentrate on the reticular portion of the ventral thalamus.

As pointed out before and shown in Figures 4 and 5, the thalamic reticular nucleus is strategically placed in the course of the axons that are going in both directions between the cerebral cortex and the thalamus. Although the positions of ventral relative to dorsal thalamus change as development proceeds, both sets of axons continue to relate to the ventral thalamic cells, and in the adult many of them give off collateral branches to the cells of the reticular nucleus. The reticular cells in turn send axons back to the thalamus, roughly to the same region from which they receive inputs. The cortical and thalamic afferents to the reticular nucleus are predominantly excitatory (but see Cox and Sherman, 1999), and the axons that go back from the reticular nucleus to the thalamus are inhibitory (summarized in Jones, 1985). Through these connections the reticular nucleus can play a crucial role in modulating the transmission of information through the thalamic relay to the cerebral cortex.

Not only do the cells within each of the functionally distinct sectors of the reticular nucleus lie in a key position in terms of their connections with pathways going both ways between cortex and thalamus, but we have seen that they also lie in a region where many of these axons undergo some of the complex interweaving discussed above. This pattern of interweaving axons gives the reticular nucleus its characteristic structure. It also contributes to important aspects of its function, since within any one sector of the reticular nucleus, connections from more than one thalamic nucleus and from more than one cortical area are established. Kölliker (1896), more than 100 years ago, recognized the crossing bundles and called the nucleus the Gitterkern, from the German word, *Gitter*, for lattice. These axon crossings and interweavings thus put the reticular nucleus in a position where the cells within any one sector can serve as a nexus, relating several different but functionally related thalamocortical and corticothalamic pathways to each other.

The thalamic reticular nucleus was for many years considered to have a "diffuse" organization, lacking the well-defined maps seen in the dorsal thalamus. Recent evidence has shown that despite the complex network that characterizes the nucleus, there are maps of peripheral sensory surfaces and of cortical areas within the reticular nucleus (Crabtree and Killackey, 1989; Crabtree, 1996; Conley *et al.*, 1991). Understanding these maps and how they relate to each other and to the maps within the main thalamocortical pathways is likely to prove a key issue in future studies of the thalamic reticular nucleus.

C. *The Overall Plan of the Next Nine Chapters*

There are many (more than 30) individually identifiable nuclei in the thalamus, and it is probable that in any one species each one has a more or less distinctive organization. Further, it is well established, and not surprising, that there are significant differences between species for any pair of homologous nuclei. For example, we shall see that the lateral geniculate nucleus of rats, cats, and monkeys has a significant population of interneurons, which establish local, inhibitory connections within that nucleus. In contrast, other thalamic nuclei, for example, the ventral posterior, have an interneuronal population that varies from almost none in the rat to quite significant numbers in the cat and monkey (Arcelli *et al.*, 1997). Some nuclei may receive their driver afferents from just one set or group of axons, whereas other nuclei receive from more than one functionally distinct set of driver or primary afferent pathways. Details of transmitters, receptors, and calcium-binding proteins differ to a significant extent from one nucleus to another, so that it may seem that a book on the thalamus must necessarily be a compendium of details about many individual nuclei. Even if such an account of the many differences among thalamic nuclei were to be limited to commonly used experimental animals, it would form a very heavy and singularly boring volume.

In the following chapters we present accounts of some of the major known structural and functional features of the thalamus. The early chapters present many of the relevant facts and start to look at interpretations, but our major interpretations are presented in detail in the later

C. The Overall Plan of the Next Nine Chapters

parts of the book. We have planned this book to be focused on questions about the functional organization of the thalamic relay in general, and we are especially interested in how this relay operates during normal, active behavioral states. As far as we can, we shall be looking for a common plan of thalamic organization that can serve as a basis for understanding any of the thalamic nuclei. Differences between nuclei can then be seen as opportunities for looking at the possible functional significance of one type of organization relative to another. Much of our discussion will be focused on the visual relay through the lateral geniculate nucleus and will extend to other sensory relays, particularly the somatosensory and the auditory relays as we look for common patterns and detailed differences. These are, at present, the best-studied thalamic nuclei, and details available for these sensory relays are often not available for most other thalamic nuclei.

The visual relay through the lateral geniculate nucleus has received considerable attention over the years. In part, this relates to the fact that we know a great deal about the organization of its input in the retina (e.g., Dowling, 1991; Rodieck, 1998) and about its cortical target (Hubel and Wiesel, 1977; Martin, 1985), so that it has been of particular interest to study the thalamic cell group that links these two. Not only has the intrinsic organization of the nucleus received detailed attention, but its reaction to varying, complex schedules of visual deprivation has taught us a great deal about the plasticity and development of thalamocortical connections (Wiesel and Hubel, 1963; Sherman and Spear, 1982; Shatz, 1994; Rittenhouse et al., 1999). In part, the interest in the visual relay relates to the intrinsic beauty of the lateral geniculate nucleus, most evident in primates and carnivores, where mapped inputs from the two eyes are brought into precise register in distinct but accurately aligned layers (Walls, 1953; Kaas et al., 1972; Casagrande and Norton, 1991). This is an arrangement that we will be exploring in later chapters to a limited extent only. Primarily, we will use current knowledge of the visual relay in the thalamus to lead to general questions about thalamic organization first in other sensory pathways and then in thalamic relays more generally.

In the next two chapters, we first consider the nerve cells of the thalamus (Chapter II), distinguishing the relay neurons from the local interneurons and looking at the several different ways in which distinct classes can be recognized within each of these two major groups. Then in Chapter III we look at the afferents that provide inputs to the thalamus, distinguishing these in terms of their structure, their origin, and, where possible, their possible functional role as drivers or modulators. In Chapter IV, we look at the intrinsic membrane properties of thalamic cells, outlining the properties of the several distinctive conductances that determine how a thalamic nerve cell is likely to react to its inputs. In Chapter V, we consider the distinct actions of different types of synaptic input, focusing on the variety of transmitters and receptors that play a role in determining how activity in any one particular group of afferents is likely to influence the cells of the thalamus.

For each of these chapters only some of the information can at present be readily related to the functional organization of the thalamus that we consider in the later chapters. Many of the points presented raise key questions about the thalamus that are as yet unanswered. We have listed some of these questions at the end of each chapter, but the reader is likely to find a great many other questions that are interesting and deserve attention. In these four chapters, we have presented evidence in some detail to indicate the range of problems that still need to be considered before anyone can claim to understand the thalamus. As new investigators are attracted to the thalamus, and we hope they will be, they will be able to look at some of these problems anew, and where we see only puzzles and unanswered questions, they are likely to look at the problems from a fresh angle and have new insights. At present, only a limited part of the knowledge that we have about thalamic cells and their connections can be interpreted in functional terms.

In the second part of the book, we introduce some of the features that are relevant to our view of what it is that the thalamus may be doing. Chapter VI explores the fact that the thalamic relay cells have two distinct response modes, which depend on the intrinsic properties and synaptic inputs discussed in Chapters IV and V. One is the tonic mode that allows an essentially linear transfer of information through the thalamus to the cortex, and the other is the burst mode that does not convey an accurate representation of the afferent signal to cortex, but instead has a high signal-to-noise ratio so that it is well adapted for spotting new signals. Chapter VII looks at the significance of maps in the brain. The extent to which maps may be essential for cerebral functions is explored as is the multiplicity of interconnected maps in thalamocortical and corticothalamic pathways. We examine the functional and connectional relationships that are produced when several maps, some mirror reversals of each other, converge upon a single thalamic nucleus or cortical area, and we look at the implications of such complex connections for understanding the basic organization of the thalamic reticular nucleus. In Chapter VIII, the distinction between first order and higher order thalamic relays is explored. The former receive their driving afferents from ascending (subcortical) pathways; the latter receive their driving afferents from the cortex and so serve as a relay in corticocortical communication, inserting thalamic functions into cortical connectivity patterns. In Chapter IX we consider criteria that will allow a distinction between afferents that are drivers and afferents that are modulators. Where we can identify the nature of the message that is being relayed in the thalamus we can see that the drivers are the carriers of that message, but in many other thalamic relays and within the corticocortical pathways, where the nature of the message is not yet understood, other criteria for identifying drivers and modulators become important. Chapter X presents an overview of some of our major conclusions, but we stress that this represents a relatively small slice of what is known about the thalamus. Many of the problems and issues raised as questions or currently unsolved problems in each of the chapters deserve close attention if we are to arrive at a more profound understanding of what it is that the thalamus does.

CHAPTER II

The Nerve Cells of the Thalamus

This chapter is concerned with the different nerve cell types that can be distinguished in the thalamus on the basis of their morphology, their position, and their connections. A basic and simple classification of cells in the dorsal thalamus distinguishes cells with axons that project to the telencephalon from those that have locally ramifying axons. These are the relay cells and the interneurons, respectively. The former have axons going through the reticular nucleus and the internal capsule to either the neocortex, the striatum or the amygdala, and we consider them first. The latter are defined as having axons that stay in the thalamus, generally quite close to the cell body. The cells of the ventral thalamus that lie in the thalamic reticular nucleus form a distinct third population; they have axons whose main terminal arbor lies in the dorsal thalamus, with some evidence for locally ramifying axons within the reticular nucleus itself.

A. On Classifying Relay Cells

1. EARLY METHODS OF IDENTIFYING AND CLASSIFYING THALAMIC RELAY CELLS

Most, probably all, dorsal thalamic nuclei in mammals have cells that project to neocortex. We shall treat the main function of the thalamus as the transmission of messages to the neocortex, and from this point of view the thalamocortical cells are almost self-evidently the most important cells. They also represent the great majority of thalamic cells, from about 70 to 99% in mammals, depending on the nucleus and the species. Cells in the relatively small group of intralaminar and midline nuclei send a more significant axonal component to the striatum, as do scattered cells in other thalamic nuclei (Powell and Cowan, 1954, 1956; Francois *et al.*, 1991; Giménez-Amaya *et al.*, 1995). In addition there are projections to the amygdaloid complex (Le Doux *et al.*, 1985) but we are primarily concerned with the pathway to the cortex, and these other pathways will not concern us further in this chapter.

We saw in Chapter I that an early method of demonstrating the cortical connections of the relay cells was to study the severe and rapid retrograde degeneration that thalamic cells undergo when their cortical terminals are damaged by local lesions (see Walker's 1938 book in particular), and it was this degenerative change that proved most useful for early studies of thalamocortical relationships. Although the method has now been superseded by others that rely on the identification of axonally transported marker molecules, it is useful to look at the way the earlier method was used and interpreted. This is of interest primarily because interpretations of the degenerative changes have had a profound and long-lasting effect on contemporary views of thalamic organization. Further, the degenerative changes that characterize most of the thalamocortical pathways provide an insight into the heavy dependence of thalamic cells on an intact cortical connection.

Small cortical lesions produce well-defined, limited segments of rapid neuronal degeneration and death with associated gliosis in a corresponding small segment of the relevant thalamic relay nucleus (see Figure 1), and larger lesions produce more extensive retrograde degeneration in more of the thalamus. Early investigators, particularly Monakow (1889) and Nissl (1913), used the method of retrograde degeneration to define relationships between the major subdivisions of the thalamus and large areas of cortex and to show that most of the thalamus is "dependent" on the neocortex in this sense. That is, destruction of all of the cerebral cortex, specifically of neocortex, produces retrograde degeneration in all of the thalamic nuclei, sparing only the thalamic reticular nucleus, cells of the epithalamus, and, to a significant extent, the cells of the midline and intralaminar nuclei (Jones, 1985).

The very dramatic retrograde changes that occur in the thalamus after cortical lesions represent an extreme form of the reaction of a nerve cell to damage of its axon. Cells in many other parts of the nervous system undergo only a mild reaction or no reaction at all after their axons are cut. The relatively modest branches that thalamocortical cells give off before they reach the cortex, mainly or entirely to the thalamic reticular nucleus (see Figure 4 in Chapter I), appear insufficient to sustain the thalamic cells damaged by cortical lesions. However, Figure 4 in Chapter I also shows that it is no longer appropriate to interpret the thalamic degeneration such as that shown in Figure 1 simply in terms of damage done to the fibers that pass from the thalamus to the cortex. Most, probably all, thalamic nuclei receive afferents back from the cortical area that they innervate, so that one is seeing a combination of transneuronal and retrograde degeneration in the thalamus. The functional details of the thalamoreticular and corticothalamic pathways are treated later. Here it is important to recognize that these pathways are relevant to understanding the thalamic cell loss produced by a cortical lesion, although it has to be

A. *Classifying Relay Cells* 21

1 mm

Figure 1
Photograph of retrograde degeneration in monkey lateral geniculate nucleus. A coronal, Nissl-stained section through the lateral geniculate nucleus of a mandrill (Mandrillus *sphinx*) to show a sector of retrograde cell degeneration produced by a restricted lesion in the visual cortex (area 17). In this animal there were several localized cortical lesions and, correspondingly, several sectors of retrograde geniculate degeneration. Dorsal is up, medial to the left. The main sector of degeneration in this section passes through the dorsal, parvocellular geniculate layers and just includes the dorsal of the ventral two magnocellular layers. Another sector of degeneration is seen in the magnocellular layers to the right (arrow). Note that a complete serial reconstruction would show each sector going through all of the geniculate layers.

recognized that the precise contribution that direct or indirect damage to any one pathway makes to degenerative changes in the thalamus is currently not defined. We know that often the thalamic cell loss is rapid and severe. It was this feature that made cell loss and shrinkage into such a useful tool for studying the connections between the thalamus and the cortex and that fixed attention onto the close link between individual thalamic nuclei and functionally or architectonically definable cortical areas (see, for example, Rose and Woolsey, 1948, 1949). The close link revealed by the lesion studies was interpreted as a "dependence" of thalamus on cortex (see particularly, Nissl, 1913; Hassler, 1964). This dependence is evidence for a trophic interaction but has often been interpreted mainly as a functional link (e.g., Macchi 1983), which of course it is, as well. In a sense this link between the degenerative change and the functional role of the pathway was the strength of the method of retrograde degeneration. But it was also a weakness, because we now know that a mild retrograde reaction, or the absence of a retrograde reaction, does not signify the absence of a functional link (Rose and Woolsey, 1958). As so often, the negative evidence is far less compelling than the positive.

This last point becomes particularly important where a thalamic relay cell in a major thalamic nucleus sends axonal branches to more than one cortical area (Geisert, 1980) or, as do some of the cells of the intralaminar nuclei, sends branches to the striatum and to a cortical area (Macchi et al., 1984). Then damage to one axon branch may not be sufficient to produce a marked retrograde change, whereas damage to both branches does so (see Figure 2). This was a view of thalamocortical relay cells first explicitly proposed by Rose and Woolsey in 1958. On the basis of the retrograde reactions visible in Nissl preparations of the medial geniculate nucleus after localized lesions of more than one area of auditory cortex, they proposed that "sustaining" projections pass from the medial geniculate nucleus to the auditory cortex. A sustaining projection was defined as one that could be cut by itself without producing any significant thalamic degeneration, but when it was cut in combination with damage to another cortical area, there were severe retrograde changes in the thalamus that the second lesion alone did not produce. This was interpreted as evidence that single thalamic cells have axons that branch to innervate more than one cortical area. Although this did not represent the only interpretation of the evidence at the time, it is now reasonably regarded as the best interpretation and represented a critical step forward in our understanding of thalamocortical relationships. Subsequently such branched axons were demonstrated for several thalamocortical pathways by the use of two distinguishable retrogradely transported markers such as those used in the studies of Geisert and Macchi et al. cited above.

The method of retrograde degeneration thus allows us to categorize relay cells in terms of the thalamic nucleus they belong to and the cortical area they connect to, and on occasion the method can show cells that send their axons to more than one cortical area. On the whole, the method was not very good at revealing the interneurons that fail to degenerate after a cortical lesion, generally because these are small and they tended to get lost among the heavy glial changes that accompany the retrograde neuronal degeneration. Initially the interneurons were recognized on the basis of Golgi preparations (see later), although incontrovertible evidence that these small cells did not send an axonal branch to the cortex had to await the advent of two techniques: reliable retrograde markers like horseradish peroxidase and the demonstration that cells not labeled

A. Classifying Relay Cells

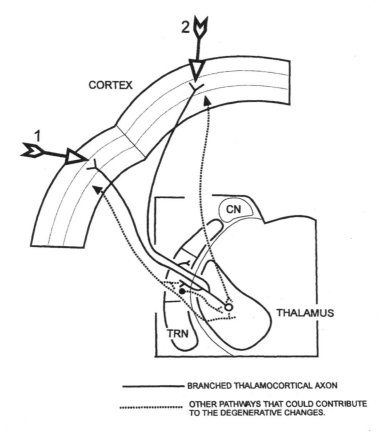

Figure 2
Schema to show a single thalamic cell projecting to two cortical areas. Damage to either of the two cortical areas produced little or no retrograde change in the thalamus, whereas damage to both areas produced severe retrograde cell degeneration. In later experiments, injections of two distinct retrogradely transported markers (arrows labeled 1 and 2) demonstrated that single thalamic cells have branching axons, suggesting that these connections rather than those represented by interrupted lines play a major role in the production of the cell changes. For further details see text.

retrogradely when injections of such markers are made into cortex are immunoreactive for GABA or GAD[1] (Penny et al., 1983; Montero, 1986). The horseradish peroxidase allows the ready distinction between relay cells and interneurons, provided that the retrograde marker has labeled a sufficiently extensive area of cortex to include all candidate relay cell axons for that sector of the thalamocortical projection. Since the interneurons of the thalamus are GABAergic whereas the relay neurons are glutamatergic, the GABA immunoreactivity of the cells that are not labeled by the horseradish peroxidase provides strong evidence about the

---
[1] Glutamic acid decarboxylase (GAD) is critical for the production of GABA and is often used as a marker for GABAergic processes.

distinction between interneurons and relay cells and has now gained sufficient acceptability that the identification of interneurons can be made on the basis of immunohistochemical methods that reveal the presence of GABA by staining either for GABA itself or for GAD, an enzyme involved in its synthesis. Other features that distinguish interneurons from relay cells are considered later and in Chapter IV (Sections A.3 and B.3) and Chapter V (Section B).

2. On General Problems of Cell Classification

Often in the past, relay cells have been classified on the basis of a single variable such as cell size, type of axonal or dendritic arbor, axon diameter or conduction velocity, or on the basis of two apparently loosely linked variables. Since each quantifiable parameter can vary continuously, it is important to ensure that one is dealing with two distinguishable populations rather than two parts of one continuous population, or (worse still) a continuous population whose extremes are described, measured, and stressed, while intermediate values are ignored or arbitrarily assigned to the "populations" represented by one or the other extreme, as has happened with earlier studies of thalamus.[2]

If a classification is based on only one variable (like size) and if this can be represented as a single bell-shaped function, one should generally reject claims for distinct classes even though cells at the high and low ends of the distribution may play somewhat different functional roles. If the relevant parameter shows two quite distinct peaks in a population histogram, or if it is a characteristic that shows no intermediate forms, such as having a particular calcium-binding protein or not having it, having terminals in layer 4 of cortex or not having them, then of course one is on surer ground. Also, it is possible that two or more single parameters vary continuously, but when plotted against each other, clear clustering can be seen, indicating two or more distinct classes.

An excellent demonstration of such clustering is the morphological study of retinal ganglion cells by Boycott and Wässle (1974) that was reanalyzed by Rodieck and Brening (1983) who found that soma size forms a single continuum, but when this was plotted against retinal position, the data points were clearly separated into two clusters. Figure 3 shows their results, and Figure 4 shows a comparable study of cells in the lateral geniculate nucleus. Here the dendritic arbors are analyzed in terms of dendritic intersections with "Sholl" circles (see legend), and it is clear that when the distribution of these intersections is plotted against cell body size, cells that have receptive field properties characteristic of retinal X cells are readily distinguished from those that have the properties of retinal Y cells (see legend for Figure 3). As we shall see, it is entirely reasonable to expect that for thalamic nuclei in general, when appropriate sets of variables are studied, dendritic arbors will contribute significantly

---

[2] Walsh's (1948) discussion of the "Giant cells of Betz" in motor cortex is entertaining, instructive, and pertinent to this point.

A. Classifying Relay Cells

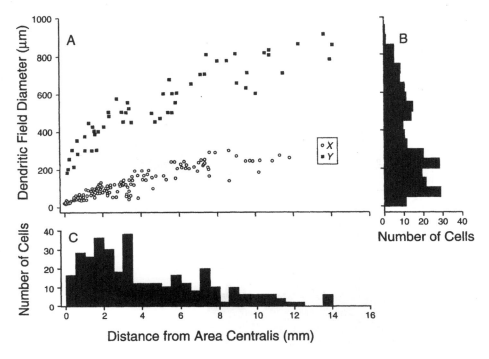

Figure 3
Measures of distance from the area centralis within retina and diameter of dendritic arbor for retinal ganglion cells of the cat. (A) Relationship of the two variables. Note the clear separation of the two classes (called X cells and Y cells) within this scatterplot. (B) Histogram of dendritic arbor diameters for cells in A. (C) Histogram of distances from the area centralis for cells in A. Note that with the single parameters of B and C, it is not possible to discern more than a single cell population.

to defining functionally distinguishable relay cell classes, but exactly how the shape of dendritic arbors relates to the function of thalamic cells is still not defined. For any proposed classification, the more independent parameters one can measure and plot in n-dimensional space, the more likely one is to find evidence for discontinuities in their distribution and, thus, justification for distinct classes if they exist.

Whereas it is possible to demonstrate that two or more classes exist within a population if the parameters measured form clusters of values, one cannot use such a method to prove that a cell population forms a single class, because the *appropriate* parameters may not have been studied. Again, the negative evidence can say nothing about the possible presence of important distinctions. In general, when one is dealing with distinct populations one expects to find that several parameters vary independently, and such independent variation will not only strengthen the logical basis of a classification but it may also make it more interesting in terms of the functional implications.

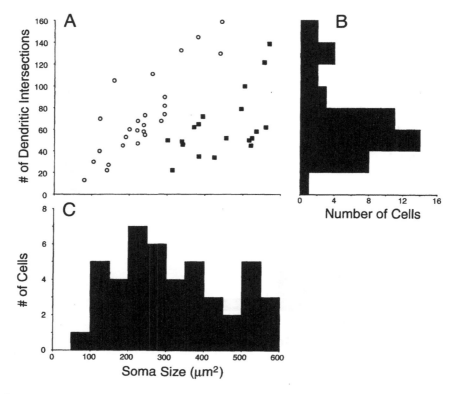

Figure 4
Measures of soma size and number of dendritic intersections for relay cells of the cat's lateral geniculate nucleus; conventions as in Figure 3. Dendritic intersections derive from Sholl ring analysis as follows. A two-dimensional reconstruction was made of the neuron onto tracing paper, a series of concentric circles spaced at 50-μm intervals was centered on the soma, and the number of intersections made by the dendrites with these rings was counted. Note again that the clear separation of two classes that could, in separate experiments, be shown to have the functional properties of X cells and Y cells is seen in (A) with two variables, but is lost when a single variable is measured (B and C). For the functional distinction between the X and the Y cells see legend for Figure 7.

3. On the Possible Functional Significance of Cell Classifications in the Thalamus

The possible functional significance of differences between any two classes of relay cell in the thalamus can be looked at in two distinct ways. These two ways are generally implicit in past studies of relay cells but have rarely been made explicit or clearly distinguished from each other. One is that the different types of relay cell have different integrative functions, modifying receptive field properties or the nature of the message in characteristically distinct ways. The other is that all relay cells do more or less the same thing in the transfer of information from afferents to cortex but differ either because different demands have to be met, in terms perhaps of impulse frequencies, conduction velocities, or cortical distribution patterns, or because of differing patterns of modulatory or gating actions that they can be exposed to. That is, on this second view, differences among relay cells represent either the nature of the traffic that is transmitted or the characteristics of the modulation. They do not reflect differences in integrative functions. This second view appears to receive experimental support in the lateral geniculate nucleus of the cat, because where functionally distinct afferent pathways can be defined in terms of the different receptive field properties that they transmit (the X and the Y pathways mentioned above, see Figure 3), they use relay cells that can be distinguished from each other on the basis of many or all of the features listed, and yet they do not appear to be concerned to any great degree with modifying receptive field properties. In the visual pathway, one finds that the receptive field properties change quite significantly at each relay from retinal receptor to the higher visual cortical areas, *except* in the lateral geniculate nucleus. There is, to our knowledge, no clear evidence to suggest that the thalamus is concerned with the sorts of transforms that change receptive field properties in the other visual relays. It appears that the major definable function of thalamic relay cells is the transfer of a particular set of receptive field properties *without* significant modification of those properties. Instead, in the geniculate relay "gating" functions or "modulatory" functions change the way in which the message is transmitted without changing its content. We discuss these functions in Chapter IX and argue that they are characteristic for the thalamus in general, although evidence for most other thalamic relays is less clear-cut.[3] At present there is no reason to think that any thalamic relay is concerned with the transformation of the message that it receives. If so, then it may prove to be a mistake to look for the "integrative" actions of the thalamus. So one has to ask what, on the second view, do differences in cell size, axon diameter, dendritic arbor, etc., of relay cells signify?

These differences may relate to the characteristics of the message transmitted, to the way in which the message is distributed, to the amount of convergence or divergence in the pathway, or to the way in which the message is modulated or gated. Thick thalamocortical axons may transmit to several cortical areas or innervate more cells in any one cortical area and may require larger cell bodies to sustain more axon terminals. Different cell sizes and dendritic architectures may relate to different patterns of modulator input. There may also be differences in the frequency with which the "gating" functions of the thalamus are active. For any one transthalamic route, one of these functional properties may dominate over others, but on this view, the basic function of the thalamic relay, the transmittal of messages from its driving pathways to cortex with minimal modification of information content, would itself remain

---

[3] Merabet et al. (1998) have described some cells in the lateralis posterior (LP)-pulvinar complex of the cat that respond selectively to the direction of moving targets and have suggested that this may reflect novel receptive field elaboration at the level of thalamus from nonselective afferents. However, the authors acknowledge that it is also possible (and we think likely) that these receptive fields arise in cortex or midbrain and are not generated in thalamus. For the critical experiment to demonstrate the thalamic origin of these receptive field properties it would be necessary to show that none of afferents to LP-pulvinar cells had these properties, a difficult experiment in view of the several afferents.

invariant. Whatever it is that the thalamus does, whether it is one gating function or several, the relay cells in one class can then all be thought to share aspects of this function or group of functions.

That is, it becomes important to look at differences among relay cells in order to define how the relay of information through the thalamus is modulated or gated in particular ways for particular thalamocortical pathways. One should regard the differences between relay cells as related, for example, to the different ways in which one pathway might be more readily brought into or out of an attentional focus. One might expect a relay concerned with details of patterned inputs to differ from one for nociceptive afferents or from one dealing with rapid movements or sudden changes in auditory or visual inputs.

The main aim, then, of a classification of relay cells will be to establish features that characterize cells with functionally different properties. Although traditionally the most important characteristic of a thalamic relay cell concerns the type of information that it transmits to cortex, we have argued that this, strictly speaking, is not a property of the cell but of its input. That is, it is probable, indeed highly likely, that two thalamic relay cells that are indistinguishable on the basis of any of the features given previously, except perhaps for the cortical destination of their axons, will be responsible for relaying quite different messages. For example, a cell in the lateral geniculate nucleus may have precisely the same morphology, synaptic relationships, postsynaptic receptors, cable properties, voltage-dependent membrane conductances, etc., as one in the medial geniculate nucleus, and if so, we would consider these to be of the same class even though they relay different messages (i.e., visual versus auditory).

For any comparison of relay cells that one cares to make, whether the comparison be of cells within a nucleus or of cells among different nuclei, one can assert that differences in structure and synaptic connectivity will relate to functionally significant differences. Conversely, where relay cells show the same features, no matter whether they are in one thalamic nucleus or another, dealing with auditory, visual, somatosensory, or other afferents, they will have functionally similar relay functions. However, at present we do not have sufficient evidence for a clear statement about exactly what the functionally relevant features are. Ideally one would like to be able to relate the intrinsic functional characteristics of cells, that is, their membrane properties, their receptors, transmitters etc. that are considered in Chapters IV and V, to the morphological features that are the focus of this chapter. We shall see that there are examples where the light and electron microscopic appearance of dendritic structures relates to the known synaptic connectivity patterns and thus to certain relay functions of some thalamic cells,[4] but generally we cannot look at a cell, classify it on the basis of the structural details of its perikaryon, or dendrites, and arrive at functionally telling conclusions. This is an important task for the

---

[4] This is the distinction between the functionally distinct X cell and the Y cell relays through the cat's lateral geniculate nucleus, illustrated in Figure 4, and briefly described in the legend for Figure 7.

4. CLASSIFICATIONS OF RELAY CELLS BASED ON DENDRITIC ARBORS AND PERIKARYAL SIZES

One of the earliest classifications of thalamic relay cells was proposed by Kölliker (1896; and see Figure 5) on the basis of cell body sizes and dendritic arbors as shown in Golgi preparations. These features have been widely used in many studies since, and we start with them in order to explore the extent to which they allow distinctions between relay cells before considering other features of relay cells that are likely to prove functionally more significant.

Figures 5A–5D illustrate the two major cell types that Kölliker recognized across several thalamic nuclei of cat, rabbit, and human. He called these "Strahlenzellen" (radiate cells, A and C) and "Buschzellen" (bushy cells, B and D). The former are larger, have a more angular cell body, are multipolar, and have dendrites that branch dichotomously, so that the number of dendritic branches increases gradually with increasing distance from the cell body. The bushy cells have somewhat smaller, rounded cell bodies and their dendrites branch so as to resemble a paintbrush (Pinsel). That is, these dendrites give off many branches close to each other so that the number of dendritic branches increases rather suddenly with increasing distance from the cell body. These cells can be multipolar, triangular, or bipolar. Kölliker reported that the bushy cells would often have a paler appearance in the Golgi preparations than the radiate cells. Varying terminologies have been used in more recent accounts, but these two basic cell types are recognizable in Golgi preparations in several thalamic nuclei (Morest, 1964; Guillery, 1966; Winer and Morest, 1983; LeVay and Ferster, 1977; see Figure 6), although the pale appearance of the bushy cells is generally not mentioned. Where cells have been filled by intracellular markers such as horseradish peroxidese or biocytin (Friedlander et al., 1981; Stanford et al., 1983; Bartlett and Smith, 1999; and see Figure 7), one sees a more complete picture of the dendritic arbor, but the distinction between stellate cells and bushy cells is also apparent.

These distinctions among relay cells have not been identified in all thalamic nuclei, are sometimes blurred because there are intermediate types (Guillery, 1966; Pearson and Haines, 1980), and strict criteria for distinguishing between cells that represent two ends of a continuum have only rarely been applied. In addition, there is some confusion because observers have used different terminologies and seem to have used different criteria. In the visual and auditory pathways of the cat the bushy cells are smaller, have smaller cell bodies, and commonly have oriented, bipolar dendritic arbors, which the radiate type do not show, and this is in accord with Kölliker's description. The two cell types have been called "stellate" and "bushy" or "tufted," respectively, in the medial geniculate nucleus of the cat (Winer, 1985), and confusingly the cells that were

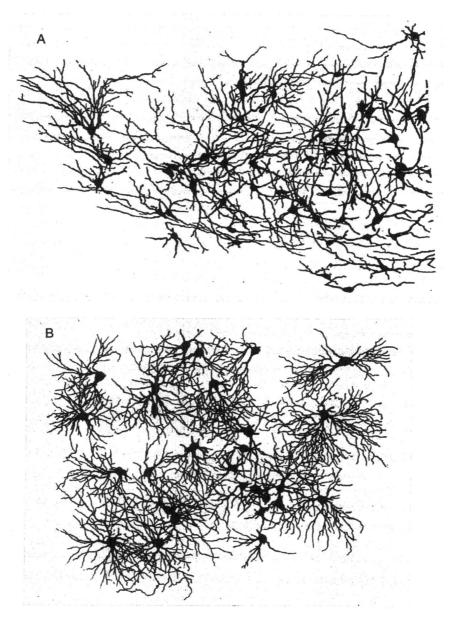

Figure 5
Thalamic cells illustrated by Kölliker (1896). (A) A group of the radiate cells. (B) A group of the bushy cells from the thalamus of a rabbit. (C and D) Individual radiate and bushy cells respectively, C, from a cat and D, from a human thalamus. Golgi method.

described as class 1 (radiate) and 2 (bushy), respectively, in the cat lateral geniculate nucleus (Guillery, 1966) were later called type 1 and type 2 by LeVay and Ferster (1977) and then type 2 and type 1 (respectively) in the cat ventral posterior nucleus by Yen *et al.* (1985). However, Yen *et al.* claim that in the cat somatosensory pathways tufted (bushy) cells are larger

A. Classifying Relay Cells

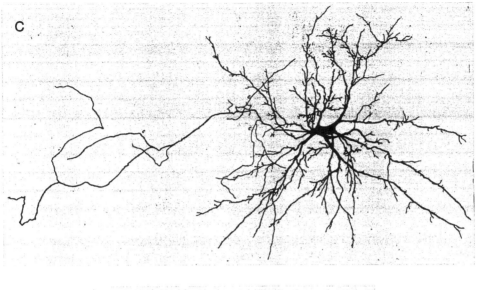

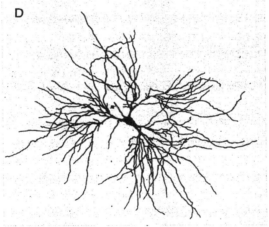

Figure 5 (continued)

than the radiate type. Peschanski *et al.* (1984), Ohara and Havton (1994), and Havton and Ohara (1994) found no distinctions among relay neurons in the ventral posterior nucleus of the rat, cat, or monkey, but a difference between bushy and radiate cells was described by Pearson and Haines (1980) for this nucleus in the bush baby. Possibly there is a species difference or perhaps the cell types in the ventral posterior nucleus deserve further study.

We have pointed out that quantitative analyses of one or more parameters cannot prove that a cell population forms a single class. The type of quantitative analysis used in the lateral geniculate nucleus of the cat by Sherman and colleagues (Friedlander *et al.*, 1981; Stanford *et al.*, 1983), who recorded dendritic intersections with spheres of increasing size centered around the perikaryon (the so-called Sholl-circles: see Figure 4), may provide a useful tool for identifying cell classes in other nuclei.

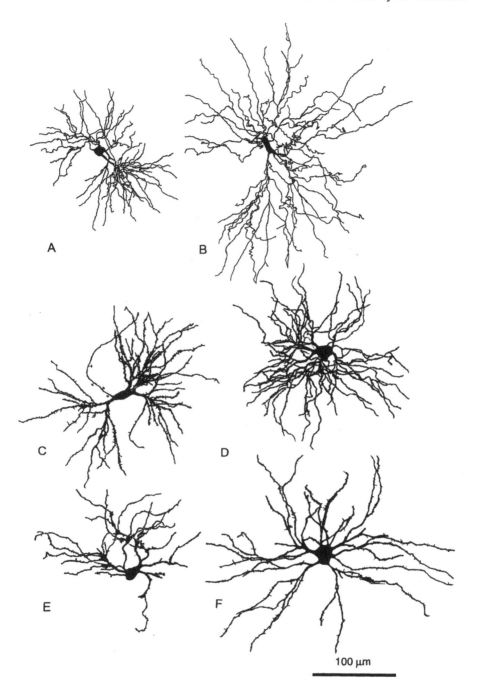

Figure 6
Three bushy (A, C, E) and three radiate (B, D, F) thalamic cells. A and B, cells from the medial geniculate nucleus of a rat that have been filled by intracellular injection of biocytin. C and D, from Golgi preparations of the medial geniculate nucleus of a cat. E and F, from Golgi preparations of the lateral geniculate nucleus of a cat.

A. Classifying Relay Cells

Bartlett and Smith (1999) used this method in the medial geniculate nucleus of the rat and found that only bushy cells could be identified in the ventral division of the nucleus but that the two cell types are distinguishable in the magnocellular division, suggesting that there may be an important difference between relay nuclei, with some containing only one type of relay cell and others having distinguishable stellate and bushy cells.

It is evident that defining distinct cell types on the basis of dendritic arbors has not been straightforward. However, there are enough hints in the literature for a reassessment of Kölliker's original account to be worthwhile. In the lateral geniculate nucleus of the cat, the more radiate type (class 1 of Guillery, 1966) has dendrites that appear to cross laminar boundaries relatively freely, whereas the bushy type (Guillery's class 2) has dendrites that tend to be confined to the lamina within which the cell body lies, suggesting that the two classes differ in the way that their peripheral dendritic segments relate to their afferents. Further, in the cat, the bushy cells are generally functionally identifiable as X cells, and the radiate cells are Y cells.[5] Close to the site where the primary dendrites of the bushy cells characteristically give rise to several secondary branches, a number of prominent rounded dendritic appendages, described by Szentágothai (1963) as "grape-like appendages," tend to be grouped (Guillery 1966; Friedlander et al., 1981; Stanford et al., 1983; and see Figure 7). These are postsynaptic specializations that relate to the retinal afferents and to synapses formed by interneurons in complex "triadic" synaptic arrangements within "glomeruli" (Szentágothai, 1963) that are described in Chapter III. That is, these relationships suggest that the bushy cells establish complex synaptic connections with interneurons that the radiate cells lack.

However, this difference, although it may apply to the cell classes found in the lateral geniculate nucleus of the cat and the ventral posterior nucleus of the bush baby (Pearson and Haines, 1980), is not likely to apply to the medial geniculate nucleus of the rat, where Bartlett and Smith (1999) have described bushy and radiate cells but where interneurons are absent or very rare (Winer and Larue, 1988; Arcelli et al., 1997), as are also the grapelike appendages. The possibility that the dendritic morphology relates to the way in which the dendrites receive contacts from local neurons is appealing and is one that merits exploration in other thalamic nuclei, but perhaps the most frustrating conclusion of even this line of thought is that we know very little about the functional significance of the local connections established by interneurons and reticular cells. The problem remains: which properties of thalamic cells relate to the nature of the message that is being transmitted and which relate to the ways in which the message is modulated on its way to the cortex, and how do these relate to the morphologically identifiable differences?

The only conclusion one can reasonably come to at present is that relay cells differ in the patterns of their dendritic structures, that this

---

[5] The functional distinctions between these cell types are considered in the legend for Figure 7.

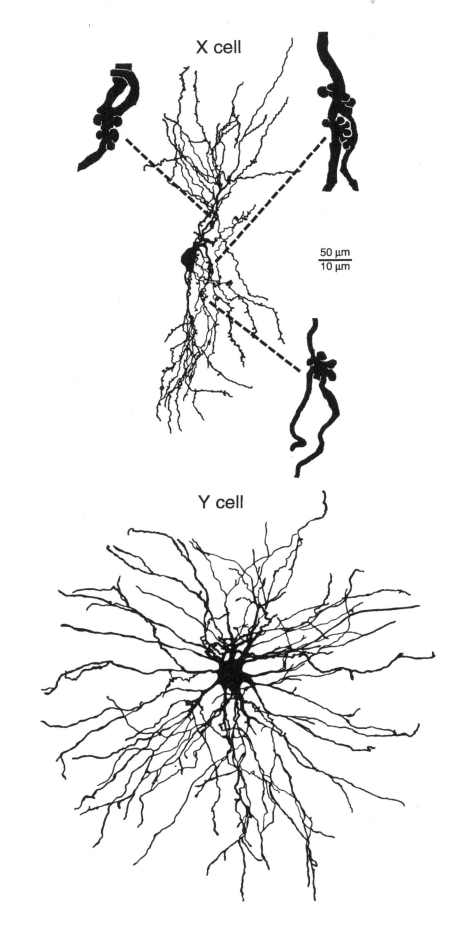

A. Classifying Relay Cells

relates to perikaryal size, and that some of the characteristic patterns seen in one thalamic nucleus can also be seen in others. A recent comparison of relay cells in the ventral posterior and pulvinar nuclei of monkey (Darian-Smith et al., 1999) did not focus on distinctions between relay cell types in either nucleus but did find that detailed quantitative comparisons showed no difference between these two nuclei. As a tool useful for distinguishing relay cell types, dendritic arbors and cell sizes have provided suggestive evidence for more than 100 years but have so far not provided crucial evidence about a functionally significant classification. One reason for this is that classifying cells is more complex than is often recognized, which was our reason for starting this section with a discussion of classifications.

So far, we have considered the two major classes of relay cells. A third, relatively small type of relay cell that is not readily identifiable as either bushy or radiate also has to be recognized and will figure in the discussion that follows. In the lateral geniculate nucleus of the cat, Guillery (1966) described these as "class 4" cells, and in the following they correspond to the "W cells" of carnivores and to the "koniocellular" relay cells of primates. In terms of their dendritic morphology they have been described so far only in the cat (Stanford et al., 1983), and they can be more readily identified on the basis of their laminar position in the lateral geniculate nucleus or the cortical arborizations of their axons.

5. LAMINAR SEGREGATIONS OF DISTINCT CLASSES OF GENICULOCORTICAL RELAY CELLS

Other criteria for distinguishing thalamic cells include the nature of the afferents they receive, the particular characteristics of their receptive field

---

Figure 7
Tracings of an X and a Y cell from the A-laminae of the cat's lateral geniculate nucleus. Compared to X cells, Y cells have larger receptive fields at matched retinal eccentricities, faster conducting axons, better responses to visual stimuli of low spatial and high temporal frequencies but poorer responses to low temporal and high spatial frequencies; also, Y cells respond to higher spatial frequencies with a nonlinear doubling response, whereas X cells show excellent linear summation to all stimuli (for details, see Sherman and Spear, 1982; Shapley and Lennie, 1985; Sherman, 1985). The cells were identified physiologically during *in vivo*, intracellular recording, and then horseradish peroxidase was passed from the recording pipette into the cell bodies. With proper histological processing the horseradish peroxidase provides a dense stain, allowing visualization of the entire somadendritic morphology. The dendritic arbor of the X cell has the tufted pattern, is elongated, and is oriented perpendicular to the plane of the layers, whereas the Y cell dendrites show a stellate distribution with an approximately spherical arbor. The X cell also has prominent clusters of dendritic appendages near proximal branch points. These are hard to see in the cell reconstructions, so three examples are shown at greater magnification, with dashed lines indicating their dendritic locations. Data from Friedlander et al. (1981).

properties, and the sizes and patterns of distribution of their axons and axonal terminals in the cerebral cortex. In the mammalian lateral geniculate nucleus all of these features are distinguishable and the cells are segregated into distinct laminae on the basis of one or several of these features, almost as though they had been designed to illustrate discussions of thalamic cell classifications for a book such as this (Figure 8). The functional or developmental reasons for this laminar segregation are not clear, but different mammalian species show quite different patterns of laminar segregation, providing a useful approach to looking at ways of classifying thalamic relay cells.

We have briefly introduced the distinction between geniculate X cells and Y cells in cats to look at the possible significance of the classifications that can be based on the dendritic arbors and on the appearance and size of the cell body. In the lateral geniculate nucleus of carnivores and primates, three functionally distinct classes of relay cell have been recognized, and these may well contain subclasses (reviews: Sherman and Spear, 1982; Sherman, 1985; Casagrande and Norton, 1991; Hendry and Calkins, 1998; Hendry and Reid, 2000). Each cell type receives afferents from a functionally distinct type of retinal driver afferent, has axons with

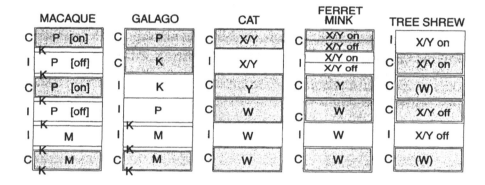

Figure 8
Schema to show the laminar distribution of cell types in the layers of the lateral geniculate nucleus of five different species (see Kaas *et al.*, 1972; Sherman and Spear, 1982; Casagrande and Norton, 1991; Hendry and Calkins, 1998). The shaded boxes show layers innervated by a crossed pathway from the eye on the opposite side (C = contralateral), and the unshaded boxes show an uncrossed pathway from the eye on the same side (I = ipsilateral). The magnocellular (M), parvocellular (P), and koniocellular pathways are shown for macaque and Galago; the X, Y, and W pathways are shown for cat, ferret, mink, and tree shrew. The on-center and the off-center afferents are indicated by on and off for the ferret and mink and are shown as [on] and [off] for the macaque to indicate that there is some disagreement about this result (see text). Note that there are often further subdivisions of the parvocellular layers in human and other primate brains, but their functional significance is not known (Malpeli, 1996; Hickey and Guillery, 1979). See text for further details.

A. Classifying Relay Cells

distinctive axon diameters terminating in a different laminar distribution in the cortex, is characterized by different cell body sizes and dendritic arbors and, to a varying extent, depending on species, shows a characteristic distribution within the laminae of the lateral geniculate nucleus (shown schematically in Figure 8). These relay cells also show distinctive types of synaptic relationships with retinal afferents and interneurons, which are discussed in Chapter III (see Section C.7).

Each class of geniculate relay cell provides an apparently independent channel for transmission from functionally distinctive retinal ganglion cells through to the visual cortex. These are the best known examples of pathways that share a single thalamic nucleus or lamina and yet provide independent parallel relays with little or no interaction between the pathways. At this point, details of the known functional distinctions between the pathways are not relevant to the following consideration of the differences between the geniculate cells (for examples of such functional distinctions, however, see Sherman, 1985; Lennie, 1980), because the distinctions depend primarily on the properties of the retinal ganglion cells that innervate the geniculate cells. This may seem strange at first sight; it is a measure of the difficulties we have in relating the structure and connections of thalamic cells to their function in the relay.

In primates, relay cells in descending order of cell body size and of axon diameter have been identified as magnocellular (M cells), parvocellular (P cells), and koniocellular (K cells). The first two lie in the correspondingly named geniculate layers (magnocellular layers 1 and 2, parvocellular layers 3 to 6 of Old World monkeys), and the koniocellular groups lie between these layers and, to a certain extent, scattered within them (Figure 8). In the bush-baby, *Galago*, however, the koniocellular group forms distinct layers, and this provided a useful first experimental approach to these small cells in primates. In carnivores (cats, ferrets, mink), the largest [here called Y cells; see Sherman and Spear (1982) and the medium sized (X cells)] are almost entirely mingled with each other in the major layers of the lateral geniculate nucleus (layers A and A1), whereas the smallest cells (the W cells) are found in the small-celled C layers nearest to the optic tract. Other parts of the carnivore nucleus show some segregation of Y and W cells from X cells. Again, cell body size relates to axon diameter.

The extent to which the apparently functionally independent pathways are or are not separated from each other topographically is illustrated in Figure 8 and is worth some further attention. It appears that a laminar segregation of the relay cells is not needed either to keep functionally distinct parallel pathways separate or to allow them to have terminals that show a clear laminar separation in the cortex (see next section, A.6).

The rules that govern how relay cells are grouped in the thalamus to form different cell nuclei or nuclear groups are not understood. The problem is particularly clearly focused in the layers of the lateral geniculate nucleus. We know that there is little or no mingling of relay cells receiving from left or right eye, but beyond this separation of monocular inputs into separate layers, there seems to be no rule that applies across species for deciding how different relay cell classes are distributed in the laminae. Figure 8 shows a striking example of this inconsistency of relay cell distributions in cat and ferret, which generally have similar geniculate structures.[6] In cats, the X- and Y-systems share the A laminae, and there is no evidence for any significant functional interaction between the two pathways on the way to cortex, except that a small minority of cells (<5%) seem to have inputs mixed from X and Y retinal axons (Sherman and Spear, 1982; Sherman, 1985). For each of these two pathways there is a so-called on-center and an off-center system (see Rodieck, 1998, for details of the two types of retinal ganglion cell that provide the afferents), and these are also mingled in the A laminae of cats. However, in ferrets (and mink), which have a geniculate relay basically similar to that of the cat, with the X and Y pathways mingled as in a cat, the on-center and off-center pathways lie in separate geniculate laminae (LeVay and McConnell, 1982; Stryker and Zahs, 1983) and have separate cortical terminations. There is no evidence that the function of these pathways is significantly different in cat and ferret (or mink). In Old World monkeys, some evidence (Schiller and Malpeli, 1978; but see Derrington and Lennie, 1984) suggests that the on-center parvocellular relay cells are partially separated from the off-center relay cells in distinct geniculate layers, but this is so only for central vision. For peripheral parts of the visual field the on-center and off-center cells share a geniculate lamina, and here, again, there is no hint of a functional correlate. A comparison of carnivores and monkeys shows that the W pathway in carnivores has its cells in a separate layer, but that in the monkeys (but not the bush babies) the koniocellular pathways have their cells scattered among the M and the P cells.

6. THE CORTICAL DISTRIBUTION OF SYNAPTIC TERMINALS FROM RELAY CELL AXONS

Thalamocortical relay cells differ in the nature of their terminal arbors and in the way in which they distribute their terminal arbors to different cortical layers and cortical areas, and these differences relate to the different classes of relay cell definable by other criteria and introduced for the visual pathways in the previous section.

---

[6] There is no demonstrated explanation of the variety of laminar structures shown in Figure 8, but it is possible to argue that the principal developmental determinant of geniculate structure is concerned with separating left eye from right eye inputs. This separation is constant in all mammalian species that have been studied and is likely to be significantly influenced by the differing activity patterns coming from each eye at the critical stages of early development (Wong et al., 1995; Shatz, 1996). Possibly, the determinant of further laminar segregation (P cells from M cells, K cells from P cells, on-center from off-center cells) would be whether the distinct activity patterns that characterize each of these cell classes appear during the period when the geniculate cells are responsive to the differing activity patterns coming from the two eyes. That is, the developmental machinery would have been established to ensure left eye/right eye segregation, but would have "accidentally" involved any other functionally distinct cell classes, setting them up in separate layers if the timing was right.

A. Classifying Relay Cells

a. The Laminar Distribution of Thalamocortical Axons

An early account of different patterns of termination of thalamocortical axons in the cortex was published just over 60 years ago and was based on Golgi preparations. Lorento de Nó (1938) showed two types of thalamocortical axon. He called one "specific," apparently having traced the fibers from medial or lateral geniculate nucleus or from the ventral posterior nucleus.[7] These axons have dense terminals in cortical layer 4 and sparser extensions into layer 3. He traced other afferents, which he called "nonspecific," from the thalamus in the mouse, and these sent much less dense terminal branches to more than one cortical area. These ascend to layer 1 and give off a few branches to other cortical laminae on the way, mainly to layer 6. This distinction between "specific" and "nonspecific" thalamocortical axons has had an important influence on subsequent studies of thalamocortical innervation patterns. Lorente de Nó's description has been widely cited and has been loosely and optimistically linked not only to important electrophysiological observations on cortical arousal but also to more dubious speculations as to the distinctions and phylogenetic history of specific and nonspecific sensory pathways.

The nonspecific system was early linked to diffusely organized thalamocortical pathways, going from the midline and intralaminar nuclei to the cerebral cortex (Jasper, 1960; reviewed in Macchi, 1993, and Jones, 1998), and thought to be a widespread system concerned with mechanisms of arousal acting through the superficial layers of cortex. Further, an additional burden was placed on the distinction between the specific and the nonspecific pathways when an extension of older ideas concerning "protopathic" and "epicritic" sensory pathways (Bishop, 1959) was added to the conceptual structure. The protopathic pathways were considered to have generalized, nonlocalized, diffuse functions and were made up of fine axons such as those in the spinothalamic pathways. The epicritic pathways were presented as having well-localized "specific" functions and were made up of thicker axons such as those in the lemniscal pathways (see Head, 1905; Walsh, 1948). The argument was functional and phylogenetic. Fine axon pathways were thought to be phylogenetically old (see Herrick, 1948) and thick axon pathways phylogenetically new; the former carried "more precise and specific" (Bishop, 1959) messages than the latter. This mixture of fact and speculation often still underlies contemporary views on the thalamocortical pathways. The relevant facts are sparse and to a significant extent yet to be defined. It is important to recognize that our knowledge about the evolution of central neural pathways is extremely limited; the evolutionary history of neural pathways is at best derived from comparative studies of extant forms, and one has to be dubious about the idea that axon diameter can serve as

---

[7] Lorente de Nó's ability to trace such myelinated axons from thalamus to cortex is extraordinary when one realizes that these axons take an oblique course from thalamus to cortex and that myelinated axons generally do not stain well with the Golgi methods. However, his completely dogmatic statement about the thalamic origins of these axons, his generally careful approach to the Golgi material, and the subsequent experimental confirmation of significant parts of his account suggest that with sufficient care and patience even the seemingly impossible can be achieved.

a marker that allows identification of distinct functional systems in species separated by enormous spans of evolutionary time. Further, distinguishing where thick axon systems are, indeed, specific, having local sign (as defined in Chapter I, Section B.1.b) that fine axon systems lack; or defining the origin of the afferents that end in different layers of cortex remain important but still unsolved questions for most thalamocortical pathways.

Recent studies have used several different techniques to define the laminar termination of thalamocortical axons. Small lesions or small injections of radioactively labeled tracer into the thalamus have shown the laminar distribution of the degenerating or labeled axons, without showing the individual arborizations formed by the cortical terminals (Hubel and Wiesel, 1972; Harting *et al.*, 1973; Ding and Casagrande, 1977; Levitt et al., 1995). Small cortical injections of retrogradely transported tracers, limited to one or a few cortical layers (Carey *et al.* 1979; Penny *et al.* 1982; Niimi *et al.*, 1984; Ding and Casagrande, 1998) have shown which thalamic cells have axons that reach any one layer or group of layers. The most informative experiments have involved the direct tracing of thalamocortical axons after injections of anterogradely transported markers into single thalamic cells or small groups of cells or axons (Ferster and LeVay, 1978; Humphrey *et al.*, 1985a,b; Ding and Casagrande, 1997; Deschênes *et al.*, 1998; Aumann *et al.*, 1998; Rockland *et al.*, 1999). Distinct groupings of thalamocortical axons are seen terminating in cortical layers 1, 3, 4, and 6. Carey *et al.*, (1979) and Penny *et al.*, (1982) claimed that for visual and somatosensory pathways small thalamic relay cells in the respective first order nuclei (lateral geniculate and ventral posterior) innervate mainly layer 1, and large ones innervate mainly layer 4. This supported Lorento de Nó's claim that different axons innervate the different cortical layers but destroyed the idea that they come from different nuclei, "specific" nuclei for layer 4 and other sources for layer 1.

In the visual and somatosensory cortex the densest terminals are in layer 4, with a sparser spillover from this plexus into layer 3; sparser projections go to layers 1 and 6. In some other thalamocortical pathways the major afferent plexus is in layer 3 or 5 rather than layer 4 (Aumann *et al.*, 1998; Deschênes *et al.*, 1998). In the ventral posterior nucleus of rats, cats, and monkeys, generally the larger cells project mainly to layer 4 with a smaller projection to layer 6, whereas the smaller cells project mainly to layers 3 and 1 (Penny *et al.*, 1982; Rausell *et al.*, 1992; Rausell and Jones, 1991). Patterns of projection from the medial geniculate nucleus are somewhat less clear, but in the monkey, some cells project mainly to middle layers and layer 6, whereas others terminate mainly in layers 3 and/or 1 (Hashikawa *et al.*, 1991; Molinari *et al.*, 1995; Nimi *et al.*, 1984; Pandya and Rosene, 1993). Pathways from the pulvinar in the monkey go to several extrastriate cortical areas, with varying patterns of laminar distribution, primarily to layer 3 but with some input to layer 1 and deeper layers (Rockland *et al.*, 1999), a pattern resembling that described for axons going from the mediodorsal nucleus of the rat to prefrontal cortex (Kuroda *et al.*, 1993).

A. Classifying Relay Cells

In summary, three different types of thalamocortical terminal arbor patterns have been described: those in middle layers, mostly layer 4, but also layers 3 and occasionally 5; those in layer 6; and those in layer 1, occasionally extending into layer 2. Most or all cells that project to layer 4 also branch to innervate layer 6, whereas most innervating layer 3 seem to branch to innervate layer 1. A few cells innervate layer 1 exclusively. The possible functional significance of these different terminal patterns will be considered further in Chapter IX.

The cortical distribution of thalamocortical axons coming from the intralaminar nuclei is of some interest in view of the long-standing perception that they represent a "nonspecific" pathway to superficial layers of cortex, distinct from the "specific" pathways that come from first order nuclei. Evidence on the supposedly specific or nonspecific nature of this cortical projection varies considerably. Although there is evidence that the projections from the intralaminar nuclei tend to be rather widespread, and in many experiments have shown no evidence of any local sign (Jones and Leavitt, 1974; Kaufman and Rosenquist, 1985), other observations show that different members of the intralaminar group of nuclei have distinct cortical projection areas (Ullan 1985; Royce et al., 1989; Berendse and Groenewegen, 1991) and that not many of the individual intralaminar cells show evidence for axons that branch to innervate more than one cortical area (Bentivoglio et al., 1981). There is some evidence for distinct local sign in some of the cortical projection pathways (Olausson et al., 1989; Royce and Mourney, 1985). The problem was reviewed earlier by Macchi and Bentivoglio (1982), and more recently by Minciacchi et al., (1993) and Molinari et al., (1993).

It should be stressed that a major target of the intralaminar and midline nuclei, apart from the cerebral cortex, is the striatum (Jones and Leavitt, 1974; Macchi et al., 1984), and this pathway clearly distinguishes the intralaminar and midline nuclei from other thalamic nuclei. However, in terms of their cortical connections, they may prove to share many of the organizational features of other thalamic nuclei. Although there is a significant contribution from the intralaminar nuclei to layer 1 of cortex, there is also a contribution to the middle and deep layers of cortex (Kaufman and Rosenquist, 1985; Royce and Mourney, 1985; Towns et al., 1990).

A survey of the laminar distribution of thalamocortical projections going to different cortical areas from individual thalamic nuclei would take us beyond the scope of the present enquiry and would show that much of the relevant information for many thalamic nuclei and cortical areas is not yet available. The important point to be noted is that the pattern of projection from thalamus to cortex varies not merely from one nucleus to another but that, perhaps more importantly, it varies from one cell type to another within any one nucleus. This aspect of the complexity of the thalamocortical pathways is well illustrated by the visual pathways as shown in Figure 9.

For the geniculocortical pathways to area 17 of primates it has been shown that the parvocellular and magnocellular cells of the lateral geniculate nucleus project to layer 4, and most or all also branch to

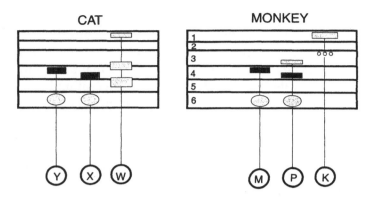

Figure 9
Schema to show the laminar distribution of geniculocortical afferents in the cat and the monkey. Based on Casagrande and Kaas (1994) and Kawano (1998).

innervate layer 6; the magnocellular axons have richer and more widespread cortical arbors than the parvocellular axons (Hubel and Wiesel, 1972; Blasdel and Lund, 1983; Florence and Casagrande 1987; Casagrande and Kaas, 1994). Within the cortex, the terminations remain distinct: the parvocellular and magnocellular projections go to different subdivisions of layer 4, the former deep to the latter. Each sends an additional component into layer 3, although this depends on the way in which the subdivisions of layer 3 and 4 are defined (see Casagrande and Kaas, 1994). The koniocellular cells have a very different arborization pattern, terminating mostly in layers 3 and 1 (Diamond *et al.*, 1985; Ding and Casagrande, 1997; Ding and Casagrande, 1998; reviewed in Casagrande, 1994; Casagrande and Kaas, 1994; Hendry and Calkins, 1998; Hendry and Reid, 2000). Most of these axons branch to innervate both layers, but a minority seem to innervate only layer 3 or only layer 1 (Ding and Casagrande, 1997, 1998). The layer 3 terminals have a limited, patchy distribution, going specifically to specialized small regions of cortex identified as having high cytochrome oxydase activity (Wong-Riley, 1979) and cells with color-opponent response properties (Livingstone and Hubel, 1984).

A similar pattern is seen in geniculocortical terminations of cats (Ferster and LeVay, 1978; Humphrey *et al.*, 1985a; Kawano, 1998). The geniculate X and Y axons terminate mostly in layer 4 with branches innervating layer 6 and within layer 4, X arbors terminate deep to Y arbors (Humphrey *et al.*, 1985a,b; Boyd and Matsubara, 1996), and the Y axons have richer and more extensive terminal arbors than do the X axons. W axons terminate mostly in layers 3 and 1. This general pattern has also been described for tree shrews, but in less detail, with one group of geniculate cells mainly innervating layer 4, and another, layers 1 and 3 (Conley *et al.*, 1984; Fitzpatrick and Raczkowski, 1990). The similarity in the gross pattern of geniculocortical terminations across such divergent mammalian representatives as primates, carnivores, and insectivores suggests that it is a general pattern for mammals.

A. Classifying Relay Cells

A revealing difference is seen when one compares the laminar arrangements in visual cortex, shown in Figure 9, with those in the lateral geniculate nucleus in Figure 8. Whereas each eye has a termination in a distinct geniculate layer, in the cortex, for any one functional type, the two eyes share a layer. The cortical axons are distributed to alternating, side-by-side patches (not shown in Figure 9), forming a part of the ocular dominance columns within one layer (Hubel and Wiesel, 1977). Further, where there is evidence for a separation of on-center from off-center pathways in distinct geniculate layers, there is also evidence that the two pathways terminate in alternating patches rather than distinct laminae in the cortex (McConnell and LeVay, 1984; Zahs and Stryker, 1988). This difference suggests that cortical layers accommodate distinct functional systems, whereas, as we have seen, geniculate layers are not readily related to particular functional specializations and may rather represent an accident or epiphenomenon of developmental forces (see footnote 6). The cortical processing required for left eye inputs is not likely to be different from that required for right eye processing, and comparably, the cortical processing for off-center and on-center pathways is likely to be very similar. In contrast to this, the functionally distinct, X, Y, W, or P, M, and K systems are likely to be making quite distinct demands on cortical processing, and so we see their separation in cortex even where they are mingled in the geniculate. This stresses that each cortical layer, and even each subdivision of a layer, is likely to have a quite distinct intrinsic functional organization, whereas in the geniculate layers (and by implication, in subdivisions of thalamic nuclei more generally) there may or may not be a laminar separation of pathways that carry different messages, and there is no necessary spatial separation of the distinct types of intrinsic thalamic circuitry acting on the relay of the messages. The clearest example of this mingling of pathways in the thalamus is the shared laminar distribution of the X and Y pathways in cats, each relating to distinct thalamic cell types and synaptic circuitry.

The difference between cortical and geniculate lamination not only throws an interesting light on the nature of the functional segregation related to the lamination, but it also demonstrates that the developmental mechanisms that lead to the production of thalamic and cortical segregation of functionally distinct pathways are likely to be fundamentally different.

b. The Distribution of Thalamocortical Axons to One or More Cortical Areas

Generally a single axon or axon branch that goes to any one cortical area branches extensively to form multiple terminations within a localized part of that area. This pattern of branching is distinct from the branching pattern described in Section A.1 (and see Figure 2) that allows an axon access to two cytoarchitectonically and functionally distinct cortical areas. Where the latter form of a dual innervation of two cortical areas relates to identifiable maps of sensory surfaces one expects both branches to innervate corresponding points on the map and, where the evidence is available, they do (e.g., Bullier et al., 1984).

Whereas the classical approach based on retrograde degeneration assigned thalamic relay cells to particular nuclei and to particular cortical areas, it is now clear that this is only a first approximation. Currently available highly sensitive methods are showing that many cortical areas are innervated by more than one thalamic nucleus and that many thalamic nuclei have cells that send axons to more than one cortical area. In some instances different cells within one nucleus send axons to different cortical areas, but in others the axonal branches of single relay cells innervate more than one cortical area. That is, relay cells within one thalamic nucleus can also be distinguished on the basis of whether they innervate one cytoarchitectonically and functionally distinct cortical area or more than one. This has been demonstrated for the first order geniculocortical pathways (e.g., Geisert, 1980; Humphrey et al., 1985b; Kennedy and Bullier, 1985; Birnbacher and Albus, 1987), somatosensory pathways (e.g., Spreafico et al., 1981; Cusick et al., 1985; but see Darian-Smith and Darian-Smith, 1993), and for higher order pathways going through the pulvinar (e.g., Lysakowski et al., 1988; Rockland et al., 1999). As methods become more sensitive and more widely used, one can expect to find more examples of multiple, complex interrelationships between thalamic nuclei and cortical areas. Here again, we will not provide a survey of thalamocortical pathways in general, but rather we will use the visual pathways to illustrate some of the relationships that have been established.

Looking specifically at the geniculocortical pathways of cats, one sees that the largest relay cells, the Y cells, are characterized by having the most richly branching thalamocortical axons. Not only does each axon innervate a well-defined sector of the visual cortex that is relatively large and has relatively many terminals compared with the X cells (Ferster and LeVay, 1978; Humphrey et al., 1985a,b), but a certain proportion of the Y cells also innervates more than one cortical area (Geisert, 1980; Humphrey et al., 1985b; Kennedy and Bullier, 1985). This is in contrast to the X cells whose axons appear to be limited to the primary visual cortex. Most evidence about the geniculocortical pathways in primates is concerned with the projection to the primary visual cortex (V1), although there is some evidence that some geniculate cells, scattered in the interlaminar zones and in the parvocellular laminae, also project outside V1 (Montero, 1986) and some certainly go to area V2 (Yukie and Iwai, 1981; Fries, 1981).

The question as to what a particular system gains (or loses) by having cells that send axonal branches to more than one cortical area rather than having two separate cell populations, each going to a different cortical area, is not resolved, except for the relatively obvious probability that a branching axon will likely transfer a truer duplicate version of relay cell activity to the two cortical areas than two separate cells would. We do not know whether in the cat the Y cells that innervate more than one cortical area differ in any other important respect from Y cells that apparently

A. Classifying Relay Cells

belong to the same morphological and functional class but innervate only one cortical area, nor do we know what the thalamocortical pathway gains by having some cells of each kind, and this seems to be true of branching axons in other thalamic relays such as the somatosensory pathways (see Spreafico et al., 1981). One could take a speculative, evolutionary view and suggest that the curiously mixed population of large cells in the cat's geniculocortical pathway represents an evolutionary stage, but one would be hard put to decide whether the branching cells are on their way in or on their way out. If there are rules that govern the distribution of such branching cells, they remain yet to be defined.

7. Perikaryal Size and Calcium-Binding Proteins

One generalization that emerges from the preceding description is that larger thalamic relay cells are most likely to innervate layers 4 and 6, whereas smaller ones tend to innervate layers 3 and 1. Recently, Jones and his colleagues (Hashikawa et al., 1991; Jones, 1998; Molinari et al., 1995; Rausell et al., 1992; Rausell and Jones, 1991) have shown that relay cells throughout the monkey's thalamus can be divided into two categories based on their immunocytochemical reactivity for parvalbumin or calbindin, two different Ca^{2+}-binding proteins. The smaller cells that project mainly to superficial cortical layers stained for calbindin, whereas the larger ones that project mainly to middle layers stained for parvalbumin (see also Hendry and Yoshioka, 1994). This is a potentially useful distinction of functionally distinct cell types, although there are two problems: one is that we do not yet know the functional implications for a cell of having one or the other calcium-binding protein, and the other is that so far the distinction only applies to primates. Since the distinction has not been seen in nonprimate forms (e.g., Ichida et al., 2000), the extent to which it can represent a basic distinction for thalamocortical systems must be doubtful. Parvalbumin-positive cells dominate the main sensory relays of the thalamus, and calbindin-positive cells dominate the intralaminar nuclei, but both cell types are found in all thalamic nuclei (De Biasi et al., 1994; Diamond et al., 1993; Goodchild and Martin, 1998; Johnson and Casagrande, 1995).

Jones (1998) has recently used this correlation of immunocytochemistry and projection pattern in primates to suggest that for the thalamus in general, two distinct projection systems are recognizable. The smaller calbindin-positive cells project mainly to superficial cortical layers, particularly to layer 1, in a diffuse pattern to form a "matrix" of thalamocortical input, whereas the larger parvalbumin-positive cells project mainly to middle cortical layers, primarily to layer 4, and do so in a specific, topographically organized fashion to form a thalamocortical "core" that has local sign. The proposal sees the core projections as bringing specific sensory information to cortex and the matrix projections as serving a role in recruitment of more widespread thalamocortical ensembles to reflect changes in behavioral state, or attention, and possibly to play a role in "binding" discharge patterns of cortical cells that represent different parts of a perceptual whole (see Singer and Gray, 1995).

At present it would seem that there is a functionally significant difference between the small cells and the large cells, which is in part expressed by the calcium-binding proteins in primates and in part expressed by the more superficial distribution of the cortical terminals of the small cells. However, if one looks at the geniculocortical pathways as an exemplar of the proposed distinction of thalamorcortical pathways, one finds that the small koniocellular cells in the monkey, which are calbindin-positive and thus should be part of the thalamocortical matrix, are not noticeably more diffuse than parvocellular and magnocellular cells. Their receptive fields, although somewhat larger, are nonetheless well focused and constrained to a small part of visual field, and the same is true for the cat if one compares the small W cells with the larger X or Y cells (Irvin et al., 1986, 1993; Norton et al., 1988; Sur and Sherman, 1982; Wilson et al., 1976). Further, as we saw earlier, although there is good evidence that the small W cells send axons to more than one cortical area (Kawano, 1998), the same has been shown for the larger Y cells (Geisert, 1980; Kennedy and Bullier, 1985). The small geniculocortical relay cells may well have some important functional characteristics in common, but they do not fit the description of a "diffusely organized" system, either in terms of the local sign of their cortical terminals or in terms of the number of different cortical areas that they innervate. Nor do the small cells all limit their terminals to layer 1, to form the "nonspecific" component of Lorente de Nó. A recent study of individual koniocellular axon arbors in cortex shows that most innervate only cortical layer 3 (a "middle" cortical layer), and most of those that innervate layer 1 do so by means of a branching axon that also innervates layer 3 (Ding and Casagrande, 1998). Thus these koniocellular cells are, generally, neither diffuse nor targeted unerringly or dominantly to layer 1.

In summary, there are a number of differences among relay cells suggestive of distinct functional roles. The position of the cells in a particular nucleus, subdivision, or layer of a nucleus and in receipt of a particular set of afferents is perhaps the simplest and the most obvious distinction, although we have seen that functionally distinct cells can share a lamina and be in receipt of functionally distinct inputs. The size of the cell and the structure and the branching pattern of the dendrites provide distinctions that may apply widely through the thalamus, but the precise functional significance of these features still remains to be defined. The distribution of the thalamocortical axons to a single functional cortical area or to more than one cortical area provides another distinction, as do the laminar distributions of the thalamocortical axons. The Ca^{2+}-binding proteins provide another variable that is likely to have functional significance. Perhaps the most important unknown about a classification of thalamic cells is whether the variable under consideration relates to the nature of the message that is being transmitted or to the nature of the modulatory influences that the message is exposed to in the thalamus.

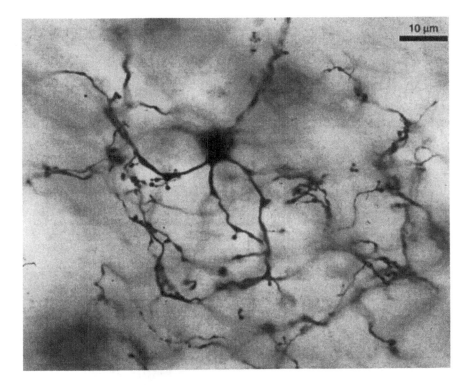

Figure 10
Interneuron from the ventral posterior nucleus of a cat. Golgi method. Provided by H. J. Ralston III.

B. Interneurons

Thalamic interneurons were described by Cajal (1911) on the basis of the Golgi method, which shows the local axons, but it was not until modifications of the Golgi methods based on aldehyde fixation, and suitable for adult tissues, were widely used that the complexity of these cells was recognized (Tömböl, 1967, 1969; Ralston, 1971; and see Figure 10). Intracellular fills (Friedlander *et al.*, 1981; Sherman and Friedlander, 1988) subsequently showed that even the best Golgi pictures gave but a limited impression of the full richness that the dendritic arbors of some of these cells could attain (Figure 11).

1. Interneuronal Cell Bodies and Dendrites

The interneurons have cell bodies that are among the smallest in the thalamus, recognizable by their size, by the fact that they cannot be retrogradely filled by tracers injected into the cortex, and because they are all GABAergic, and thus immunoreactive to GABA and GAD. The proportion of cells in any one thalamic nucleus that are interneurons varies considerably, depending on the species and on the nucleus (see Arcelli

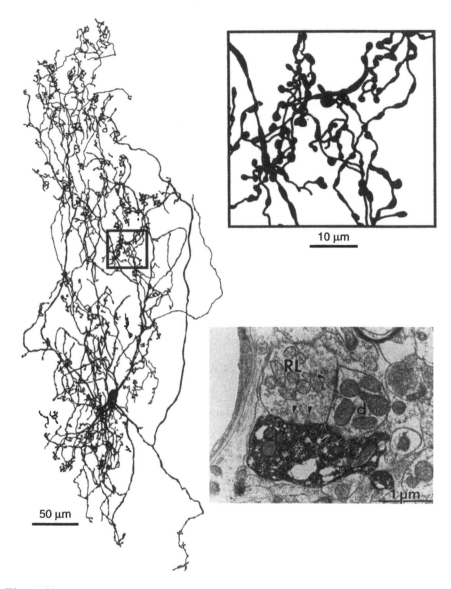

Figure 11
Interneuron from lateral geniculate nucleus of a cat. Intracellular injection with horseradish peroxidase. Redrawn from Hamos *et al.* (1985). The upper inset shows an enlarged view of axoniform dendritic terminals. The electron micrograph shows a triad in one section, with the labeled F2 terminal from the interneuron indicated by an asterisk. Arrowheads point to the three synapses of the triad: from the F2 terminal to a dendritic appendage (app) of an X relay cell, from the retinal terminal (RLP) to the F2 terminal, and from the retinal terminal to the dendritic appendage. Data from Hamos *et al.* (1985).

et al., 1997; Spreafico, 1997). In the lateral geniculate nucleus of the cat the interneurons represent about 20–25% of the total cell population (LeVay and Ferster, 1979; Fitzpatrick *et al.*, 1984), and a figure as high as 35% has been reported for the magnocellular layers of the monkey's lateral geniculate nucleus (Montero, 1986). Whereas in the rat, the lateral geniculate nucleus has 20–30% interneurons, all other thalamic nuclei in the rat, including the ventral posterior nucleus, have less than 1%. Variation between these extremes, of 20–30% for carnivores and primates and less than 1% for rats, has been reported for other species (Arcelli et al., 1997). Although 1% looks as though it represents a population that could easily be ignored, and sometimes is, two points are relevant. One is that a richly branched arbor of interneuronal processes can have several hundred terminals, so even with only 1% of the cells being interneurons, there is a distinct possibility that all cells in the nucleus could be within reach of interneuronal processes and each relay cell could be interacting with the processes of several interneurons. The second point is that, as we show in Chapter III, it has been possible to identify two distinct, parallel, independent pathways through the cat's lateral geniculate nucleus to cortex, one (the X cell pathway) having extensive connections with interneurons and the other (the Y cell pathway) lacking these connections. The relative numbers of interneurons in any one relay nucleus may reflect the extent to which interneuronal activity is relevant to the relay as a whole, or it may reflect the extent to which one or another of two or more parallel pathways through a nucleus depends on interneuronal connections. For most thalamic nuclei we do not have information about either the proportion of nerve cells that are interneurons or the extent to which those cells do or do not have very rich terminal arbors.

Figures 10 and 11 show that the terminal portions of the dendrites of interneurons differ strikingly from those of the relay cells. Whereas the dendrites of relay cells, like most dendrites in the vertebrate central nervous system generally, become thinner toward their ends and have a simple branching pattern (Figures 5 and 6), the terminal parts of the interneuronal dendrites are more complex, have a relatively rich terminal branching pattern, and form enlarged en passage and terminal swellings of the sort more commonly seen on axons in most other parts of the brain. These dendrites have been described as "axoniform" for that reason. The term is further justified on the basis of electron microscopic studies showing that the terminals contain synaptic vesicles and make specialized synaptic contact on other cells just as one expects a classical axon to do (Ralston, 1971; Famiglietti and Peters, 1972; and see Figure 11 and Chapter III). There is also recent pharmacological evidence to indicate that these dendritic terminals in fact do act to produce inhibitory postsynaptic potentials in relay cells (Cox et al., 1998). That is, although they are reasonably regarded as dendrites and serve as *postsynaptic* sites for the axons of incoming primary and other axons, many of their terminal processes and some of their en passage swellings also have the structure and function of axons and are *presynaptic*.

2. On Distinguishing Interneuronal Axons and Dendrites

At this point it is worth digressing briefly to consider the identification of presynaptic "axoniform" dendrites, a problem that is largely semantic and historic, but that still often leads to puzzlement about processes such as these, which seem to be both dendrite and axon. The problem arises primarily because Cajal proposed two important conceptual tools for analyzing the nervous system. One was the neuron doctrine and the other was the "law of dynamic polarization" (or the "law of functional polarity" in the 1995 translation of the 1911 book). Each, the doctrine and the law, provided tremendously powerful tools for analyzing the nervous system, and they served well for half a century or more; but we now know that neither the doctrine nor the law is entirely correct in its original formulation, or indeed in any formulation that fails to recognize that each is merely a rough approximation that can serve as a guide for studying the brain. The names Cajal chose, the *doctrine* and the *law*, indicate the extent to which he was using these concepts as propaganda. The fact that there has been confusion for most of the 20th century reflects his success and his authority. In the 1995 English translation, the law is introduced and stated as follows:

"Does this mean that incoming and outgoing impulses . . . pass indiscriminately from cell body to cell body, from one dendrite to another, from one axon to another, or from one of these three elements to any other? Or conversely, is there some well established and immutable rule that determines what parts of one neuron may contact another cell? Everything we know about the function of the nervous system points to the latter supposition as true, and indicates that there is in fact such a law. Our own observations have led us to define this law as follows: *A functional synapse or useful and effective contact between two neurons can only be formed between the collateral or terminal axonal ramifications of one neuron and the dendrites or cell body of another neuron.*"*

It is worth noting the "only" in the last, stressed phrase.

Although this law is a brilliant recipe for starting to analyze neuronal circuits in the vertebrate brain, it is also a quite extraordinarily limited and dogmatic statement. Cajal knew that most invertebrate neurons are unipolar, with processes that do not allow a ready distinction between axons and dendrites, and he clearly recognized that in vertebrates the granule cells of the olfactory bulb and the retinal amacrine cells lack axons, so they were, according to the law, out of the business of influencing other neurons. The dogmatism of the law led to some interesting later discussions about various neuron types that did not fit readily into the law, such as the dorsal root ganglion or the variety of neurons found in invertebrate brains (e.g., Bullock, 1959; Bodian, 1962), but there has never been any clear resolution of how to fit the law into the variety of known neuronal structures, and there is not likely to be. The variety of neurons is too great. Textbooks are generally less dogmatic about the functional distinction between axons and dendrites than they used to be, but this distinction is not one that should be lightly discarded. For many nerve cells the distinction between axons and dendrites is useful. The former is

---

* From Cajal (1995), "Histology of the Nervous System of Man and Vertebrates," with permission from Oxford University Press.

B. Interneurons

a single slender process that almost invariably lacks ribosome rosettes, conducts action potentials, is quite commonly myelinated, and terminates in presynaptic processes. The latter are generally represented by several processes for each cell, these are gently tapered, contain ribosomes, are commonly postsynaptic, and generally conduct decrementally, but as we discuss in Chapter V can perhaps also conduct spikes. More recent observations suggest that the orientation of the microtubules and the nature of the microtubule-associated proteins and of several other groups of proteins distributed differentially on the neuronal membrane add other important features for distinguishing axons from dendrites (Black and Baas, 1989; Baas, 1999; Winckler and Melman, 1999). However, there is no general law that relates axons or dendrites to their synaptic relationships. Axons can be postsynaptic and dendrites can be presynaptic, and for any neural process, whatever we decide to call it, we need to know whether it contains and can release transmitter, how the release is controlled (e.g., voltage and Ca^{2+} entry or other mechanisms), whether it has receptors that can be influenced by transmitters released near by, and what are its membrane properties that may or may not allow spike propagation. These are issues discussed in later chapters.

3. THE AXONS OF THE INTERNEURONS

Whereas most thalamic interneurons have several characteristic axoniform dendrites, they have only a single axon that is recognizable on the basis of an axon hillock and a slim initial segment and, at least in some instances, myelin, and this axon in turn leads into a long branched process that does not taper significantly and that generally branches in the neighborhood of the cell body. On the basis of the axonal ramification, Tömböl (1969) has described two sorts of interneuron, one with a locally ramifying axon and one with an axon that goes beyond the dendritic arbor but generally stays in the same nucleus or nuclear group. In the lateral geniculate nucleus this type of axon may go into an adjacent lamina, in the medial geniculate nucleus, Tömböl describes that it can go from one nuclear subdivision to another. She described such cells in the ventral posterior nucleus and the mediodorsal nucleus as well, but comparable descriptions have not appeared subsequently, apart from the interneurons described by Winer and Morest (1983 and 1984) in the medial geniculate nucleus of the cat, where they distinguished large and small interneurons and suggested that some of the interneuronal axons passed from one subdivision of the medial geniculate nucleus to another.

In Golgi preparations the distinction between the single axon and the axoniform dendrites is not always obvious, and it is not uncommon for a Golgi preparation to leave the axons unimpregnated or only partially impregnated. The possibility has been raised that interneurons may lack axons (Lieberman, 1973; Wilson, 1986), but in view of the difficulties of demonstrating the axons, this negative result needs to be interpreted cautiously. Electron microscopic studies may show the absence of a characteristic axonal initial segment arising from the cell body, but because axons can arise from dendrites, such evidence, which is based on a limited sample, leaves considerable room for doubt. Interneurons demonstrated by intracellular fills have shown axons, but this may be a highly selected sample because in these experiments the neurons are usually identified on the basis of their action potentials, and, as shown in Chapter V, the dendrites of the interneurons, acting locally, can function without action potentials. That is, axonless interneurons may not have action potentials, and if so, then the method of filling interneurons by intracellular injections that has been used in the past would favor neurons that have axons. Therefore, with the currently used methods of labeling by intracellular injection, one might expect to find no axonless interneurons, even if they exist. So, the question whether all thalamic interneurons have axons must be left open for the present. Given the available techniques, observations of interneurons that lack axons cannot be regarded as decisive evidence that such interneurons exist, and the demonstration that some interneurons have axons cannot be generalized to all interneurons.

The axonal terminals and the presynaptic dendrites of an interneuron are both GABAergic so that they may be thought of as having the same actions. However, the postsynaptic contacts made are different, and other differences in ultrastructure are noted in later chapters. Whereas the dendrites have terminals that are also postsynaptic, the axon terminals appear not to receive synaptic contacts from any other processes (Montero, 1987). Also, in the lateral geniculate nucleus the dendritic terminals innervate glomeruli mostly associated with relay X cells, whereas the axon terminals contact extraglomerular process (Wilson *et al.*, 1984; Hamos *et al.*, 1985).

4. Classifications of Interneurons

We have seen that a classification has been proposed, for interneurons with an identifiable axon, based on the territory occupied by the axon relative to the dendrites. This feature stresses the possibility that the axon and the dendrites may be concerned with different aspects of the thalamic circuits. Other differences between interneurons that have been described may well depend on the extent to which individual cell processes have been revealed by the particular method used, so that the significant range of dendritic arbors that has been described for interneurons may not provide a categorization of interneurons that can be readily interpreted in functional terms. Montero and Zempel (1985) distinguished interneurons on the basis of cell body size (see also Winer and Morest, 1983), but on this criterion alone it is not clear whether one is dealing with a continuum of cell sizes rather than two distinct cells classes. Recently Bickford *et al.*, (1999) have described two types of interneuron in the visual thalamus, both reacting positively to GAD immunostaining, but one reacting positively and the other negatively to nitric oxide synthase. The former are somewhat larger than the latter, have a simpler dendritic form, resembling a cell type described earlier by Updyke (1979) as "class V" cells. These appear to be involved primarily in nonglomerular synapses, whereas the latter are seen more commonly in relation to glomeruli.

B. Interneurons

Montero (1989) and Sanchez-Vives et al. (1996) described a somewhat special class of interneuron in the interlaminar regions of the lateral geniculate nucleus of the cat. These resemble the cells of the perigeniculate nucleus (described later) rather than the other geniculate interneurons. The perigeniculate nucleus is a specialized portion of the cat's thalamic reticular nucleus that relates particularly to the A laminae of the lateral geniculate nucleus. These interlaminar interneurons are larger than other interneurons, show a somewhat distinctive orientation of their dendrites, and have some of the physiological properties of perigeniculate or reticular cells, discussed in Chapters IV and V.

The possibility that these cells may be displaced perigeniculate cells has a clear functional implication. The interneurons of the lateral geniculate nucleus, and possibly of the thalamus in general, differ from the perigeniculate and reticular neurons because the former receive synaptic contacts from primary, driving afferents (e.g., retinal axons for the lateral geniculate cells) but the latter do not (see Chapter III). Further, the reticular neurons send their axons to regions some distance from the cell body. In this they resemble the second of Tömböl's interneuronal classes, those having axons that extend beyond their dendritic arbors. These may, therefore, have included such interlaminar interneurons. Montero (1989) found no retinal input to the interlaminar interneurons, but he was not looking at labeled retinal axons, and his survey was limited to the cell body and the proximal parts of the dendrites. Since a significant part of the retinal input to geniculate interneurons contacts the distal, axoniform dendrites, the relationship of the interlaminar interneurons to the retinal input must remain an open question. Evidence obtained from intracellular filling of relay cells and their axons in slice preparations suggests that these interlaminar interneurons do receive afferents from collateral branches of the relay cell axons, which also branch within the perigeniculate nucleus (Sanchez-Vives et al., 1996). Other studies of geniculate relay cells filled with horseradish peroxidase have shown that at least some form collaterals within the nucleus (Friedlander et al., 1981; Stanford et al., 1983). It appears that local collaterals of relay cells are present at least in some parts of the thalamus. They arise from the axon in its course through its nucleus of origin, before it enters the reticular nucleus, and they may thus provide an innervation for interneurons.

The implication, that these interneurons actually represent migrated perigeniculate cells, would need developmental confirmation, which is currently not available. Such confirmation would be feasible if there were a specific marker of perigeniculate cells that is present at an early enough stage and the result could prove of interest. Given the extent to which the segregation of functionally distinct cells in the lateral geniculate nucleus varies, so that the X cells and Y cells and on-center and off-center cells can be either intermingled or not (see Section A.5) without apparently affecting their functional role, such displaced reticular cells may well be a further sign that the precise locus of a cell in the thalamus is not an important determinant of its functional connectivity. The developmental forces that might produce a migration of cells from ventral to dorsal thalamus are undefined. Possibly these cells should be included in the next section on the reticular nucleus rather than in this one on interneurons, but at present that remains an open question.

C. The Cells of the Thalamic Reticular Nucleus

The thalamic reticular nucleus is made up of nerve cells that lie in a complex meshwork of intertwining thalamocortical and corticothalamic axons. The nucleus forms a slender shield around the dorsal and lateral aspects of the dorsal thalamus, and it is placed so that any axon passing between cortex and thalamus must go through the nucleus. Many of these traversing fibers, possibly all, innervate the reticular cells, with glutamatergic afferents that are generally excitatory. The reticular cells tend to have relatively large cell bodies and discoid dendritic arbors that lie in the plane of the nucleus (Scheibel and Scheibel, 1966; Lübke, 1993; Figure 12). Like the thalamic interneurons, they are all GABAergic and provide an inhibitory innervation to the relay cells of the thalamus (Jones, 1985; Ohara and Lieberman, 1985; Sanchez-Vives et al., 1997). There are some indications that the cells of the thalamic reticular nucleus are not a homogeneous group. The cells in the anterior part of the nucleus do not have the characteristic discoid shape seen in the main part of the nucleus, and there have been accounts that distinguish large from small cells on the basis of Golgi preparations (Spreafico et al., 1991).

Reticular cells and interneurons both send inhibitory axons back to thalamic relay cells. However, they have distinct developmental origins and lie in different parts of the diencephalon, the reticular cells in the ventral thalamus and the interneurons in the dorsal thalamus (see Chapter I). Further, there are other clear differences between the two cell groups. One is that the reticular cells do not show the striking axoniform terminal arbors on their dendrites such as those seen on many thalamic interneurons (Scheibel and Scheibel, 1996; Ohara and Lieberman, 1985; Lübke, 1993; see however below). Another difference, mentioned previously and treated more fully in the next chapter on afferents, concerns differences in the connectivity patterns of the two cell groups.

The axons of reticular cells terminate in the thalamus, where they generally form well-localized terminal arbors (Pinault et al., 1995a,b). For the major sensory modalities, the corticoreticular and reticulothalamic pathways show matching topographical maps, as do the pathways from the thalamus to the reticular nucleus. That is, limited parts of any sensory surface produce activity within limited parts of the reticular nucleus because this receives afferents from thalamic and cortical regions innervated by axons representing the same parts of the sensory surface. The reticular cells in turn send their inhibitory axons back to the same parts of the thalamus.

The main part of the thalamic reticular nucleus can be divided into "sectors" on the basis of its afferent connections with groups of thalamic nuclei and groups of cytoarchitectonically definable cortical areas (Jones, 1985; Guillery et al., 1998; and see Chapter VII). Thus one sector is related

C. The Cells of the Thalamic Reticular Nucleus

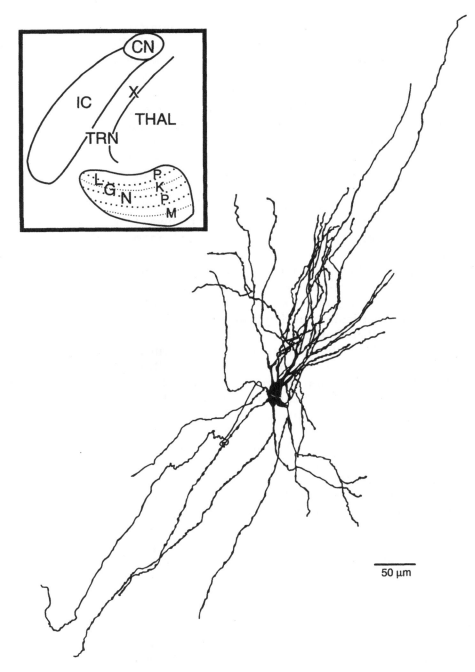

Figure 12
Intracellularly filled neuron from the thalamic reticular nucleus of Galago. The cell was filled with neurobiotin and illustrates the pattern of orientation of the dendrites in the plane of the thalamic reticular nucleus. The inset shows the position of the cell from a coronal section through the lateral geniculate nucleus (LGN) and internal capsule (IC). CN, caudate nucleus; TRN, thalamic reticular nucleus; K, M, and P indicate the layers of the lateral geniculate nucleus; X marks the position of the cell. Drawing provided by P. Smith, K. Manning, and D. Uhlrich.

to the visual pathways, another related to the auditory pathways, and there is one for the somatosensory and one for the motor pathways. However, there are no identifiable boundaries or architectonic distinctions between sectors in preparations that do not reveal the connections. Nor does it seem that the dendrites of cell bodies lying in any one sector respect the borders of the sectors. That is, the sectors of the reticular nucleus are not a morphologically distinct entity like the nuclei of the dorsal thalamus. Any one reticular sector, although generally limited in its connection to one modality or presumed functional group of pathways, can relate to more than one thalamic nucleus and, correspondingly, to more than one cortical area that is concerned with the same modality. In these connections and in the lack of clearly definable borders to the sectors there is an indication that the individual reticular cells may be less closely linked to any one functionally defined afferent pathway than are the cells of the dorsal thalamus. These topics are discussed in more detail in Chapter VII.

The extent to which local interconnectivity patterns in the reticular nucleus might play a part in the control of rhythmic discharge patterns in the thalamocortical pathways has been important for theories of reticular function, particularly as they may relate to the synchronization of reticular activity that characterizes certain forms of sleep and epilepsy (Steriade *et al.*, 1990). Although the complex axoniform dendritic appendages seen on thalamic interneurons have not generally been reported on reticular cells, some such appendages were shown light and electron microscopically in the rat by Pinault *et al.* (1997), and in addition there is evidence that reticular cells have axons that can give off local branches within the nucleus itself in cat and rat (Yen *et al.*, 1985b; Spreafico *et al.* 1988; Liu *et al.*, 1995). Electron microscopic evidence for some local circuitry established by serial synaptic junctions in the nucleus is considered in Chapter III, Section F.

When we look at the inhibitory afferents that thalamic cells receive from nearby neurons, we see that the inhibitory circuitry going to a thalamic relay neuron can come from interneuronal dendrites or axons or from thalamic reticular axons. Although the precise pattern of the inhibition in a nucleus may depend on the relative number of interneurons in that nucleus, we have seen that in a nucleus like the cat's lateral geniculate nucleus, which is rich in interneurons, there may be some cells like the X cells that have extensive connections with interneuronal presynaptic dendrites, whereas other, adjacent cells like the Y cells lack such rich connections. The extent to which connections with interneuronal axons may play a comparable role is not known, although the connectivity patterns considered in the next chapter suggest that the interneuronal axons and dendrites have distinct functions. Further, it is not known whether the axons of reticular neurons may provide a functional replacement for interneuronal axons where interneurons are lacking. From this point of view the possibility that some geniculate interneurons should be seen as displaced reticular cells could either prove very interesting if the functions are clearly distinct or rather dull if they are not.

D. Summary

We have shown that there are morphological characteristics on the basis of which distinct types of thalamic relay neuron can be recognized, and that the same may prove to be true for interneurons and also for the nerve cells of the reticular nucleus. Although relay neurons differ in several features, including perikaryal size, axon diameter, pattern of dendritic arbors, thalamic position in a particular nucleus, nuclear subdivision or lamina, and site of cortical termination in a particular cortical area or lamina, it is difficult to interpret these features in functional terms. Perhaps the key issue is to define for each variable whether it relates to the nature of the message that is being passed through thalamus to cortex or to the nature of the thalamic control (modulation, gating) to which the message is exposed in its passage through the thalamus. Some of the morphological differences that have been described in this chapter, such as the difference between interneuronal dendrites and relay cell dendrites, clearly have direct and important functional implications, although it has to be recognized that exactly what it is that the presynaptic interneuronal dendrites are doing is still rather obscure. Other differences, such as those between bushy and radiate dendrites or between axonal terminals in any one particular cortical layer remain to be explored. The most important issues that have yet to be defined include the extent to which particular characteristics of thalamic cells that have been considered in this chapter are generalizable across all (or most) thalamic nuclei and a resolution of exactly how each characteristic, whether generalizable or local, relates to the functional properties of the thalamic relay that are discussed in subsequent chapters.

E. Some Unresolved Questions

1. Are there structural features of thalamic relay cells that will lead to a functionally useful classification applicable generally across all of the thalamus?

2. How many functionally distinct types of relay cell can be identified in the major relay nuclei of the thalamus, and does the distribution of these cell types vary significantly across thalamic nuclei?

3. If the classes of relay cell reflect different functional gating properties (as opposed to integrative properties), how many distinct gating functions or classes can be defined in the mammalian thalamus?

4. What, if any, is the functional significance of regional segregations of cell classes across thalamic nuclei? Specifically, are there basic ground rules that govern the segregation of distinct functional cell types within the layers of the lateral geniculate nucleus?

5. Is there a basic difference in the organization of geniculate and cortical laminae, such that circuits having distinct functions must occupy distinct laminae and sublaminae in the cortex, whereas they can mingle within a single lamina in the thalamus?

6. Where a thalamic nucleus sends axons to more than one cortical area, are there functionally significant ground rules that govern whether the projection is established by branching axons or by two distinct populations of thalamic cells?

7. Are there several different types of thalamic interneuron and do all interneurons have axons?

8. What is the functional significance of the different calcium-binding proteins found in thalamic relay cells?

CHAPTER **III**

The Afferent Axons to the Thalamus

The thalamus sends most of its outputs to the cerebral cortex, and the messages that are received by the cortex from the thalamus form the major focus of this book. Therefore, we have to look closely at the several different sources of afferents that contribute to the messages that the thalamic relay cells pass on to cortex. Many of these afferents come from sources outside the diencephalon, and we will consider several of them individually. However, two important local afferents come from the thalamic interneurons and from the cells of the thalamic reticular nucleus. Since one or both of these afferents innervate all thalamic relay cells, depending on the nucleus and on the species, we will need to discuss not only how these more local afferents relate to the relay cells but also how each of these cell groups itself is innervated. The interneurons are scattered among the relay cells, so that all afferents going to most of the thalamic nuclei have an opportunity to innervate both relay cells and interneurons; most afferents do not limit their synaptic contacts to either relay cells or interneurons. The thalamic reticular nucleus itself shares many of its afferents with the nuclei of the dorsal thalamus. We will stress the distinction between afferents that are shared between the reticular nucleus and the nuclei of the dorsal thalamus, on the one hand, and those that go to the dorsal thalamus but do not innervate the reticular nucleus, on the other hand. However, we will have a separate section on the afferents to the reticular nucleus.

In this chapter we describe ways in which one can distinguish afferents on the basis of their light microscopic appearance, their fine structural, electron microscopic appearance, and their extrathalamic origins. We show how these features may relate to the functions of the afferents as drivers or modulators, and we look at the patterns of synaptic interconnections established in the thalamus by each of the major afferent groups.

A. A Functional Classification of Afferents to the Thalamic Nuclei

We have pointed out that we would be looking for features that can be generalized across thalamic nuclei and that we would be using the visual pathways as our prime example of thalamic organization in order to make comparisons with other relays where possible. When we look at the afferent axons that go to innervate thalamic cells, the visual pathways show a crucial distinction between two types of afferent, which is also identifiable in other first order nuclei receiving sensory afferents. This is the distinction between "drivers" and "modulators" that we introduced in Chapter I and discuss in detail in Chapter IX. In this and subsequent chapters we argue that this distinction will prove critical to a functional analysis of all thalamic relays. Essentially, the drivers carry the message that is passed on to cortex, whereas the modulators can affect the way in which this message is transmitted but seem to have little or no effect on the contents of the message itself. In the visual relay in the lateral geniculate nucleus we can recognize the retinal afferents as the drivers because they carry the crucial visual information that is relayed to the cortex, and there are good reasons for thinking that all other afferents going to the lateral geniculate nucleus, with the possible, but currently unproven, exception of the tectal afferents, are all modulators. There is equally strong evidence for thinking of the axons that come from the medial lemniscus and the spinothalamic tract as the drivers for the ventral posterior nucleus and of those that come from the inferior colliculus as the drivers for the ventral part of the medial geniculate nucleus. For most of the other nuclei of the thalamus there is much less direct experimental evidence to help identify drivers and modulators, but we will treat the afferents that come from the cerebellum and from the mamillary bodies to the thalamus as being essentially like the lemniscal afferents. Our view of afferents will lead us to two important proposals: One is that we can use the morphological characteristics, synaptic relationships, and functional properties of the drivers in some well-studied thalamic nuclei to identify the drivers in nuclei about which we currently have very much less information, and the second is that not all afferents are equal.

The first proposal will form a major part of this and some of the later chapters. It leads to the conclusion that by identifying the drivers for any one thalamic nucleus, the origin of the main functionally significant input sent by that nucleus to its cortical area can be defined, and in this way useful light can be thrown on the function of those thalamic relays and cortical areas that are currently not well understood. Each thalamic nucleus may be capable of a variety of functional modes, but its essential function in awake behaving animals is to pass information received over its driving afferents on to its own cortical area, and we regard this as a crucial feature of thalamic organization. Further, the functional capacities of the receiving cortical area depend critically on the nature of the messages that are carried to the thalamus by the driving afferents. We think of the visual cortex as visual because of the retinal messages that reach it

A. A Functional Classification of Afferents to the Thalamic Nuclei

by way of the lateral geniculate nucleus, and correspondingly, auditory and somatosensory cortex acquire their major functional characteristics through the driving auditory and somatosensory afferents that end in the thalamus. There is no known neocortical area that lacks a thalamic input, and we suggest that for any one cortical area, no matter how high it may be in the hierarchy of corticocortical pathways often thought to be responsible for the processing of sensory information (van Essen et al., 1992), the functional characteristics of the driving afferents that supply its thalamic relays will provide a powerful clue regarding the functional significance of that cortical area.

One can ask: What would be the consequence if, for example, visual pathways were directed to auditory cortex? The issue is not as purely theoretical as might appear at first. Sur et al. (1988; and see Pallas et al., 1994; Angelucci et al., 1997), working with young ferrets, have redirected retinofugal axons into auditory thalamus after removing the visual cortex and superior colliculus. The auditory thalamus then receives messages from the retina and sends them to the auditory cortex, where cells in the experimental adults respond to visual instead of auditory stimuli. Recent experiments (Melchner et al., 2000) in which such misrouting was unilateral show that behavioral responses are appropriate for the stimulus (visual or auditory), not for the cortical area receiving the message.

Although the point about the key and functionally dominant nature of the driving afferents that go to a thalamic nucleus seems simple and obvious when one is looking at thalamocortical pathways whose major role is well understood, like the visual or auditory pathways, it is worth stressing here because the point is far less obvious when one is looking at pathways that go through thalamic nuclei whose function is not understood. We shall argue that it is critical to define the driving afferents for any nucleus before one can hope to understand what it is that some of the more mysterious thalamic nuclei may be passing to their cortical areas. That is, we can expect to learn more about nuclei like the medial dorsal, lateral dorsal, pulvinar, or intralaminar nuclei (see Chapter I) if we can define the origin and the functional nature of their *driving* afferents. To do this, we need to look in detail at the structural and functional characteristics of *known* driving afferents in other nuclei so that we can make comparisons between nuclei and recognize the same functional class of afferents in each. For this reason we consider the characteristics of the driving afferents in considerable detail in this and later chapters. There is one caveat to be noted, which is discussed more fully in Chapter II, Section B.2.a, and in Chapter VIII. This is that, although we discuss afferents in terms of first and higher order *nuclei* it is probable that many thalamic nuclei will prove to contain a mixture of first and higher order *circuits*; that is, there may be some cells or some regions of a nucleus that receive cortical drivers, and others that receive subcortical drivers. Considering the connections in terms of two distinct types of nucleus is undoubtedly an oversimplification, but it provides a useful introductory distinction.

The second proposal, that not all afferents are equal, naturally derives from the first and warns that classifications of thalamic nuclei, based on the origin or the total number of any one group of afferents, are not likely to prove instructive unless the functional nature of the afferents has been taken into consideration (see e.g., Macchi, 1983). It has been a general practice to classify thalamic nuclei in terms of the ascending afferents that they receive from lower centers and to regard the inputs from cortex as categorically different, not contributing to this classification. We shall treat afferents in terms of their morphological relationships and of the effects that they are likely to have on relay cells, no matter what their origin, and in the following account we will recognize ascending drivers as well as drivers that come from the cerebral cortex. Similarly, modulators will be seen to have cortical as well as subcortical origins.

B. Afferent Axon Types as Seen Light Microscopically

1. Light Microscopy of Ascending Afferents and of Cortical Afferents in First Order Nuclei

A useful clue for distinguishing drivers from modulators is likely to be whether they send a branch to innervate the thalamic reticular nucleus. On the basis of currently available evidence, the ascending driver afferents to first order thalamic nuclei innervate relay cells and interneurons but do not innervate the cells of the thalamic reticular nucleus. The afferents from the cerebral cortex that go to first and higher order nuclei fall into two distinct groups: one has branches going to the reticular nucleus and the other does not. Both innervate relay cells and interneurons. First order relays receive axons from cortex that also send a branch to the reticular nucleus. We regard these as modulators. Higher order relays receive afferents from both groups, those that send branches to the reticular nucleus and those that do not. That is, the distinction between the corticothalamic axons that do and those that do not innervate the reticular nucleus is likely to represent a fundamental distinction between corticothalamic drivers, which by definition are innervating only higher order relays, and modulators, which innervate all thalamic nuclei. The importance for modulators of the reticular innervation is discussed further in Chapters VIII and IX.

The earliest morphological distinction between afferent axons recognized in *first order* thalamic nuclei like the lateral and medial geniculate nuclei and the ventral posterior nucleus was based on Golgi preparations from adult animals. These demonstrated that the ascending driving afferents and descending corticothalamic axons had very different terminal ramifications (Szentágothai, 1963; Guillery, 1966; Jones, 1985). The former, coming from the retina, the inferior colliculus, or the medial lemniscus, are sometimes generically referred as "lemniscal" afferents because they resemble each other closely, and the latter also share a common basic structure in each of these nuclei.

Figures 1 to 4 show the light microscopic appearance of these two groups of afferents in first order thalamic nuclei as they appear in Golgi preparations or after individual axons have been filled by intracellular markers such as phaseolus lectin, horseradish peroxidase, or biocytin (see

B. Afferent Axon Types as Seen Light Microscopically

Bolam, 1992, for an account of some of these methods). Within the first order nuclei the distinction between the lemniscal afferents (Figures 1 to 3) and the cortical afferents (Figure 4) is strikingly clear and consistent, and the difference does not depend on the method (Robson, 1983; Murphy and Sillito, 1996). However, it is not yet clear exactly how this marked morphological difference relates to the functional organization of these afferents.

To illustrate what one might expect to gain from understanding the structural characteristics of an afferent axon, it is useful to look at other well-characterized axonal terminal patterns, such as those in the cerebellum, where the morphology of the terminal arbor can be clearly related to the way in which a particular class of axon establishes its synaptic relationships (Eccles *et al.*, 1967). The synaptic portions of basket cell axons are shaped to match the shape of the relevant receptive surface of the Purkinje cells, the climbing fibers match the branching pattern of their postsynaptic dendrites, and the parallel fibers show a distribution of synaptic swellings that relates closely to the contacts established as they cross the dendritic trees of the Purkinje cells. Unfortunately, although the illustrated differences between the two axon types in the thalamus are seen wherever they have been studied, in many different nuclei and in several different mammalian species, and although we know something about the different synaptic relationships established by each class, these provide only elusive clues to relate the morphological characteristics to the connectional patterns of these axons. In spite of this somewhat surprising ignorance, it is reasonable to assume that when one sees axons resembling those in Figures 1 to 3 there is a shared pattern of connectivity, which differs from that of axons like those of Figure 4, no matter where in the thalamus we are looking; to a limited extent fine structural studies of synaptic connections and functional studies of the pathways confirm this.

It is important to note that all of the corticothalamic afferents, both driving and modulating, are regarded as glutamatergic and that most or all of the known driver afferents to first order thalamic nuclei are also

---

Figures 1–6 on the following pages show two distinct types of afferent axons to first order (Figures 1–4, pp. 64–67) and higher order (Figures 5 and 6, pp. 69 and 70) nuclei. Figures 1–3 show ascending driver afferents (type II axons) and Figure 4 shows corticothalamic afferents (modulatory, type I axons) terminating in first order nuclei. Whereas the former have well-localized terminal arbors with relatively large *en passant* and terminal swellings and do not send branches to the reticular nucleus, the latter have more widespread terminal arbors, with small, scattered, terminal side branches and a few *en passant* swellings. They are often seen to send branches to the reticular nucleus (not shown in the figures). Figures 5 and 6 show the type I and type II axons that provide the same two sorts of afferent to higher order nuclei, both types here coming from the cortex, but only the former having branches to the reticular nucleus (not shown in the figures).

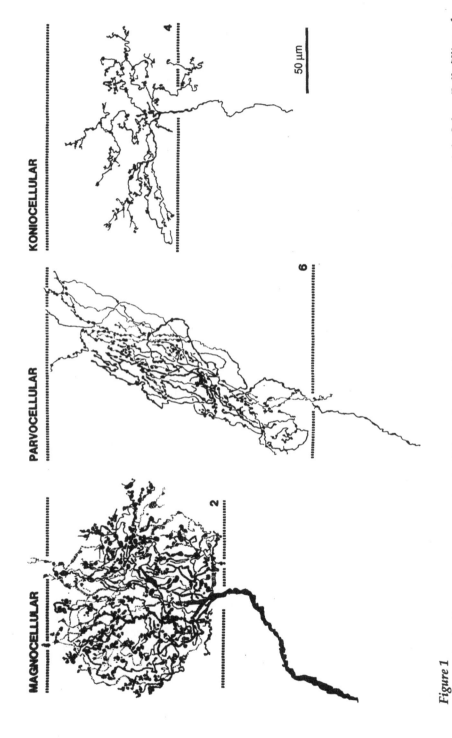

Figure 1
Three retinal afferent axons terminating in different layers of the lateral geniculate nucleus in an adult *Galago*. Bulk filling of retinogeniculate axons with horseradish peroxidase.

B. Afferent Axon Types as Seen Light Microscopically

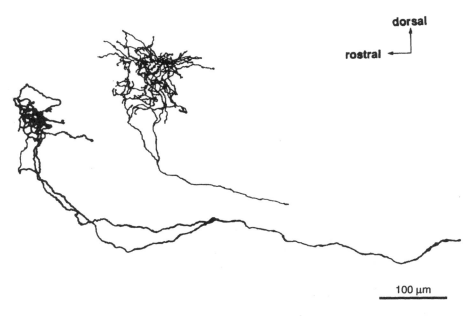

Figure 2
Two ascending driver afferents terminating in the medial geniculate nucleus of an adult ferret. Bulk filling of axons in the inferior colliculus with horseradish peroxidase.

thought to be glutamatergic (de Curtis et al., 1989; Scharfman et al., 1990; McCormick, 1992; Salt and Eaton, 1996; Vidnyánszky et al., 1996b). Other afferents, which include cholinergic, serotonergic, noradrenergic, and histaminergic afferents, are not considered to be drivers and are discussed in Section B.4. In addition there are several groups of GABAergic afferents: these include the afferents from the thalamic reticular nucleus going to each of the relay nuclei; afferents from globus pallidus and substantia nigra to the ventral anterior, ventral lateral, and center median thalamic nuclei (Penney and Young, 1981; Balercia et al., 1996; Ilinsky et al., 1997); afferents from the pallidum and basal forebrain structures to the mediodorsal nucleus (Churchill et al., 1996); afferents from the pretectal nuclei to the lateral geniculate nucleus (Cucchiaro et al., 1991, 1993) and from the inferior colliculus to the medial geniculate nucleus (Peruzzi et al., 1997). For reasons discussed in Chapter IX, most or all of these are likely not to be drivers and are considered separately.

a. Ascending Driving Afferents in First Order Nuclei

These were called type II axons on the basis of Golgi preparations from the lateral geniculate nucleus (Guillery, 1966). We have seen that they have well-localized terminal arbors and are characterized by generally quite thick, richly branched axons with large and rather heterogeneous terminal structures, some relatively simple bulbous swellings, others more complex, irregular, or "flowery" in structure when viewed at

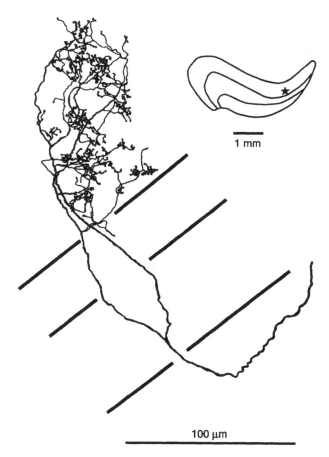

Figure 3
A retinogeniculate X axon terminating in layer A of the lateral geniculate nucleus of a cat. The axon was filled intracellularly with horseradish peroxidase and is shown in a coronal section. The star in the small inset shows the location of the terminal arbor within the lateral geniculate nucleus.

higher light microscopic magnifications. These terminals have been demonstrated by intracellular fills of single axons for the lateral geniculate nucleus (Bowling and Michael, 1984; Sur *et al.*, 1987), for the medial geniculate nucleus (Pallas and Sur, 1994), and for the ventral posterior nucleus (Jones, 1983). Figure 1 shows that these axons vary somewhat in thickness and in the details of their terminal structure, but they are always clearly recognizable. On the basis of the electron microscopic accounts that are discussed below one would expect the afferents that reach the thalamus from the cerebellum and from the mamillary bodies to resemble the terminals illustrated in Figure 1, but to our knowledge comparable light microscopic illustrations of these afferents have not been published. All of the driving afferents going to first order nuclei show

B. Afferent Axon Types as Seen Light Microscopically

local sign; that is, they are a part of a pathway that forms an orderly map of a sensory surface or a lower relay.

b. Descending Cortical Afferents in First Order Nuclei

Within the first order nuclei the corticothalamic axons are generally thinner and have long thin branches that run for considerable distances, with some *en passage* swellings and many short, stubby, terminal side branches, each of which ends in a tiny single swelling (see Figure 4). These short side branches are an absolute hallmark of these corticothalamic axons and were aptly described as "drumstick-like" side branches by Szentágothai (1963) for the cat lateral geniculate nucleus. These axons were called type I by Guillery (1966). Each such corticothalamic afferent in a first order nucleus distributes primarily to a relatively limited region

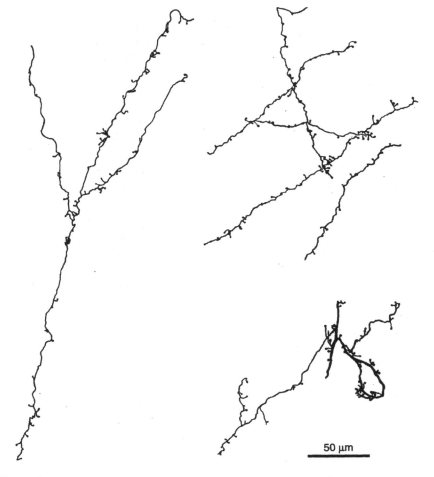

Figure 4
Corticogeniculate (type I) axons from the A layers of the lateral geniculate nucleus of a cat; Golgi method.

of the nucleus and has a few more scattered terminals that go somewhat further afield (Murphy and Sillito, 1996; Darian Smith *et al.*, 1999). In addition there is commonly a branch to the thalamic reticular nucleus. The terminals of the type I axons are generally less tightly grouped and less well localized than are the terminals of the type II axons (see Figures 1 to 3). There is some evidence (e.g., Murphy *et al.*, 2000) that some of the type I corticothalamic axons project back to the thalamic nuclei and to the subdivisions of the thalamic nuclei from which the cortical area in which they arise receives thalamocortical afferents. However, such evidence is only available for relatively few thalamic nuclei, and the central dense focus with a surrounding sparser zone that characterizes the type I projection from cortex suggests that this may not be a truly reciprocal arrangement of connections. Cortical areas that send type I axons to thalamic regions from which they receive no input are mentioned in Section G.2.

Where direct evidence is available, as for the ventral posterior and medial and lateral geniculate nuclei (e.g., Gilbert and Kelly, 1975; Ojima *et al.*, 1996; Ojima 1994; Bourassa *et al.*, 1995; Deschênes *et al.*, 1994), the type I axons have been shown to arise from pyramidal cells in layer 6 of cortex and, where their action has been tested, they have been seen to act as modulators (see below and Chapters VIII and IX). Although these axons vary somewhat in thickness, they show more homogeneity of structure than do the lemniscal terminals; axons like these are remarkably similar in appearance not only from nucleus to nucleus but also from species to species. It should be recognized that it is the shape of the terminal arbor more than the thickness of the axons that represents the crucial distinguishing feature between type I and type II axons. We have mentioned that there are some relatively fine type II axons with small terminals (for example, the retinal afferents to W cells in the lateral geniculate nucleus of the cat or to the koniocellular layers in the monkey) and, especially in the rat, there can be significant overlaps of size for the two axon and terminal types. The relative uniformity of the terminal pattern of the type I axons is their most striking feature. The major variation in their structure is that within the terminal distribution of even a single one of these axons the spacing of the characteristic side branches can vary greatly (see Figure 5), suggesting that the extent to which any one branch is focused on a small part of the terminal zone is likely to depend on some currently undefined local conditions. In our own material (R.W.G., unpublished observations), obtained after small injections into cortex, we have looked to see whether the spacing of the side branches is denser near the major focus of termination, becoming sparser in the peripheral branches, but have not been able to demonstrate such a relationship.

2. LIGHT MICROSCOPY OF CORTICAL AFFERENTS IN HIGHER ORDER NUCLEI

In contrast to the single, characteristic type of corticothalamic axon seen in first order nuclei, there are two distinct types of afferent reaching

B. *Afferent Axon Types as Seen Light Microscopically*   69

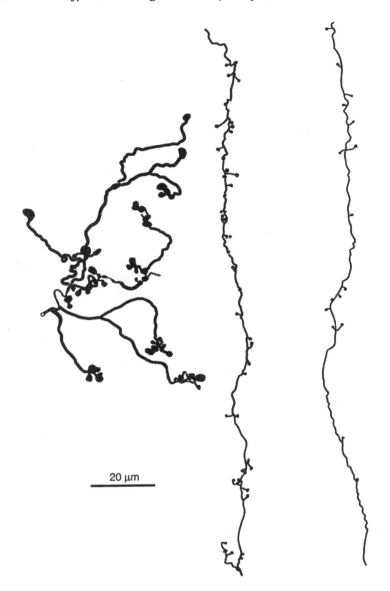

Figure 5
The terminal parts of three corticothalamic axons in the lateral posterior nucleus labeled by biotinylated dextran amine. The type II axon is shown on the left and two type I axons on the right. Note: the drawing shows the type I axons as thicker than they were in the section, in order to allow adequate reproduction of the thinner parts of the axon. From an unpublished experiment by S. L. Feig and R. W. Guillery.

higher order nuclei from the cortex (Figures 5 and 6). One looks just like the afferents that pass to first order nuclei (type I), and the second resembles the ascending, lemniscal afferents that terminate in the first order nuclei (type II). Corticothalamic axons that look essentially like those

70 CHAPTER III *The Afferent Axons to the Thalamus*

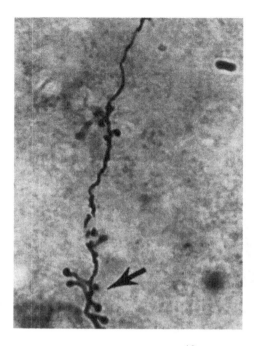

10 μm

Figure 6
A type I (E type) corticothalamic axon from the pulvinar of a macaque monkey. The axon was filled by an injection of Phaseolus lectin into cortical area TE. Previously unpublished figure provided by Dr. K. Rockland.

described earlier for the first order nuclei were described in the macaque pulvinar as E-type axons by Rockland (1996). She recognized them as equivalent to type I axons described in other nuclei and also described a second type of corticothalamic axons, which she called R-type axons. These are like the type II or lemniscal axons seen in first order thalamic nuclei. Similar axons had earlier been traced from pyramidal cells in layer 5 of the cerebral cortex to the posterior nucleus (Bourassa *et al.*, 1995), the lateral posterior nucleus (Deschênes *et al.*, 1994), and the magnocellular division of the medial geniculate nucleus of the rat (Ojima 1994). From this evidence and from experiments in which cortical cells have been retrogradely labeled after injections of horseradish peroxidase into thalamic nuclei (e.g., Gilbert and Kelly, 1975; Abramson and Chalupa, 1985) it is reasonable to conclude that the type II corticothalamic axons arise from pyramidal cells in cortical layer 5 and, further, that they can be branches of long descending corticofugal axons going to the midbrain or pons (Deschênes *et al.*, 1994). In contrast to this, we have seen that the type I axons with the characteristic drumsticklike side branches arise from pyramidal cells in cortical layer 6, and it is probable, but not yet proven, that such axons are to be found in every thalamic nucleus.

B. Afferent Axon Types as Seen Light Microscopically

It should be noted that the two types of pyramidal cell contributing to any one thalamic nucleus are not necessarily found in the same areas of cortex. For example, in the cat, area 17 contributes a layer 5 (type II) input to the lateralis posterior nucleus but essentially no layer 6 (type I) input and sends a layer 6 input but no layer 5 input to the lateral geniculate nucleus (Gilbert and Kelly, 1975; Abramson and Chalupa, 1985). Both of these nuclei, the lateral geniculate nucleus and the lateralis posterior nucleus, receive layer 6 inputs from several other visual cortical areas, some of which also send layer 5 inputs to some of the same extrageniculate regions of the thalamus (Updyke, 1977, 1981; Abramson and Chalupa, 1985).[1]

There is evidence for the visual pathways (Kalil and Chase, 1970; Schmielau and Singer, 1977; Geisert et al., 1981; Bender, 1983) and for the somatosensory pathways (Diamond et al., 1992) that receptive field properties of thalamic relay cells in first order relays, which receive cortical afferents from layer 6 only, are qualitatively unaffected by inactivation of cortex, whereas in the higher order relays that receive afferents from layer 5 as well, cortical inactivation abolishes the receptive field responses. That is, the layer 5 afferents are drivers, whereas the layer 6 afferents are modulators. We shall regard this distinction between layer 5 and layer 6 afferents as being generalizable to all of thalamus, treating this as a working hypothesis and recognizing that its validity has not been as widely checked as one might wish.

We have stressed the probable importance of the innervation of the thalamic reticular nucleus. The layer 5 afferents to the thalamus differ from the layer 6 afferents in not sending a branch to the thalamic reticular nucleus (Bourassa et al., 1995; Deschênes et al., 1994). Although we regard this as an important distinction, there have not been many critical studies, and negative evidence about branches that may be fine and hard to fill or trace must be accepted as tentative. We know of one report of a layer 5 cell that sent a short branch into the thalamic reticular nucleus (Rockland, 1996), but the course of the branch was not followed far in the relevant illustration nor was its course in the reticular nucleus described. Where known ascending driver afferents have been traced by modern methods, there appear to be no examples of branches going to the reticular nucleus. We shall treat the lack of a reticular branch as a characteristic of all driver afferents but recognize that negative evidence is weak and that, like most laws about the nervous system, this one could well be broken, possibly even before this book is published.

In general, the functional properties of the layer 5 cells that send terminals to the thalamus have not been studied specifically (see also Chapter VIII, Section B.3). Those that are branches of axons also going

---

[1] A recent observation (Feig and Guillery, 2000), that both type I and type II corticothalamic axons appear to innervate small blood vessels in the regions close to their terminal thalamic arbors, raises two interesting possibilities that will need further study. One is that these axons may play a role in controlling the blood supply to the regions that they innervate, and the other is that they may alter the permeability of vessels in those regions, producing modulatory actions quite distinct from those produced by transmitter release from the axon terminals.

more caudally to the midbrain or pons are likely to be transmitting information to thalamus that represents a copy of at least a part of the descending output of the relevant cortical area, and this will then be the message that the higher order nucleus passes on to another cortical area. For understanding a higher order corticothalamic system the response properties of the layer 5 driver afferents should prove particularly instructive, but although layer 5 pyramidal cells have been studied, especially in visual and motor pathways, details about the functional properties of the ones that send branches to the thalamus are few.

So far we have presented the first and higher order nuclei as distinguishable structures and have argued that the distinction depends on whether the nucleus receives driver afferents (type II) from subcortical pathways or from the cerebral cortex. However, it is possible that several thalamic nuclei may represent a mixture of the two types of circuit, having some afferents from lower centers contributing to first order circuits and others coming from the cortex and contributing to higher order circuits. This is discussed further in Chapter VIII. For the purposes of this chapter we treat all of the nuclei shown with a heavy outline in Figure 2 of Chapter I as first order nuclei and all others as having significant higher order components.

3. TOPOGRAPHICAL ORGANIZATION OF CORTICOTHALAMIC AFFERENTS

There is evidence from early fiber degeneration studies and from more recent labeling studies (Guillery 1967a; Bourassa *et al.*, 1995; Murphy and Sillito, 1996; Murphy *et al.*, 2000) that the corticothalamic axons that go to the first order nuclei all show a strict topographic organization that matches the thalamocortical projection relatively closely (see Figure 7 and compare with Figure 1 of Chapter II). When single axons are traced, individual type I axons show more scatter than do the individual type II axons in the corresponding part of a thalamic nucleus. This has been demonstrated for a comparison of corticogeniculate and retinogeniculate axons for the visual pathways by Murphy and Sillito (1996) in the cat's lateral geniculate nucleus. It has also been shown for the two types of cortical afferent, one coming from layer 6 and one from layer 5 for the somatosensory pathways in rat and monkey (Bourassa *et al.*, 1995; Darian-Smith *et al.*, 1999), and for the visual pulvinar (Rockland, 1998). This relationship may to some extent reflect the fact that the former, the type I axons, distribute to the peripheral segments of relay cell dendrites (see below), whereas the latter distribute to proximal dendritic segments. However, some of the scatter of the type I axons extends rather further than the average length of a dendrite, and in the geniculate relay one finds that where retinal afferents are limited to a single set of monocularly innervated layers, most cortical afferents distribute to adjacent layers, one receiving from each eye. Both sets of axons dealing with any one

B. Afferent Axon Types as Seen Light Microscopically

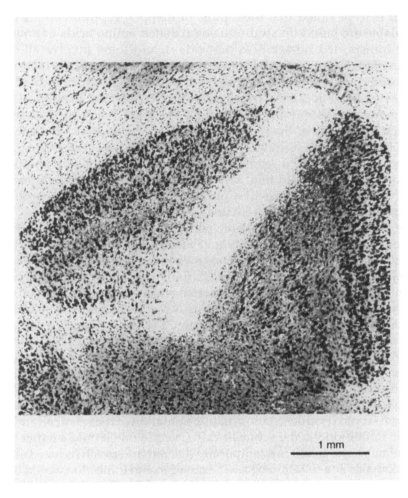

Figure 7
A coronal section through the lateral geniculate nucleus of an owl monkey, showing a "pencil" of autoradiographic label passing through all of the geniculate layers. Radioactive proline had been injected into a small zone of the primary visual cortex (area 17) on the same side. The proline was incorporated into proteins or polypeptides in the cortical cells and transported along the corticothalamic axons to the lateral geniculate nucleus. The radioactivity in the terminals of the labeled corticothalamic terminals shows up on the photographic emulsion with which the section was coated, and in this darkfield micrograph appears as a bright streak. From a photograph taken by Dr. Jon Kaas.

receptive field position are thus focused on much the same population of relay cells, although the type Is extend further afield than do the type IIs.

The corticothalamic axons that go to higher order thalamic nuclei also show a well-defined topographical organization where this has been studied (e.g., Updyke, 1977, 1981; Bourassa *et al.*, 1995; Groenewegen, 1988), although the details are less clear than they are for first order nuclei. It is to be noted that most plots of mapped representations that are available are based on studies using tritiated amino acids or anterogradely transported horseradish peroxidase, and these involve all corticothalamic axons, the type II (drivers) as well as the type I (modulators). Studies of individually filled axons show the degree to which individual axon terminals are localized to a limited part of the total projection but do not generally display the whole map. There are many cortical areas for which the details of the map of the thalamic projection have not yet been studied and, from the evidence available (e.g., Tusa et al., 1978, 1979; Tusa and Palmer, 1980), it seems probable that, in general, the mapping of sensory surfaces would tend to be lost as processing passes through a hierarchy of cortical areas. The possibility that some other, currently undefined feature is mapped, remains largely unexplored.

The fact that the corticothalamic pathway, made up of the type I axons, innervates relay cells directly and also indirectly through the thalamic reticular nucleus or interneurons means that the pathway can excite relay cells by the direct connection and inhibit them by way of the reticular or interneuronal connection. The actual effect on relay cells of activating these cortical afferents depends critically on the details of connectivity at the single cell level, and this is illustrated in Figure 8 for two variants among many more that are possible. In the version schematically shown in Figure 8A, a single corticothalamic axon branches to innervate a reticular cell or an interneuron (*cell 2*) and relay cell (*cell b*), and the reticular cell or interneuron contacts the same relay cell. This is an example of *feedforward inhibition,* and the result would be that strong activity in the corticothalamic axon would produce monosynaptic excitation of the relay cell and disynaptic inhibition via the reticular cell. Overall, one would predict that this scheme would result in a temporally discrete and much reduced effect of corticothalamic activation because excitation and inhibition would both be present to offset each other in the same relay cells. The schema in Figure 8B is radically different. Here, the corticothalamic axon branches to innervate reticular *cells 1* and *3* and relay *cell b. However,* reticular *cells 1* and *3* do not contact relay *cell b* but instead contact its neighbors (*cells a* and *c*). This would not produce feedforward inhibition; rather, this would tend to excite a central relay cell, or a small group of cells, and inhibit neighboring cells. This can also be regarded as a form of lateral inhibition. The schema of Figure 8B would thus produce zones of quite strong excitation and inhibition that are topographically organized, and these would tend to form center/surround patterns.

The schemas of Figures 8A and 8B could only produce distinct outcomes if corticothalamic activity were modulated with a very fine grain. For instance, thalamic cells would respond to large-scale inactivation or activation of cortex in much the same way with either schema. The one relevant study, which suggests that Figure 8B is the more plausible alternative, was published by Tsumoto et al. (1978). They excited a small cluster of cells in layer 6 of area 17 in cats with iontophoretic application of glutamate while recording from relay cells of the lateral geniculate nucleus. They found that if the receptive fields of the relay cells overlapped

B. Afferent Axon Types as Seen Light Microscopically

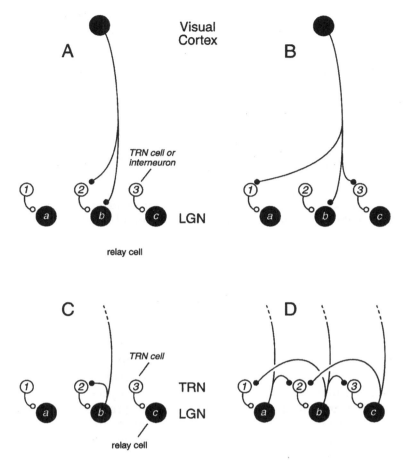

Figure 8
Schematic representation of possible patterns of interconnections of thalamic relay cells with reticular cells and corticothalamic type I axons. See text, pp. 74 and 81–82 for further details.

or were within about 1 degree of those of the excited layer 6 cell cluster, the geniculate relay cell response was increased; if the receptive fields of the geniculate cells were offset by about 1 to 2 degrees from those of the cortical site, the geniculate cell responses were reduced; if the receptive field offset was more than 2 degrees, there was no effect of cortical glutamate application on geniculate cell responses. Whether this observation can be generalized to favor the circuitry of Figure 8B for all cortico-reticulo-thalamic examples remains to be determined, and a combination of Figures 8A and 8B may also exist.

4. OTHER AFFERENTS TO THALAMIC NUCLEI

Other afferents include those coming from the brain stem (see Figure 9), which represent a significant proportion of the total number of

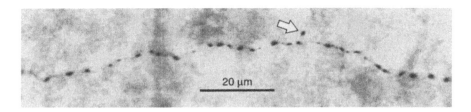

Figure 9
Photomicrograph of part of terminal arbor of axon labeled from the parabrachial region. This view is from the A-laminae of the cat's lateral geniculate nucleus, and the anterograde label placed into the parabrachial region is the lectin, *Phaseolus vulgaris leucoagglutinin*. The synaptic boutons are evident mostly as *en passant* swellings, with an occasional terminal found at the end of a short side branch (arrow).

synapses, up to perhaps 30% or more in the lateral geniculate nucleus of the cat (Erişir *et al.*, 1997). Cholinergic axons from the brain stem have been described in rats and cats (Raczkowski and Fitzpatrick, 1989; Hallanger *et al.*, 1987; Cucchiaro *et al.*, 1988). The density of the distribution of these axons varies from one thalamic nucleus to another (Raczkowski and Fitzpatrick, 1989; Hallanger *et al.*, 1990). Cholinergic axons coming from the parabrachial nucleus of cats have been described as fine, branching, beaded terminal axons, some of which have a relatively localized zone of termination within one or two functionally related thalamic nuclei. In the lateral geniculate nucleus they go mainly to the A layers of the lateral geniculate nucleus, whereas in other parts of the thalamus they show a less localized and more widely distributed terminal arbor (Uhlrich *et al.*, 1988).

Other afferents from the brain stem include serotonergic axons from the dorsal raphé nucleus (Gonzalo-Ruiz *et al.*, 1995; Kayama *et al.*, 1989; de Lima and Singer, 1987), histaminergic axons from the hypothalamus (Uhlrich *et al.*, 1993), and noradrenergic axons from the region of the parabrachial nucleus (de Lima and Singer, 1987; Morrison and Foote, 1986). The origins of the latter are mixed with those of the cholinergic afferents in some species and separated into a well-defined locus ceruleus in others (de Lima and Singer, 1987; Morrison and Foote, 1986). An interesting difference between two cholinergic inputs to the lateral geniculate nucleus is that the main one, coming from the parabrachial nucleus also contains nitric oxide synthase, suggesting that these cells release nitric oxide as a possible neuromodulator, whereas a smaller component from the parabigeminal nucleus does not (Bickford *et al.*, 1993).

There are also certain ascending pathways from the brain stem that appear to be specific to particular thalamic nuclei. There is a GABAergic input from the inferior colliculus to the medial geniculate nucleus (Peruzzi *et al.*, 1997) and for the lateral geniculate nucleus there are GABAergic afferents from the pretectum (Cucchiaro *et al.*, 1991). Other afferents go from the superior colliculus primarily to the C-laminae of the cat or K-laminae of the primate (Harting et al., 1986, 1991), and there is a cholinergic input from the parabigeminal nucleus in the midbrain, which also goes mainly to the C-laminae of the cat or K-laminae of the primate (Harting et al., 1991).

5. Afferents from the Thalamic Reticular Nucleus to First and Higher Order Nuclei

Axons from the reticular nucleus to the thalamus have been demonstrated by electrophysiological methods in vivo (Sefton and Burke, 1966; Ahlsén et al., 1984, 1985) or in slice preparations (von Krosigk et al., 1993) and have also been shown on the basis of anterograde labeling of single reticulothalamic axons (Yen et al., 1985b; Uhlrich et al., 1991; Pinault et al., 1995a,b; Cox et al., 1996). The reticulothalamic pathway is GABAergic (Houser et al., 1980; Ohara and Lieberman, 1985) and so far as is known, all thalamic nuclei receive these GABAergic afferents from the thalamic reticular nucleus. Although many of the illustrations of these reticulothalamic axons do not show the details of the terminal morphology, the most common picture shows beaded axons with few side branches (Uhlrich et al., 1991; Cox et al., 1996; Figures 10 and 11), and in this they are readily distinguishable from the driver afferents and from the type I corticothalamic afferents considered earlier. However, reticulothalamic axons may not have a uniform structure. They differ in the degree to which their terminals are tightly localized or widespread, and they may also differ in the detailed structure of their terminals. Cox et al. (1996) described a range of terminal patterns in the ventral posterior nucleus of the rat, with some showing a very small, well-localized terminal plexus, and others showing a more widespread, but still localized arbor. The latter produce a weaker inhibitory action on the thalamic relay cells of the ventral posterior nucleus than do the former (Cox et al., 1997; Figures 10A and 10B); these may represent extremes of a continuum rather than distinct cell classes. Whereas Yen et al. (1985b) and Uhlrich et al. (1991) describe relatively fine beaded axons in first order thalamic nuclei; Pinault et al., (1995b) describe a more complex pattern of terminal branches not unlike that of driver afferents described earlier, suggesting that there may be a range of reticulothalamic axon terminal patterns that vary in their terminal localization and their terminal structure.

Earlier methods seemed to show a rather widespread and diffuse projection for the reticulothalamic axons (e.g., Yen et al., 1985b), but there is now good evidence for a topographically organized pathway going from the reticular nucleus to the major first order nuclei. Locally limited injections of retrograde tracers into some of these thalamic nuclei have produced localized zones of labeled cells in the reticular nucleus (Crabtree and Killackey; 1989; Crabtree, 1992a,b; Loszádi, 1995) and small fills of one or a few reticular cells with an anterograde tracer can reveal individual axons with locally ramifying branches within one thalamic nucleus, which can be either first or higher order (Uhlrich et al., 1991; Cox et al., 1996; Pinault et al., 1995a,b; Pinault and Deschênes, 1998a). In some

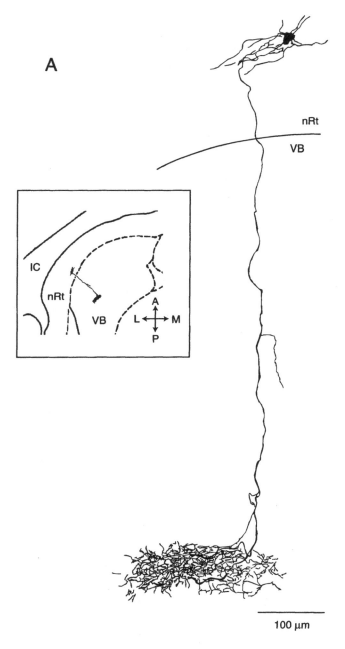

Figure 10
(A) Reticulothalamic axon with a closely clustered terminal arborization in the ventral posterior nucleus of a rat. In this and in B the inset shows the localization of the cell and the terminal arbor in a horizontal section through the thalamus.

instances the innervation of a higher order nucleus is formed by a branch of an axon that is also innervating a first order nucleus (Pinault *et al.*, 1995a; Pinault and Deschênes, 1998a,b; Crabtree, 1996). There is some evi-

B. Afferent Axon Types as Seen Light Microscopically

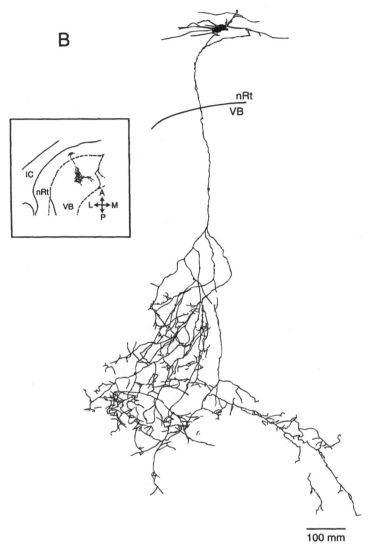

Figure 10 (continued)
(B) Reticulothalamic axon with a relatively diffuse terminal arborization in the ventral posterior nucleus of a rat. Details as in (A).

dence that the topographic map of sensory surfaces is less clearly preserved in the reticulothalamic pathways of higher order thalamic nuclei than for first order nuclei (see Crabtree, 1996; Conley *et al.*, 1991; Conley and Diamond, 1990, and see Chapter VII).

Uhlrich *et al.* (1991) described axons that pass from the perigeniculate nucleus to the A laminae of the lateral geniculate nucleus (see Figure 11). They showed that these axons generally have two branches, a medial one distributing to a very narrow column of geniculate cells that must correspond to a tiny part of the visual field and a lateral branch with a broader but still rather limited distribution. This branch shows two types of local sign. In the first place, it distributes to geniculate laminae receiving from

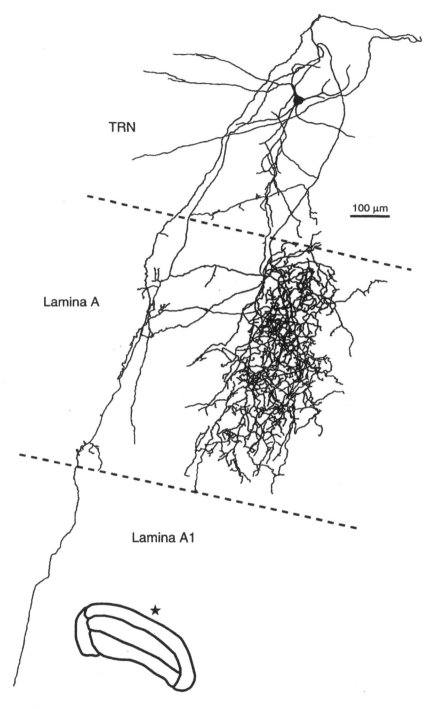

Figure 11
An afferent axon from the perigeniculate nucleus of the cat passing to the A layers of the lateral geniculate nucleus, showing a slender medial branch and a lateral branch with broader distribution. The axon was filled with horseradish peroxidase and is shown in a section cut in a coronal plane.

B. Afferent Axon Types as Seen Light Microscopically

left or right eye in accordance with the dominant ocular input to the parent perigeniculate cell, and second, it distributes to a column of geniculate cells that corresponds roughly to the position of the receptive field of that perigeniculate cell.

On the evidence available to date, where the reticulothalamic pathway shows local sign, it appears to be in a topographically reciprocal relationship with the thalamoreticular pathway. That is, the part of the thalamic nucleus receiving local afferents from the reticular nucleus sends branches of its thalamocortical axon back to roughly the same part of the reticular nucleus. However, Pinault and Deschênes (1998b) labeled cells in the thalamic reticular nucleus of the rat so that they could trace the axons of these cells into the thalamus, and at the same site they labeled the axon branches that thalamocortical relay cells sent to the same region of the reticular nucleus. The thalamic relay cells themselves were labeled retrogradely in these preparations. The results showed that these retrogradely labeled thalamic relay cells lay outside, but close to, the limited volume of the relevant thalamic relay nucleus within which the terminal arbor of the labeled reticular cell ramified. This is an elegant and, so far as we know, unique morphological demonstration of the sorts of relationships that are schematically represented in Figure 8. However, there is some evidence from intracellular recording from geniculate relay cells of cat and ferret that action potentials in a relay cell are sometimes followed disynaptically by an IPSP (Lo and Sherman, 1994; Kim and McCormick, 1998), implying a direct feedback (as in Figure 8C). It is probable that both types of connection occur, and at present there is significant doubt about the details of the functional connectivity established between the thalamic reticular nucleus and dorsal thalamus.

Currently there is much of the same ambiguity for the reticular connections as described earlier for the cortico-reticulo-thalamic connections, and again, the functional significance of the different types of connection can be appreciated by the schematic representation in Figures 8C and 8D. In Figure 8C, the relay *cell b* contacts reticular *cell 2* that projects back to innervate *cell b*. This is a classical example of feedback inhibition and means that strong activity in a relay cell will be somewhat suppressed after two synaptic delays. The scheme drawn in Figure 8D is more complex and more interesting. Here, relay *cell b* contacts reticular cells *1* and *3*, but not reticular *cell 2*. However, cells *1* and *2* do not contact *cell b*, but rather contact its neighbors (*cells a* and *c*); and reticular *cell 2*, which does innervate *cell b*, is in turn innervated by *cells a* and *c*. Consider what this means when a relay cell becomes highly active: activity in *cell b* would produce disynaptic inhibition (via reticular *cells 1* and *3*) of its neighbors, *cells a* and *c*; the reduced activity in the neighboring relay cells means that reticular *cell 2* is less activated, and thus its inhibitory contribution to *cell b* is reduced. This scheme thus yields *feedback disinhibition*, precisely the opposite of the scheme illustrated in Figure 8C. Clearly, we cannot understand how the thalamo-reticulo-thalamic circuit functions until we know whether Figure 8C, Figure 8D, or some combination best describes the functional connectivity. We have seen that there is some evidence for the connections shown in Figure 8C and that the circuit shown in Figure 8D may be present and may, indeed, prove dominant. Until we can determine which circuit, Figure 8C or 8D, best represents the dominant thalamoreticular interactions, this most important part of thalamic functioning must remain an enigma. In Chapter VII we discuss different topographical arrangements linking the reticular nucleus to the thalamus, and it is possible that the details of connectivity patterns discussed previously may vary depending on the nature of the topographical link.

6. CONNECTIONS FROM INTERNEURONS TO RELAY CELLS

The interneurons are in receipt of most, perhaps all, of the afferents that we have considered so far, but from the point of view of afferents received by the relay cells, the interneurons are themselves also a source of important afferent connections. We have seen in Chapter II that there are two types of connection established from interneurons to relay cells, one coming from the interneuronal axons and the other from the dendrites. It is probable that these two GABAergic inputs to relay cells are functionally distinguishable, and this issue will be explored further in Chapter V. The synaptic connections that are established by the interneurons have been demonstrated by electron microscopic studies, and these are considered in Section C.

7. THE HETEROGENEITY OF THE SEVERAL AFFERENT SYSTEMS

So far, we have grouped the afferents by their source, such as one cortical layer, sensory pathway, or brain stem nucleus. This reflects the extent to which each source gives rise to a more or less characteristic type of axonal terminal in the thalamus. However, it does not imply that afferents coming from any one source are homogeneous: we know that the retinal afferents coming from different ganglion cell types known as W, X, and Y (see Chapter II) have morphologically distinguishable types of terminal arbors in the lateral geniculate nucleus (Sherman, 1985); we have suggested that reticulothalamic axons may vary in their structure; we present evidence in Section E.2 for some significant possible heterogeneity in corticothalamic axons in terms of their origin and function; and we have presented evidence in Chapter II that there may be more than one type of interneuron. Information about the afferents that any one relay cell or relay cell type in any one particular thalamic nucleus in any one species receives can be sketched in terms of the groupings we have presented, but details have to be defined on a more precise basis if an accurate view of the input to the relay cells is being sought.

C. Electron Microscopic Appearance of the Afferent Axon Terminals and Their Synaptic Relationships

1. GENERAL DESCRIPTION

With the electron microscope one can see that a "generic" structure is present in essentially all thalamic nuclei. If one is looking at a relatively

C. Electron Microscopic Appearance of the Afferent Axon Terminals

low magnification at a section of a thalamic nucleus, one sees primarily dense bundles of myelinated axons that represent the many afferent and efferent axons, blood vessels, glial cells, neuronal cell bodies with relatively few synaptic contacts on their surface, many dendritic profiles that receive a few more but still not many synaptic contacts, and then patchy areas of closely grouped synaptic profiles, called "glomeruli" (Figures 12 and 13A), where several different sorts of structure appear to be making specialized synaptic contacts with each other. Within the glomeruli, which are considered in more detail later, and also in the regions between the glomeruli there are several different types of synaptic terminal profile, most of which have now been identified in terms of their origins by experimental studies. Some contain round synaptic vesicles, called R type profiles, and some contain flattened or pleomorphic vesicles. We shall refer to these as flat vesicles even though the vesicle shapes are often irregular or undefined (see Gray, 1969) and the profiles that contain such vesicles will be called F-type profiles. The shape of the vesicles is determined by the vesicular contents and the osmolarity of the solutions used in the early processing of the tissue (Valdivia, 1971), and it appears from immunohistochemical studies that in the thalamus, as in many other parts of the mammalian brain, the F profiles are all GABAergic, whereas many of the profiles containing round vesicles are glutamatergic.

Two main types of R profile have been recognized on the basis of a number of variables, including profile size. They fall readily into large and small categories (see Figures 12 and 13A) and have been called RL (Round vesicles and Large terminals) and RS (Round vesicles and Small terminals), respectively. These two classes also differ on the basis of the types of synaptic junctions they make, of their extrathalamic origins, and in some instances in the neurotransmitters they use. The relevant literature is discussed more fully later. The RL profiles represent the driver afferents; they are characteristically seen within the glomeruli, although they can also be found in extraglomerular regions. The RS afferents represent a mixed population including modulator afferents from cortex and from the brain stem. Virtually all of the cortical RS terminals are extraglomerular, and all of the RS terminals within glomeruli are of brain stem origin, although brain stem terminals are also found outside of glomeruli. Two types of F profile are also recognized, F1 and F2. The former represent the axons of cells in the thalamic reticular nucleus, the axons of local interneurons and for some thalamic nuclei, other extrinsic GABAergic extrinsic afferents, whereas the latter represent the axoniform dendritic processes of interneurons (see Chapter II). They are not distinguishable on the basis of size (see Figure 14), but the former are never postsynaptic to another process, whereas the latter commonly are.

Before considering the synaptic contacts established by these various profile types, it is useful to look at the basis of their classification and naming. Problems concerning classifications in general were discussed in the last chapter, and the same issues apply whether one is classifying nerve cell bodies, synaptic terminals, or synaptic vesicles. Given that most successful and functionally significant classifications will depend on several variables, an obvious, but not widely recognized, problem of

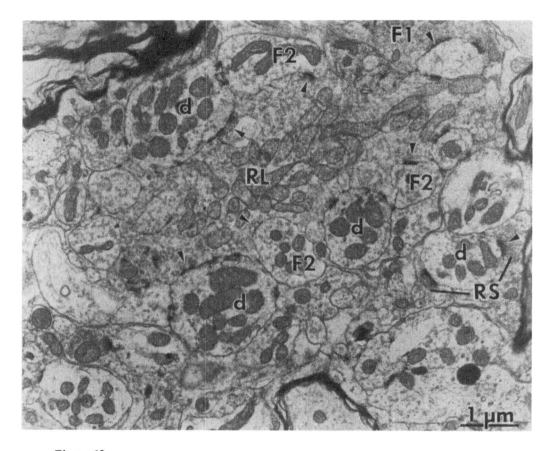

Figure 12
Electron micrograph of a glomerulus with a large central RL terminal from an A-lamina of the lateral geniculate nucleus of a cat. Some of the synaptic junctions are identified by arrowheads. F1, F2, RL, and RS identify the major types of axon terminal (see text) and "d" identifies some of the dendritic profiles. A schematic interpretation of the major profiles identifiable in this micrograph is shown in Figure 13A. Note that this particular glomerulus shows very little evidence for the encapsulation by sheets of astrocytic cytoplasm that is often regarded as a typical feature of thalamic glomeruli. See text for more details. Electron micrograph prepared by Sue Van Horn.

naming arises. The traditional approach to the naming problem as applied to axons and their terminals was to identify the origin and the termination of a particular group and then name them accordingly as spinothalamic, retinogeniculate, etc. However, in an electron microscopic study one is initially faced with the problem of identifying axons on the basis of their fine structure and synaptic relationships. There may be good experimental evidence that a particular terminal type defined in this way comes from a particular source, and then it would seem sensible to name those terminals in accordance with that knowledge. Currently we are close to being able to name all RL terminals as driving afferents

C. Electron Microscopic Appearance of the Afferent Axon Terminals

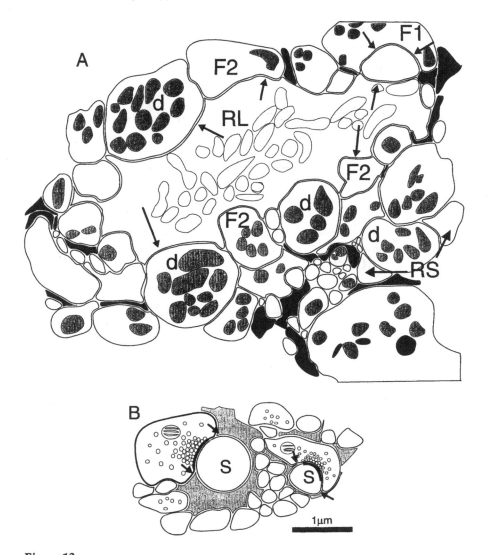

Figure 13
(A) Interpretative schema to identify the major profiles in Figure 12. The abbreviations are as for Figure 12, and the arrows indicate the positions of the synaptic junctions that are marked by arrowheads in Figure 12. The mitochondria are shown as pale in the central RL profile and as dark in all other profiles. Three structures interpreted as lysosomes are shown in a darker, almost black shade in the unlabeled dendritic profile at the lower right of the figure. Astrocytic profiles are shown in black. See text for further details. (B) A comparable schematic representation for the cerebellar molecular layer of synapses made onto Purkinje cell spines (S) to show the close relationship of the astrocytic cytoplasm (gray) to these synaptic junctions. Based on Figure 5.9 of Peters *et al.* (1991).

for their particular nucleus so that such a terminal in the lateral geniculate nucleus is reasonably treated as a retinal terminal. But it could be coming from the tectum or possibly, on the basis of the light microscopic evidence discussed earlier, there may yet be a few that have a different origin. So

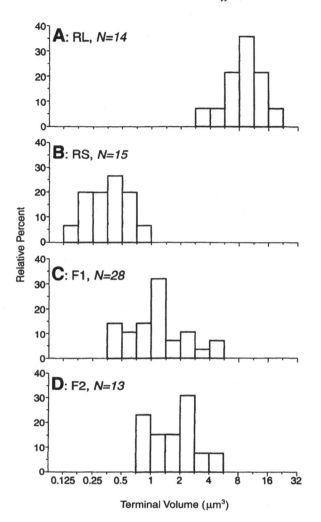

Figure 14
Histograms showing the volume, computed from serial sections, of representative population of terminal types from the A-laminae of the cat's lateral geniculate nucleus. Data from Van Horn *et al.*, 2000.

we continue with the more neutral term RL. In general, whenever a particular morphological class is likely to include afferents from more than one source, as is the case for the RS terminals, which have at least two distinct origins in cortex and brain stem, it is best, in accounts that do not identify the origin, to work with a neutral terminology.

It has to be stressed that a number of characteristics other than vesicle shape (Roundness) and terminal size (Largeness) contribute to our recognition of the RL terminals, and this is also true for all of the other terminal types. We stress that the terminology should be regarded as neutral and interim. The R and F and the L and S should be treated as labels for terminals that combine several distinct morphological features, some of

C. Electron Microscopic Appearance of the Afferent Axon Terminals

which are not part of the name. These are dealt with more fully later and include synaptic relationships, mitochondrial appearance and distribution, types of contact made, etc. Terminal size alone, vesicle shape alone, and even the combination of the two provide a very limited basis for classification. Some RL terminals are smaller than others, and in single sections some RS terminal profiles can be as large as some of the smaller RL terminal profiles. However, as a population, RL terminals are significantly larger than RS terminals.

2. THE RL TERMINALS IN FIRST ORDER NUCLEI

These are a prominent feature on the basis of their size, although numerically they represent a relatively small proportion of the total synaptic profiles or contacts. Axon terminals that have the characteristics of RL terminals are shown in Figures 12, 15, and 16 and have been identified as the driving afferents in the lateral geniculate nucleus (Szentágothai, 1963; Colonnier and Guillery, 1964; Guillery, 1969a; Peters and Palay, 1966), the medial geniculate nucleus (Jones and Rockel, 1971; Morest, 1975; Majorossy and Kiss, 1976), and the ventral posterior nucleus (Ralston, 1969, Jones and Powell, 1969; Ma et al., 1987a,b); Ohara et al., 1989). The cerebellar axons that innervate the ventral lateral nucleus have also been identified as forming RL terminals (Harding 1973; Rinvik and Grofova, 1974a,b; Ilinsky and Kultas-Ilinsky, 1990; Kultas-Ilinsky and Ilinsky, 1991) as have the mamillothalamic afferents to the anterior nuclei (Somogyi et al., 1978). These axons correspond closely to the type II axons defined light microscopically, that is, to the driving afferents. They commonly have characteristically pale mitochondria, and so have in the past been called RLP terminals. This feature is likely to have a functional correlate, but at present there is no information about what this might be. Since the pale mitochondria are not evident in all preparations, we shall use the simpler term, RL. The RL terminals can present profiles 2 to 3 μm or more in diameter; they almost invariably contain a group of mitochondria and often make more than one specialized synaptic or nonsynaptic junction in any one section. Glomeruli generally contain one, occasionally more than one, such profile, and so one can regard them as a characteristic part of the glomeruli.[2] However, RL profiles are also seen outside glomeruli, especially in nuclei that contain few interneurons.

RL profiles make asymmetric synaptic junctions[3] on stem dendrites,

---

[2] The related question as to whether every glomerulus contains at least one RL terminal is harder to answer because to our knowledge this has not been systematically investigated. However, in a nonsystematic survey of glomeruli in the A laminae of the lateral geniculate nucleus of cats carried out in SMS's laboratory, every glomerulus encountered had an RL terminal.

[3] Gray (1959) characterized synaptic junctions as symmetric or asymmetric on the basis of the relative thickening of the postsynaptic specialization that most probably relates to the nature of the receptor at the junction. The presynaptic thickening is roughly the same across all synapses in the brain. "Asymmetrical" means that the postsynaptic thickening is much more pronounced than the presynaptic, and "symmetrical" means that the thickening on both sides is similar. Colonnier (1968) showed that asymmetrical junctions are generally associated with round vesicles in the presynaptic process and that symmetrical junctions are associated with flat vesicles.

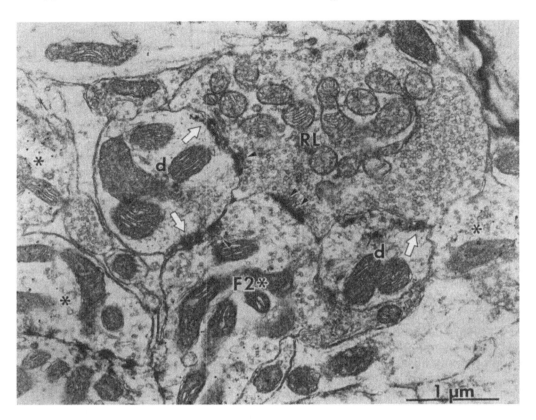

Figure 15
Electron micrograph to show a triad from an A-lamina of the lateral geniculate nucleus of a cat. In this section the GABA immunoreactive profiles are identifiable by the very small colloidal gold particles and have been marked by asterisks. The synaptic junctions are shown by arrowheads. The retinal RL terminal is presynaptic to the F2 terminal at the synaptic junction marked by two arrowheads and is presynaptic to the dendrite (d) at the synapse marked by one arrowhead. The F2 terminal, in turn, is presynaptic to the same dendrite at another synapse marked by a single arrowhead. Several filamenous contacts (see text) are also shown in this figure (white arrows). They are seen adjacent to synaptic contacts and can be distinguished from synaptic contacts because they show no accumulation of presynaptic vesicles and, on the postsynaptic side, show dense structures that relate closely to a web of intermediate filaments. Electron micrograph prepared by Sue Van Horn.

on the grapelike dendritic appendages (see Chapter II) and occasionally on cell bodies. They also synapse on one group of the F profiles (the presynaptic dendrites or F2 terminals; Figures 12, 15, and 16). The dendritic profiles, other than the F2 dendritic terminals and rare interneuronal stem dendrites, that the RL terminals contact are among the larger dendritic profiles in the nucleus, indicating that these contacts, including those formed in the glomeruli, are relatively close to the cell body of

C. Electron Microscopic Appearance of the Afferent Axon Terminals

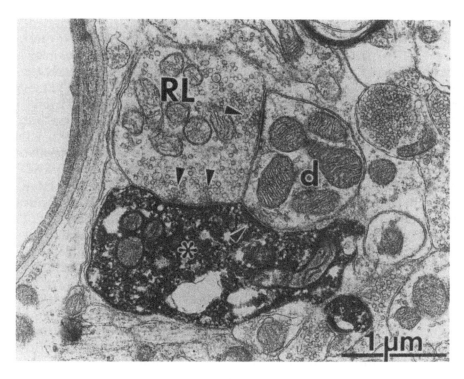

Figure 16
The interneuronal profile that is shown as an inset of Figure 11 in Chapter II is shown again to illustrate the basic structure of a triad. The labeled interneuronal F2 profile is filled with horseradish peroxidase and is identified by the asterisk. The retinal axon terminal (RL) is presynaptic to the interneuronal profile (shown by two arrowheads) and is also presynaptic to the dendrite (d) at the single arrowhead. The interneuronal profile, in turn, is presynaptic to the same dendrite. Electron micrograph prepared by Sue Van Horn.

the relay cells. Figure 12 shows a typical glomerulus from the lateral geniculate nucleus of a cat. Some of the dendritic profiles (labeled "d" in the figure) that are postsynaptic to the central RL terminal are likely to be grape-like appendages, and others appear to be dendritic stems. The two types of dendritic profile have not been separately identified in this figure because the certain identification would require analysis of serial sections.

There is some variation in the part of the relay cell to which these terminals relate; not only is there variation within any one nucleus of the thalamus, but there is also a systematic difference between nuclei and even between geniculate layers (Kultas-Ilinsky and Ilinsky, 1991; Feig and Harting, 1994). For example, in the magnocellular and parvocellular layers of primates, the RL terminals are often found close to the cell body, whereas they are further from the cell body in the koniocellular layers (Feig and Harting, 1994; Guillery and Colonnier, 1970). RL terminals are never postsynaptic to any other process. In addition to these synaptic junctions the

RL terminals establish some desmosome-like contacts with adjacent dendrites. These are not associated with synaptic vesicles (see Figure 15) and so are regarded as nonsynaptic. They are associated with groups of intermediate filaments that accumulate on either side of the junction and for this reason have been called filamentous junctions (Guillery, 1967b; Lieberman and Spacek, 1997). They may represent an unusual adhesive junction (Peters et al., 1991). Their functional role has not been explored, but they provide a useful clue for distinguishing RL from RS terminals.

The RL profiles, either within glomeruli or within a nonglomerular region, often establish "triadic" junctions. A triad is formed where an RL profile is presynaptic to two adjacent profiles, one a dendrite of a relay cell, and the other an F2 profile. The F2 profile in turn is then presynaptic to the same dendrite (see Figures 15 and 16). This is the classical form of the triad and, in most of the examples that have been documented for thalamus, involves an RL terminal as the common presynaptic element. Triads that involve RS profiles as the common presynaptic element are considered in the next section.

Since the RL terminals contact F2 terminals, which are the axoniform dendritic processes of interneurons, and also contact the dendrites of the relay neurons, they are presynaptic to both types of thalamic cell. However, whereas interneurons receive the retinal driver contacts primarily on the (presynaptic, axoniform) distal dendritic appendages, the relay cells receive their afferents on the proximal dendrites. RL terminals do form some contacts on the stem dendrites of interneurons (Hamos et al., 1985; Montero, 1991; Weber et al., 1989; Wilson et al., 1996), and functionally these two types of contact are likely to have quite distinct actions (Cox et al., 1998; Cox and Sherman, 2000, and see Chapter V). The synaptic contacts on the presynaptic dendrites may well act locally on individual dendritic processes, whereas the contacts on stem dendrites nearer the cell body are more likely to produce action potentials capable of discharging the axon and invading at least parts of the dendritic arbors of the interneurons.

3. THE RL TERMINALS IN HIGHER ORDER NUCLEI

Already in 1972 (Mathers, 1972) it was shown that in the pulvinar of the monkey corticothalamic axons having the characteristic RL (driver) structure degenerate after lesions of visual cortex, and this was later confirmed for the pulvinar and lateralis posterior nucleus of the cat and monkey (Ogren and Hendrickson, 1979; Feig and Harting, 1998) and the squirrel (Robson and Hall, 1977). More recently Schwartz et al. (1991) showed in the mediodorsal nucleus RL terminals that came from frontal cortex. Kultas-Ilinsky et al. (1997) report that some RL terminals in the ventral anterior nucleus of the monkey have a cortical origin, and Hoogland et al. (1991), who traced axons marked by intracellular label from somatosensory cortex of the mouse to the posterior nucleus (PO), have demonstrated that some of them form type II terminals in light microscopic terms and have then identified these as large RL terminals within that higher order thalamic nucleus. Comparable observations on cortical afferents to the dorsal division of the medial geniculate nucleus have also been reported by Bartlett et al. (2000). However, apart from these two correlated light and electron microscopic studies, the evidence that the light microscopically labeled axons coming from layer 5 and described in Section B.2 are, indeed, RL axons in terms of their electron microscopic structure and relationships is generally indirect. We cannot yet be certain that the layer 5 axons in higher order nuclei always form RL terminals, nor can we assert that the afferents coming from layer 6 cells are always RS axons. The indirect evidence is highly suggestive (see e.g., Schwartz et al., 1991, for the mediodorsal nucleus), but so far it is only circumstantial. At present it is the combination of evidence obtained from retrograde labeling studies, electron microscopic degeneration, and anterograde labeling studies combined with light microscopic tracing of individual axons that serves to make a strong indirect case for this conclusion. We will treat a corticothalamic afferent that forms RL terminals as representing a higher order circuit and will expect that these are type II afferents that arise from cortical layer 5 cells in all instances. It is of interest, then, to note that the report of corticothalamic RL terminals in the ventral anterior nucleus, reported by Kultas-Ilinsky et al. (1997) and the more recent report of some type II corticothalamic terminals in the ventral posterior nucleus (Rouiller et al., 1998; Darian-Smith et al., 1999) suggest that these nuclei must be regarded as having some higher order circuits, a detail that does not appear in Figure 2 of Chapter I but is discussed further in Chapter VIII.

4. THE RS TERMINALS

These are smaller than the RL terminals in single sections and in reconstructions (Figures 14A and 14B). Occasionally a section will include only a small part of an RL, which then appears to be in the size range of the RS terminals, so that serial sections or other criteria may have to be used for definite identification. The vesicles are generally more closely packed in the RS than in the RL terminals and the RL terminals make multiple junctions in single sections (Figure 12), whereas the RS terminals do not. The RS terminals do not show filamentous junctions and where the differential appearance of the mitochondria is evident, the RS terminals have dark mitochondria whereas the RL terminals have pale ones. RS terminals are somewhat more likely to be found outside the glomeruli than in them, and where they occur in the glomeruli, they contribute fewer synapses than do the RL terminals. In terms of the overall visual impression and of the volume occupied, the RL profiles dominate in the glomeruli, where RS profiles are seen more rarely. However, since the RS terminals are significantly smaller than the RL terminals, they are less likely to be cut, and the visual impression cannot be taken to represent the numerical relationships, which are considered separately.

There are two major sources of the RS terminals: the cortex and the brain stem. The RS terminals that come from the cortex represent the *en passant* synapses and the terminal portions of the "drumsticklike side branches" of the type I axons described in Chapter II. We have seen that these come from layer 6 pyramids in the cortex, although an additional layer 5 origin cannot be completely excluded at present. These provide the corticothalamic modulatory afferents, and for a while this was the only demonstrated origin of the RS terminals. More recently it has become clear that many of the RS terminals, especially the larger ones and those in the glomeruli, are the terminals of ascending cholinergic axons from the brain stem (Erişir et al., 1997).

In the lateral geniculate nucleus of the cat the smaller, cortical RS terminals contact relatively small, peripheral dendritic profiles of relay cells and the dendritic shafts of interneurons (Wilson et al., 1984; Weber et al., 1989; Montero, 1991; Erişir et al., 1997). The larger, brain stem cholinergic terminals make their contacts closer to the relay cell body (Erişir et al., 1997; Wilson et al., 1984). Both types of RS terminal (from cortex and brain stem) contact the dendritic shafts of interneurons, but those from brain stem are the major type of RS terminal contacting the distal axoniform processes (F2 terminals) of the interneuronal dendrites, with few if any cortical RS terminals contacting these (Erişir et al., 1997; Vidnyanszky and Hamori, 1994).

When RS profiles that come from the brain stem are selectively labeled in the cat's lateral geniculate nucleus (Figure 17), it is possible to see that they often form a type of triad. This involves two different RS terminals, from two branches of a common brain stem axon, making the contacts that are comparable to the contacts made by the single RL in the classical triad (Erişir et al., 1997). We shall refer to this as a "pseudotriad." RS terminals of cortical origin do not show this pattern of synaptic contacts.

The terminology that has been used in the past to describe RS terminals and our own use needs brief consideration (Ide, 1982; Erişir et al., 1997). As mentioned earlier, in many preparations the RS profiles have dark mitochondria, whereas the RL profiles have pale mitochondria. This led to the introduction of a further distinction between the larger and the smaller RS profiles so that RSD profiles were distinguished from RLD profiles (the "D" referring to dark mitochondria), and the latter were recognized as clearly distinct from the RLP profiles, which are generally even larger than the largest RLD profiles and which also differ in the other respects considered previously. However, since mitochondrial appearance is somewhat variable and since it has been shown that in the lateral geniculate nucleus the RSD and RLD profiles form a continuum and cannot be distinguished from each other without specifically labeling one or the other (Erişir et al., 1997), it continues to be useful to refer to all of these as RS. Also, as is indicated in Figure 14, there is no overlap between RS and RL terminal sizes, at least for the A-laminae of the cat's lateral geniculate nucleus (Van Horn et al., 2000). Further, there are terminals in the reticular nucleus that are thought to come from the thalamus (Ohara and Lieberman, 1985; Ide, 1982; Williamson et al., 1993); these terminals are larger than most of the RS terminals in the thalamus, and these have also been called RLD in the past (see Ide, 1982). We shall be

C. Electron Microscopic Appearance of the Afferent Axon Terminals

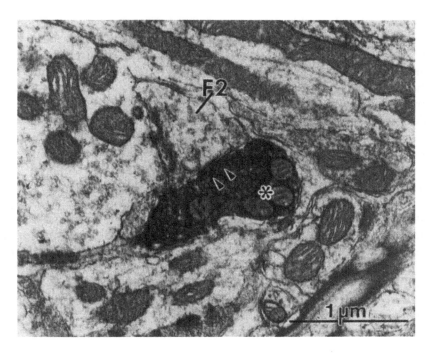

Figure 17
Labeled RS profile (*) contacting an F2 terminal in the A-laminae of the cat's lateral geniculate nucleus. The synapse is indicated by arrowheads. The labeling is with an antibody directed against brain nitric oxide synthase (BNOS), a precursor for the neuromodulator, nitric oxide (NO), suggesting that NO is present in and released from the terminal. One significance of the BNOS labeling is that all terminals in the neuropil of the cat's lateral geniculate nucleus that label for BNOS also label for choline acetyltransferase, indicating that the terminals are also cholinergic, and it is known that the sole source of such terminals is the parabrachial region of the brain stem (Bickford *et al.*, 1993; Erişir *et al.*, 1997). From Erişir *et al.* (1997).

using the simple RS term throughout for the terminals in the dorsal thalamus, identifying axon terminals by other criteria where this is possible and useful.

5. THE F TERMINALS: F1 and F2

Two types of F profile, not varying greatly in size, have been called F1 and F2 profiles. They are often difficult to distinguish, although the distribution of the vesicles is rather more irregular in the F2 profiles. As mentioned earlier, the main, critical difference is that the F1 profiles can be presynaptic at a synaptic junction but are never postsynaptic, whereas the F2 profiles can be both presynaptic and postsynaptic. The F2 profiles occasionally contain ribosomes (Ralston, 1971; Famiglietti and Peters, 1972). This evidence, and the evidence from serial reconstructions

(Rapisardi and Miles, 1984; Wilson *et al.*, 1996; Hamos *et al.*, 1985), identifies the F2 profiles as parts of interneuronal dendrites, occasionally the shafts, more commonly the axoniform, *dendritic* appendages. The F2 profiles make a major contribution to the glomeruli in the thalamus, and they are the only vesicle-containing synaptic terminals so far described in the thalamus that are postsynaptic.

Reticulothalamic axons form medium-sized terminals with flattened vesicles that belong to the F1 class. In the ventral posterior and lateral geniculate nuclei of the rat they terminate as symmetrical, extraglomerular, axodendritic synapses (Montero and Scott, 1981; Peschanski *et al.*, 1983). Similarly, in the cat, where they mainly contact relay cells (Bickford, Van Horn, and Sherman, unpublished), they terminate predominantly outside the glomeruli, primarily (about 90%) on the peripheral dendritic segments that receive the cortical RS terminals and only to a limited extent (about 10%) on the more proximal dendritic segments receiving the retinal RL afferents (Cucchiaro *et al.*, 1991). This raises the possibility that the thalamic reticular nucleus may be more effective at modulating cortical than retinal inputs on relay cells. Further, we show in Chapter IV that relay cells express many voltage-dependent properties in their dendrites and that some favor peripheral dendritic segments. For example, the voltage-dependent, T-type Ca^{2+} channels described in Chapter IV seem to be concentrated on peripheral dendrites. It is likely, therefore, that the reticular inputs and also the cortical RS inputs may be especially effective in controlling these channels (see also Chapter VI).

Other sources for F1 terminals in the lateral geniculate nucleus include the axons of interneurons and the GABAergic axons that innervate the lateral geniculate nucleus from the pretectum (Montero, 1987; Cucchiaro *et al.*, 1993). In general, the morphological distinctions among these F1 terminals that arise from different sources are elusive (but see Montero, 1987).

6. Quantitative Relationships

The relative distributions of synapses from the various terminal types have not been defined throughout the thalamus generally, either in terms of their distribution on major postsynaptic sites or in terms of their relative numbers. We have indicated that the RL terminals synapse on relatively large dendritic profiles and on F2 profiles. These synapses are often in the form of triads and in glomeruli. Many of the RS and F1 profiles relate to more distal sectors of the dendrites and some of the F1 and RS profiles also lie in the glomeruli. Axosomatic synapses are relatively rare, but synapses from RL and F profiles do occur on cell body profiles.

Relative numbers are hard to interpret in functional terms because a numerically large input may be modulatory with a relatively subtle or finely graded effect, whereas a powerful driver input if it is critically located can have a dominant effect even though numerically it forms only a small proportion of the terminals. Relative numbers for the lateral geniculate nucleus of the cat provide a general guide, but there is likely

C. Electron Microscopic Appearance of the Afferent Axon Terminals

to be considerable variation among nuclei and species. Figure 18, based on results from Wilson *et al.* (1984) and Van Horn *et al.* (2000) for the geniculate A layers, schematically shows the numerical distributions of synapses onto relay cells in the lateral geniculate nucleus of the cat. (As noted below, the distributions on interneurons would be somewhat different.)

Several earlier electron microscope studies (e.g., Guillery, 1969b; Weber *et al.*, 1989; Montero, 1991; Erişir *et al.*, 1998) documented the relative distribution of synapses from various terminal types in the lateral geniculate nucleus, but these have generally not taken account of sampling biases that lead to an overrepresentation of large and an underrepresentation of small structures. Van Horn *et al.* (2000) applied corrections for this bias

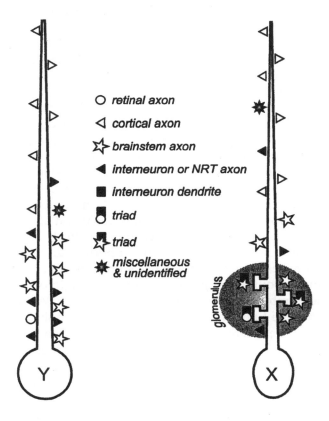

Figure 18
Schematic view of synaptic inputs onto an X and a Y cell of the cat's lateral geniculate nucleus. For simplicity, only one, unbranched dendrite is shown. Synaptic types are shown in relative numbers. Note that, for both cell types, cortical synapses occur on distal dendrites and brain stem and retinal synapses are found on proximal dendrites; there is no overlap of these two zones. The main difference between the X and the Y cell is that the X cell has numerous triadic inputs in glomeruli from F2 terminals and either retinal or brain stem terminals. These typically occur on dendritic appendages. The Y cell is essentially devoid of such triadic inputs and glomeruli.

and have used GABA immunohistochemistry in the lateral geniculate nucleus of the cat to distinguish postsynaptic profiles that were relay cells (i.e., GABA−) from those that were interneurons (i.e., GABA+). They found that, for relay cells, 7% of the synaptic contacts are from RL terminals, 31% are from F terminals, and 62% are from RS terminals; for interneurons, the values are 47, 24, and 29% from RL, F, and RS terminals, respectively. These numbers can be further broken down by the estimate that roughly half of these RS terminals are from corticogeniculate axons, and nearly all the rest are from parabrachial cholinergic afferents (Erişir et al., 1997).[4] Thus corticogeniculate and parabrachial afferents each provide roughly 30% of the synapses on relay cells and roughly 15% on interneurons. The uncorrected data for interneurons and relay cells on which these figures (from Erişir et al., 1998) are based are similar to those reported by Montero (1991), but the corrections increase the numbers for the smaller RS profiles and reduce the numbers for the larger RL profiles.

In a related study (Weber et al., 1989) of the cat's lateral geniculate nucleus, corticogeniculate axons were labeled with ³H-proline for identification by electron microscopic autoradiography on sections that were treated for GABA immunocytochemistry so that interneuronal processes could be distinguished from relay cell processes. This study concluded that cortical terminals ended predominantly on the dendritic shafts rather than the vesicle-containing F2 terminals and were three times more likely to contact interneurons than relay cells. The first conclusion is in accord with the result of Erişir et al. (1997), but the second conclusion cannot be reconciled with the figures of Van Horn et al. (2000), cited above on the basis of what is known about ratio of interneurons to relay cells (Fitzpatrick et al., 1984; Montero and Zempel, 1985) and about numbers of synapses that contact the relay cells (Wilson et al., 1984).

It is important to note that the critically important driver input from retina has the smallest number of synapses onto relay cells. This reinforces the caution about judging the functional strength of an input on the basis of the numbers of synapses that it contributes. We shall argue this point again more fully in Chapter IX. One further point about these figures is that they represent numbers of synapses as identified in terms of electron microscopic specializations. They do not represent numbers of terminals because in the cat's lateral geniculate nucleus, a single RL terminal commonly makes about nine distinct synaptic contacts, F terminals make about three, and RS terminals make only one (Van Horn et al., 2000). A further point about counts of synaptic junctions is that some afferents function without synaptic specializations. The histaminergic input to the lateral geniculate nucleus from the hypothalamus is an example (Uhlrich et al., 1993; Wilson et al., 1999). This may operate by releasing histamine from sites that show no structural specializations along the terminal course of the axon, with the transmitter diffusing to affect target cells in the vicinity.

---

[4] The parabigeminal afferents mentioned earlier innervate only the C layers.

7. THE GLOMERULI AND TRIADS

The complexity of a glomerulus can vary greatly and is much greater in the lateral geniculate nucleus of a cat or monkey than it is in, for example, the medial geniculate or the ventral posterior nucleus of a rat. To some extent we shall see that the complexity of glomeruli relates to the presence of interneurons, since interneuronal processes provide all of the F2 profiles and possibly some of the F1 profiles in a glomerulus. However, this may not be a general rule. Kultas-Ilinsky and Ilinsky (1991) claim that in the ventral lateral nucleus of the monkey about 25% of the nerve cells are interneurons but that glomerular formations and triads are rare in this nucleus. A detailed comparison of this nucleus with one like the lateral geniculate nucleus of the same species could prove rewarding in showing exactly what are the critical features of the interneuronal structure, the distribution of interneuronal dendrites, or the synaptic connectivity of the relevant processes that produce the reported difference. Within the glomeruli, the RL terminals often lie centrally (Figure 12) among many other synaptic profiles that are also in contact with each other. Although they vary considerably from one thalamic nucleus to another in their complexity, these glomeruli are a characteristic feature of many thalamic nuclei, and it is reasonable to ask about the possible functional significance of these quite striking synaptic arrangements. That is, given that the glomeruli are such a visually impressive and characteristic feature of thalamic synaptic arrangements, we need to ask what it is about the glomerular structures that differentiates them from other synaptic groupings and that may be relevant to the way in which these particular synaptic arrangements function.

The glomeruli have in the past been described as regions of closely grouped synaptic processes that are wrapped in slender sheets of astrocytic cytoplasm. The implication of this description focuses on the possible function of the astrocytic sheets that appear to be wrapping the glomeruli and leads one to look at them as though they may provide a barrier that prevents transport into or out of the glomeruli. However, an effective diffusion barrier requires either a quite significant wrapping, or else some significant reduction of the extracellular space in general, the strategic placement of tight junctions, or a specialization of extracellular material. There is no evidence for any of these in the regions surrounding the glomeruli. And the astrocytic ensheathment is often incomplete. It may be more instructive to regard the glomeruli as regions that are essentially free of any astrocytic processes (see Figures 12 and 13A). It is as though there had been a developmental process that cleared the astrocytic cytoplasm out of the area of the glomeruli and removed it to the glomerular periphery where the appearance of a wrapping is produced. This alternative view of the glomerulus, as a zone free of glia, has two important consequences. One is to help counter the idea that the astrocytic sheets that wrap around glomeruli may actually serve as diffusion barriers, preventing the passive diffusion of materials out of (or into) the glomeruli. The second consequence arises from a comparison with nonglomerular synapses, such as are seen in many other parts of the brain, for example, in the cerebral cortex, the molecular layer of the cerebellar cortex, or the nonglomerular parts of the thalamus. The relationships are shown schematically in Figure 13B and can be contrasted with Figure 13A. Note that in the nonglomerular synapses there is almost invariably a tongue of astrocytic cytoplasm on one or both sides of the synaptic cleft, whereas in the thalamic glomeruli, and also in the cerebellar glomeruli where mossy fibers contact granule cells, the synaptic clefts are all at some distance from the astrocytic cytoplasm.

In view of the known functions of astrocytes, which include ion transport and transmitter uptake at synaptic junctions (Bacci et al., 1999; Pfrieger and Barres, 1996), the absence of astrocytic processes from the glomeruli may reflect functional properties not shared by extraglomerular synapses. For example, it can be proposed that in the glomerular synapses K^+ ions and some transmitters are allowed to accumulate in the immediate, extracellular environment of the synaptic junctions and that they are not removed until they have reached the glial sheets that surround the glomeruli. Earlier evidence (Angel et al., 1967) that repetitive stimulation of brain stem or visual cortex produced an increase in the excitability of optic nerve axons, which was initially interpreted as presynaptic inhibition acting on the retinogeniculate axons, is now more readily interpreted as an extracellular accumulation of K^+ ions because we know that the retinogeniculate axons are not postsynaptic to any other processes that could produce the postulated presynaptic inhibition. It may be relevant that in the cerebellar glomeruli, which, as indicated above, also lack juxtasynaptic astrocytic processes there is evidence that at junctions having multiple sites, but not at those with single sites, transmitter release can lead to changes suggesting a functionally significant accumulation of extracellular transmitter (Silver et al., 1996). As the functional interactions between nerve cells and astrocytes at synaptic junctions become more clearly understood, a new consideration of the relationships seen in the glomeruli will almost certainly prove worthwhile.

We have noted that a common feature of the primary afferents is their synaptic involvement in three closely associated synapses that form a triad. These triads are common within glomeruli, often relate to the grape-like dendritic appendages of relay cells described in Chapter II, but are not an essential part of all thalamic relays. Thus, when one compares the retinogeniculate X and the Y pathways that were introduced in Chapter II, the innervation of X relay cells commonly involves such triads and glomeruli, but that of Y relay cells does not. This relates to the observation that the X relay cells are more likely to have grape-like appendages at their primary branch points than the Y relay cells. In the ventral posterior complex of the rat, axons that come from the dorsal column nuclei relate to triads, thus resembling the X pathway, but spinothalamic axons do not and may be more like the Y pathway (Ma et al., 1987a). The functional significance of this difference is still not clear but will be discussed further in Chapter VI, which deals with the properties of synaptic inputs.

Axon terminals having the fine structural features of the driver afferents are seen in all of the major thalamic nuclei that have been studied. However, as we have indicated, the complexity of the glomeruli varies from very complex, with many interconnected presynaptic and postsynaptic profiles, to a simple triad, and in nuclei that have few or no interneurons triads are correspondingly rare or absent. Although for RL terminals, their participation in triads varies, and although they vary in size and in the preferred postsynaptic surface that they contact (close to cell body, or beyond the first branch point), we shall regard all relatively large axons having round synaptic vesicles, making multiple asymmetrical synaptic and filamentous contacts on proximal dendrites of relay cells, and not forming the postsynaptic element of any synaptic junction as driver afferents in the thalamus. We shall treat them as the drivers of the relay neurons in that thalamic nucleus, particularly if they relate to glomeruli and/or triads and if they are shown to be glutamatergic. The relationships to astrocytic cytoplasm of the simpler synaptic arrangements have not been studied, but the absence of astrocytic cytoplasm from the region of these synapses could also prove of diagnostic significance and may be of functional interest when we have a better knowledge of the role of the astrocytes. Exactly which of these several features are critical for the production of the "driver" function of passing ascending messages on to cortex with relatively little modification is not clear. We are inclined to argue that probably most of these features may be relevant for understanding the driver function.

D. *Afferents from Interneurons and Reticular Cells*

Interneurons provide a numerically major source of inhibitory afferents to relay cells in some but not in all thalamic nuclei. For example, in cats and monkeys, in the thalamic nuclei that have been studied (Arcelli *et al.*, 1997) 20–30% of the nerve cells are interneurons. This is also true for the lateral geniculate nucleus of the rat, but the other thalamic nuclei of the rat have less than 1% interneurons. We have seen that they can contribute to the synaptic circuitry of the thalamus by means of two different types of process, their dendritic presynaptic processes, which are F2 terminals, and their axon terminals, which are F1 terminals. The axons of interneurons resemble the axons of reticular cells. Both have medium-sized terminals that contain flattened vesicles and make symmetrical contacts. These terminals are the F1 profiles and have been traced to glomerular and extraglomerular synaptic sites (Montero, 1987; Ohara *et al.*, 1980; Montero and Scott, 1981). It is important to note that in contrast to the interneuronal dendrites, which form the F2 terminals, participate in triads, and are mostly seen within the glomeruli, the F1 profiles are never postsynaptic to another process. The dendrites of interneurons ramify within about 200 μm of the cell body, and generally, where axons have been demonstrated, they relate to the same volume of the relevant thalamic nucleus and presumably to the same group of relay cells. However, the two types of interneuron, briefly discussed in the last chapter, those described by Tömböl and the putative displaced reticular cells, may prove an exception with axons that ramify beyond the region of the dendritic arbors.

E. GABA Immunoreactive Afferents

Afferents that are immunoreactive for GABA or GAD and that come from extradiencephalic sources have only been mentioned briefly so far. We have very little idea about their function, and we consider them here in terms of the features we have discussed for other afferents. They are a mixed group that includes the axons going from the globus pallidus and pars reticulata of the substantia nigra to the ventral lateral/ventral anterior cell group and to the center median nucleus (Kultas-Ilinsky *et al.*, 1997; Balercia *et al.*, 1996) as well as afferents from the inferior colliculus to the medial geniculate nucleus (Peruzzi *et al.*, 1997) and from the pretectum to the lateral geniculate nucleus (Cucchiaro *et al.*, 1993). There are reasons, discussed in Chapter IX, for thinking that these are likely to be modulators rather than drivers, but direct evidence is not available at present.

We have argued earlier that known drivers lack a branch to the thalamic reticular nucleus, whereas known modulators have such branches. However, since the reticular branches provide an indirect inhibitory innervation for the thalamic cells, the ground rules for the GABAergic axons, which are themselves inhibitory, may well be different. The pretectal afferents send a branch to the reticular nucleus (Cucchiaro *et al.*, 1993), but the axons from the inferior colliculus probably do not send a branch there, since studies of the ascending pathways from the inferior colliculus have failed to show any innervation of the reticular nucleus (Kudo and Niimi, 1980; Le Doux *et al.*, 1985), and the evidence for the globus pallidus and substantia nigra leaves some room for doubt. The pallidal and nigral cells do send GABAergic afferents to the thalamic reticular nucleus (Paré *et al.*, 1990; Cornwall *et al.*, 1990; Gandia *et al.*, 1993; Asanuma, 1994), which at first sight suggests that these afferents will prove to be modulators. However, Asanuma's study showed that the projection to the reticular nucleus arose in the external segment of the globus pallidus, and showed no component to the motor thalamus coming from this part of the globus pallidus. In contrast to this, Sidibe *et al.* (1997), who demonstrated the projection from the internal segment of the globus pallidus to the thalamus, showed no projection to the thalamic reticular nucleus, although there was a projection to the zona incerta, adjacent to the reticular nucleus. Functionally such an incertal connection cannot be treated as a reticular component, so it would appear that the pallidal cells that send axons to the reticular nucleus do not innervate thalamus, and that, conversely, the pallidal cells that innervate the thalamus do not innervate the reticular nucleus. Comparable evidence is not available for the nigral afferents to the thalamus, but there is no evidence currently available to show that the cells that provide the thalamic afferents themselves have branches innervating the reticular nucleus.

Other morphological evidence about these axons does not take the problem much further. Electron microscopic studies show that the pallidal axons can form the presynaptic components of a triadic junction (Kultas-Ilinsky et al., 1997; Balercia et al., 1996) and that at least some relate to primary branch points of dendrites. That is, they show some resemblance to the RL axons. However since the cholinergic afferents that come from the brain stem and that are not likely to be drivers also show such a relationship, we cannot regard this as a useful criterion for determining the function, driver or modulator, of these axons.

The most interesting question regarding these extradiencephalic GABAergic afferents is whether they can act functionally as drivers or whether they all operate as modulators (see Chapter IX). We have suggested that their morphology and connections are not decisive on this issue. The problems that arise if an inhibitory afferent that innervates a spiking cell is to act as a driver are considered more fully in Chapter IX. The point is made that whereas for a nonspiking postsynaptic cell, either an EPSP or an IPSP can convey the same information, as commonly happens in the retina; for a cell that passes on information as action potentials, the postsynaptic effect of an IPSP differs significantly from that of an EPSP and is unlikely to meet the requirements for a functional driver afferent. For the reasons discussed in Chapter IX we regard inhibitory, GABAergic inputs to thalamus as unlikely candidates to be drivers and suggest they are all modulators.

F. Afferents to the Thalamic Reticular Nucleus

Afferents to the thalamic reticular nucleus come from the nearby thalamic nuclei and from the cortex. These are collateral branches of the thalamocortical and corticothalamic axons that link individual thalamic nuclei with functionally related groups of cortical areas. The thalamic reticular nucleus can be divided into a number of distinct sectors (Jones, 1985), each related to a group of thalamic nuclei and their cortical areas, and each of the major sensory modalities relates to one sector as do the motor pathways and the mamillothalamic pathways. Both groups of axons, from thalamus and from cortex, are glutamatergic (Eaton and Salt, 1996; Kharazia and Weinberg, 1994) and make asymmetrical synaptic junction on the dendrites of the cells in the reticular nucleus (Ohara and Lieberman, 1985; Williamson et al., 1993).

Other afferents reach the thalamic reticular nucleus from several sources. These include GABAergic axons from the external segment of the globus pallidus (Asanuma, 1994) and from the substantia nigra (Paré et al., 1990), which terminate in a rostral part of the reticular nucleus that probably corresponds to the sector of the nucleus also receiving afferents from the motor cortex. GABAergic afferents also come from cells scattered in the basal forebrain, and these have a more widespread distribution in the reticular nucleus without contributing to the dorsal thalamus (Asanuma and Porter, 1990; Asanuma, 1989; Bickford et al., 1994). In addition, there are GABAergic afferents from the pretectum that project to the lateral geniculate nucleus and that send branches to the visual sector of the reticular nucleus (Cucchiaro et al., 1993). Similarly, the cholinergic afferents that go to the thalamus from the parabrachial region of the brain stem also send branches to the thalamic reticular nucleus (Uhlrich et al., 1988). The fine structure of the cholinergic afferents to the thalamic reticular nucleus has been studied in rats (Hallanger and Wainer, 1988). Asanuma (1992) has described noradrenergic afferents to the reticular nucleus, and Manning et al. (1996) described histaminergic afferents coming from the hypothalamus and distributing to the visual sector of the reticular nucleus and the perigeniculate nucleus.

Within the reticular nucleus there are possibilities for local connections between neurons. Dendrodendritic junctions have been described in cat and monkey (Deschênes et al., 1985; Ohara, 1988; but see Williamson et al., 1994). Pinault et al. (1997) described dendrodendritic synapses in rats and noted that local, intrareticular branches of reticular cell axons could form the postsynaptic component of a synapse, so establishing axoaxonal contacts. Such local connections are likely to play a role in generating rhythmic activity in the reticular nucleus (see Chapter V, Section B.3) and are likely to add further to the complexity of the interconnections established within the nucleus.

It can be seen that most of the afferents to the reticular nucleus are shared either with a limited part or with all of the dorsal thalamus. The balance of these inputs in the final action that they can have on the thalamic relay cells and thus on the messages that reach the cortex depends in the first place on whether they are making local connections to a limited part of the reticular nucleus or global connections to most or all of it, in the second place on the details of the way in which these several pathways distribute their connections to the relay cells, the interneurons, and the reticular cells, and in the third place on how the relevant thalamic and reticular cells are interconnected with each other (see Figure 8 and Sections B.3 and B.5). These are still areas that need further exploration.

G. *Some Problems of Synaptic Connectivity Patterns*

1. THE RELATIONSHIP OF TWO DRIVER INPUTS TO A SINGLE THALAMIC NUCLEUS: DOES THE THALAMUS HAVE AN INTEGRATIVE FUNCTION?

There is an important question about the driving afferents that has so far received limited attention for most thalamic nuclei. Where a cell group receives driving afferents from more than one source, one needs to know how the two groups of afferents relate to each other, to the individual thalamic glomeruli, and to their postsynaptic neurons. The problem is again well illustrated in the lateral geniculate nucleus of the cat, where the retinogeniculate X and the Y pathways have a common terminal site in one set of layers but show no significant interaction. With only rare examples of mixing, X cells have glomerular synapses, Y cells have extraglomerular synapses, and any one geniculate cell serves as a relay for either the X or the Y pathways, not both (Wilson et al., 1984; Hamos et al., 1987). Comparable problems arise in the thalamic nuclei that receive both first and higher order afferents, and also in the ventral posterior nucleus, where spinothalamic and lemniscal terminals overlap. There is some indication that these two pathways may share postsynaptic cells in rats (Ma et al., 1987a), but for the monkey the evidence is still not in, although in general there is good reason to think that the two pathways engage different cell types in adjacent but interdigitating parts of the nucleus (Krubitzer and Kaas, 1992). The nature of the relationships established by two (or more) driver pathways within a thalamic nucleus is likely to prove crucial to understanding the functional organization of that nucleus. If functionally distinct relay pathways do not share cells in the transmittal of the messages carried by the drivers, then one can reasonably regard the thalamic relay in general as comparable to the geniculate relay in producing no changes in the receptive field properties on the retinogeniculocortical pathway. That is, the thalamic relay is concerned with the transmittal of a message, not with integration of messages. Where one finds that two pathways share a relay, as possibly in the ventral posterior nucleus of the rat, one would need to look closely at how those two inputs interact functionally in the relay, in terms of the relay cell responses, and also in terms of their synaptic arrangements on the relay cell dendrites and, where relevant, within the glomeruli.

2. THE COMPLEXITY OF MULTIPLE LAYER 6 AFFERENTS TO THE THALAMUS AND SOME RELATED PROBLEMS

An unsolved problem about the corticothalamic pathway to first order thalamic nuclei concerns the possibility that these axons do not represent a homogeneous population. We have discussed the dangers of creating distinct classes on the basis of a few criteria that have not been shown to have a clear bimodal or multimodal distribution. However, the opposite danger, of looking for a single function where several functions are contributing to a system like the corticothalamic system, has also to be recognized. Katz (1987) described two types of cell in cortical layer 6 of area 17 projecting to the lateral geniculate nucleus in cats. These differed in the richness of their dendritic arbors and local cortical collaterals. In primates it has been shown that distinct populations of cortical cells project to different layers of the lateral geniculate nucleus (Fitzpatrick et al., 1994; Casagrande and Kaas, 1994) and Bourassa and Deschênes (1995) showed that cells in the upper parts of cortical layer 6 of area 17 in the rat project to the lateral geniculate nucleus, whereas cells in the lower parts send axons to the lateralis posterior nucleus and the lateral geniculate nucleus, with single axons innervating both nuclei. Tsumoto and Suda (1980) described three types of cortical cell projecting to the lateral geniculate nucleus of cats, basing these types on axonal conduction velocities and receptive field properties. Interestingly, they found that one type, having intermediate values for conduction velocity, is absent from the monocular segment of the nucleus, which led them to suggest that this particular type of corticothalamic cell may be specifically concerned with binocular interactions, a function that was earlier proposed by Schmielau and Singer (1977) for corticogeniculate axons and that would fit well with the observation that many corticogeniculate axons pass through the layers in the direction of the lines of projection and can have terminals in adjacent layers that receive retinal innervation from different eyes (Guillery, 1967a; Robson, 1983; Murphy and Sillito, 1996).

As the possible functions of the corticothalamic pathway from layer 6 are considered, it may be important to bear in mind that there are a great many of these axons in most (probably all) thalamic nuclei and that they may span a range of connectional patterns and functions that will only be fully appreciated on the basis of further studies of these pathways. That is, the precise origin in particular cortical cytoarchitectonic areas of the layer 6 corticothalamic afferents has only been defined to a limited extent for some thalamic nuclei, and as the functional significance of these axons becomes clearer, so there will be an increasing need for information for each thalamic nucleus about precisely which cortical areas send layer 6 afferents to that nucleus and how these several inputs are related to each other. It will be important to define where two thalamic nuclei share a cortical afferent, as do the first and higher order visual relays in the study of Bourassa and Deschênes (1995), and it will be equally important to know where distinct corticothalamic afferents relate to a single thalamic relay cell or group of relay cells. That is, this is a system that shows both divergence and convergence. For example, we know that the lateral geniculate nucleus of the cat receives layer 6 afferents from cortical areas 17, 18, and 19. There is significant overlap of these three inputs, but the extent to which inputs from two different cortical areas impinge on the same geniculate cell or the same parts of any one geniculate cell is unknown. Areas 17 and 18 distribute to all geniculate layers, with the layer 18 afferents possibly being somewhat more focused on intralaminar regions, whereas area 19 sends its axons to the C layers almost exclusively in the lateral geniculate nucleus and also sends a significant layer 6 component to the lateralis posterior nucleus (Updyke, 1977; Murphy *et al.*, 2000). There are additional afferents to parts of the lateral geniculate nucleus from other visual cortical areas (Updyke, 1981), but we know nothing about how these several afferents interact in the control of geniculate activity.

The anterior thalamic nuclei receive RS afferents not only from the cingulate cortex, which is the area of cortex to which the relay cells project, but also from the hippocampus (Somogyi *et al.*, 1978). The cortical projection to the mediodorsal nucleus as described by Négyessi *et al.*, (1998) is of interest because there is a bilateral corticothalamic pathway. The uncrossed pathway is relatively large and includes axons that have

G. Some Problems of Synaptic Connectivity Patterns

RS terminals as well as some axons that have RL terminals. In contrast, the smaller crossed projection includes only the RS type of terminal. For the center median nucleus it has been reported that injections of retrograde tracers into the nucleus label cortical pyramidal cells in layer 5 but not in layer 6 of the motor cortex (Catsman-Berrevoets and Kuypers, 1978; Royce, 1983). This suggests that there should be a significant RL or type II input to the center median nucleus from motor cortex with no significant type I or RS input. However, fine structural studies of this pathway (Harding, 1973; Balercia et al., 1996) show a predominant RS input. Possibly the layer 6 cells in motor cortex are resistant to the retrograde marker used, perhaps most of the RS terminals that come from the motor cortex come from layer 5 cells, or possibly the RS terminals that were described were mainly parts of larger RL terminals and would have been identified as such had the studies used serial sections. The issue remains unresolved.

Experiments designed to test the action of corticothalamic afferents coming from visual cortex have generally focused on one or two cortical areas (17, 18) in anesthetized animals (Kalil and Chase, 1970; Richard et al., 1975; Schmielau and Singer, 1977; Baker and Malpeli, 1977; Sillito et al., 1994; Cudeiro and Sillito, 1996) and have demonstrated generally subtle effects of cortical cooling, destruction, or stimulation. We argue in later chapters (VIII and IX) that the effects produced by manipulations of these modulatory pathways may well be elusive and are perhaps more likely to be relevant to responses of an awake behaving animal than an anesthetized one. It is reasonable in the first instance to search for the action of corticothalamic afferents that come from primary receiving areas like area 17, but in the long run we will need to know how each of the several areas giving rise to a type I innervation acts, and what happens when several cortical areas are sending concurrent afferents to a thalamic relay.

The nature of the modulatory influence that the layer 6 cells exert in the thalamus is discussed in Chapter IX, but if one is to understand the circumstances under which this modulatory influence acts, then one needs to know about the activity patterns of the relevant layer 6 cells. Relatively little information is available on this score, and generally it is recorded in terms of receptive field properties of the layer 6 cells in anesthetized animals. An important point is that the functional significance of at least some of the corticothalamic axons may be entirely lost in an anesthetized animal. Their function is most likely to be exerted in a conscious animal that is maintaining attention on one particular part of its sensory environment or is in the process of switching from one part to another. For instance, Tsumoto and Suda (1980) reported cortical cells that were silent in their anesthetized preparations and that could only be identified on the basis of antidromic stimulation from the thalamus. To address this requires detailed study of the firing patterns of these layer 6 cells under different conditions of visual stimulation *and* behavioral state, and the firing patterns seen under these conditions may well be quite different from those observed in the classical receptive field studies.

We have earlier pointed out that the corticothalamic axons coming from layer 6 send collateral branches to the thalamic reticular nucleus. However, it is not known whether each of these corticothalamic axons has such a branch, or whether only some of them do. A preparation that reveals a group of corticothalamic axons passing through the reticular nucleus and continuing on to the thalamus shows that the plexus in the reticular nucleus is well localized and relatively dense, but in its total extent it is modest compared with that developed in the thalamus (Murphy and Sillito, 1996). An intuitive, nonquantified view would suggest that either each reticular branch is rather modest, or that only some of the corticothalamic axons passing through the nucleus have reticular branches.

H. Summary

The most important distinction among afferents to the thalamus is likely to be that between drivers and modulators. For some of the afferents to first order nuclei the functional distinction is clear: the type II axons that have RL terminals are drivers, whereas the type I axons with RS terminals are modulators. The drivers are represented by ascending afferents to first order relays and by corticothalamic afferents from layer 5 to higher order relays. The modulators are corticothalamic afferents from layer 6 and other components coming from subcortical origins. We have proposed that this functional distinction applies throughout the thalamus and have explored differences in the distribution and the synaptic connectivity patterns of the afferents with relay cells and with interneurons. The generality of this proposed distinction for the thalamus remains to be tested, and in a later chapter (Chapter IX) functional tests for distinguishing drivers from modulators are proposed. The possibility that there are several types of modulator having distinct functions remains largely unexplored. An appreciation of thalamic circuitry must include the patterns of connectivity of several axonal groups that include candidate modulatory axons from brain stem and basal forebrain and the important afferents that come from the thalamic reticular neurons and from the interneurons. The presynaptic dendrites of the interneurons provide yet another source of input to many, but not to all, thalamic relay cells.

A survey of thalamic afferents and of the major synaptic connections that they establish in the thalamus shows, in the first place, that the details of the synaptic connections, and the possibilities for interactions between drivers and modulators, are extremely complex. In the second place, such a survey shows us how little we really understand about what the circuitry of the thalamus is like. If we focus on the way in which afferents from different cortical areas, different cortical layers, or different cortical cell types relate to each other, to other afferents, and to particular thalamic relays, we find that most of the facts that are critical for a functional appreciation of these pathways are not yet available. The same is true if we look for the details of synaptic interconnections in any one

H. Summary

thalamic nucleus, particularly in a higher order nucleus. The introduction of the electron microscope and, more recently, the development of a variety of techniques for tracing axons individually or in bundles have provided a great deal of new information about the thalamus, but perhaps the most important conclusion that can come out of this chapter is the literally abysmal degree to which we lack key information about details of thalamic organization, information that is technically obtainable but that will require a clear sense of what are the important questions and a significant amount of painstaking work.

I. Some Unresolved Questions

1. How is the distinctive morphology of the type I and type II axons related to the developmental or functional constraints that they obey?

2. What are critical features by which one can unmistakably recognize a driver afferent: (a) the structural features; (b) the pharmacological and physiological features? (See also Chapter IX for a further view on this question.)

3. Can driver afferents be generally characterized as not having a reticular branch?

4. Can one define for each thalamic nucleus, and consequently for each corresponding neocortical area, one or more set(s) of driving afferents; and can one use this information to clarify the functional role of each thalamocortical pathway?

5. Are there thalamic cells that receive inputs from two functionally distinct drivers and that thus serve as thalamic integrators?

6. Are there rules that define how the corticofugal axons arising in layers 5 and 6 of any one cortical area are distributed in the thalamus?

7. What are the functional consequences of having groups of synaptic processes arranged within glomeruli, and how important are the glial relationships in producing these consequences?

8. What is the function of the triads?

9. Do all corticothalamic axons from layer 6 innervate relay cells and thalamic reticular cells and interneurons, or do some just innervate one or two of these cell classes?

10. Are any of the extradiencephalic GABAergic afferents drivers? Or are they all modulators?

11. What are the detailed functional relationships established by cortical, reticular, and interneuronal synapses on the relay cells?

CHAPTER IV

Intrinsic Cell Properties

One of the key questions that needs to be answered is how thalamic circuitry affects the relay of information to cortex. There are likely to be several different ways in which thalamic circuitry can act on the relay, and each of these will involve many interrelated properties of the relay, including the nature of the various driver and modulator inputs, the properties of synaptic transmission from these afferents to the thalamic neurons, and finally the intrinsic properties of these neurons, because these properties dictate how synaptic inputs will be integrated to control the cell's firing properties. We begin here with a consideration of these intrinsic neural properties, and later in Chapters V and IX we shall consider the nature of synaptic inputs to these cells.

A. Cable Properties

In the general case of a classical neuron, with an output limited to an axon, the only signal that can be relayed for the sort of distances (i.e., > 1 cm) required by thalamocortical axons requires the generation of conventional Na^+/K^+ action potentials. Such axonal signaling is used by thalamic relay cells, reticular cells, and probably most interneurons. We need to understand how synaptic potentials generated in the dendrites, where nearly all synaptic inputs are found, affect the soma or axon hillock, the site where action potentials are usually generated.[1]

A few decades ago, our view of a neuron was that it more or less

---

[1] In some neurons in other parts of the brain, there is evidence that action potentials can be initiated in the dendrites, but this issue has not been addressed for thalamic cells. Also, some neurons have axons that emanate from proximal dendrites rather than the soma, but in thalamic relay cells and interneurons, axons generally arise from the soma.

linearly sums excitatory and inhibitory synaptic inputs to create a net voltage change at the axon hillock, where firing of action potentials is initiated. This was thought to be all that was needed to predict the output of the neuron. That is, when postsynaptic potentials were summed to depolarize the cell past its firing threshold, an action potential would be initiated, or if the cell were already firing, the further depolarization would increase its firing rate. Conversely, if the summed postsynaptic potentials produced a net hyperpolarization, the cell's firing rate would drop or cease altogether. We now know that this concept of a linearly summing postsynaptic potential is an oversimplification because membrane properties show highly nonlinear, voltage-dependent properties. Nonetheless it is a first step to understanding how inputs to a cell can be integrated to control the cell's output.

Even with the simplifying assumption that postsynaptic potentials sum linearly, the problem of how this can be analyzed by the investigator is not trivial. Imagine that a synaptic input at some specified dendritic site locally changes an ionic conductance in the postsynaptic membrane, allowing current in the form of charged ions to flow into (or out of) the cell at that site. How does this event change the membrane potential at the soma or axon hillock? This will depend on the flow of current that is initiated and on how it spreads through the cell. The actual pattern of current flow is mostly related to the complex, three-dimensional architecture of dendritic arbors. Not all the current will flow directly to the soma because some will flow in the opposite direction toward more distal dendrites. As each dendritic branch point is passed in either direction, the current flow will divide from the active branch into adjacent branches. Thus if we start with the current flowing toward the soma, once a branch point is reached, the current will divide, some heading down the parent branch toward the soma, and some heading down the other daughter branch(es) away from the soma. This process is repeated as other branch points are reached, regardless of the direction of current flow toward or away from the soma, until either the soma or a dendritic ending is reached. Finally, because the membrane enclosing the dendrites and soma is itself somewhat permeable to electric current, some of the current will flow out of all parts of the dendrites and the cell into the extracellular spaces.

A hydraulic analog often helps to explain this phenomenon. Think of the dendritic tree as an intricately branched rubber hose with tiny holes all along the surface. These holes make the hose slightly leaky, and thus water will leak in much the same way that current will leak across the neuronal membranes. If you inject water at some specific "dendritic" site, that will increase the water pressure, which is the hydraulic analog of increasing the voltage difference across the membrane of the neuron. This leads to water flow in all directions throughout the branched hose, with some of it leaking out through the surface holes. A more formal term for electrical "leakiness" as applied to cell membranes is "conductance": the greater the conductance the more current (or water) will flow across the membrane. The amount of leakiness or conductance matters: the less the conductance (i.e., with fewer or smaller holes), the larger the amount of

A. Cable Properties

injected current that will reach the soma and axon hillock. Now, to further complicate the determination, imagine that the amount of membrane conductance (to electric current) or leakiness (to water) varies in a complicated fashion with time. To make matters even worse (but not quite hopeless), imagine finally that multiple sites along the dendritic tree (or hose) can have current (or water) injected with variable temporal interrelationships. This is the analog of different synapses becoming active with a variety of temporal interrelationships. The problem is to determine the amount of extra electric current (or water) that will appear at the soma, because this is the ultimate determination of the pattern of action potentials that will ensue.

To begin solving this difficult problem, neurons may be modeled as passive cables (Jack *et al.*, 1975; Rall, 1977), and the postsynaptic potentials can be thought of as being conducted electrotonically through the dendritic arbor, and particularly from synaptic sites to the soma or axon hillock. Such modeling remains the chief linking hypothesis between cell shape, the distribution of synaptic inputs, and the efficacy of synapses in the production of postsynaptic action potentials. It has thus proven a useful first step in assessing the impact of synapses at different dendritic locations. One must be mindful of the many assumptions and simplifications typically used in cable modeling because these limit the validity of the model. For one example, various parameters that are often impractical to measure in neurons, such as electrical resistance of the cytoplasm and electrical resistance and capacitance of the membrane, can be estimated from measurements made in other cells (Jack *et al.*, 1975). Although this assumption has not been explicitly tested for neurons, the cable values computed with these parameters seem to be within reasonable bounds. For another, these parameters are typically, but not always, assumed to be uniform spatially and temporally, but there is evidence to the contrary based, among other factors, on nonuniform distribution of voltage-sensitive ion channels, the opening or closing of which alters membrane conductance. However, membrane conductance can be viewed as just another cable parameter, and these dynamic alterations can be taken into account in cable modeling. Finally, most cable models assess only the effects of an isolated synaptic event and do not attempt to compute the various spatial and temporal combinations of multiple synaptic activations. For a real neuron, the electrical properties of the membranes or cytoplasm may not be distributed uniformly throughout, these values almost certainly change with time (see later), and complex spatiotemporal combinations of synaptic activation are common. Here again, the effect of active synapses is to change membrane conductance locally, and this can be accounted for in cable modeling.

1. CABLE PROPERTIES OF RELAY CELLS

Several experiments have described the cable properties of thalamic relay cells from *in vitro* slice preparations, but these tend to differ from the more physiological properties of the *in vivo* preparation for two reasons (Holmes and Woody, 1989; Bernander *et al.*, 1991; Destexhe *et al.*, 1996,

1998a). First, slices often tend to cut off significant portions of the dendritic arbors of relay cells, leaving a cell with only a partial arbor. Although the cut ends of dendrites appear to close, thus repairing major leaks caused by the cuts, this partial arbor has less surface membrane area and thus less overall leakage or membrane conductance than does the intact cell. Second, and probably more significantly, compared with the *in vivo* condition, there is much less spontaneous activity in most *in vitro* slice preparations, and thus much less synaptic activation of a recorded cell. Less synaptic activation means much less membrane conductance because most active synapses result in opening ion channels, thus making the postsynaptic cell more leaky. Notice that both of these differences act to make cells appear less leaky to electric current *in vitro*, and this affects the final estimation of cable properties. It is probably more physiologically appropriate to consider cable properties of thalamic cells as determined from *in vivo* experiments. However, although it is true that *in vivo* experiments, in principle, offer a more physiological estimate of cable properties, the fact that recordings are generally of higher quality and easier to control *in vitro* somewhat negates these arguments.

The only thalamic relay cells that have been formally tested for cable modeling during *in vivo* recording are those of the lateral geniculate nucleus in cats (Bloomfield *et al.*, 1987). Two of the major factors that determine cable properties of neurons are dendritic geometry and the electrical properties of the cytoplasm and membranes of the cell. Because the morphological variations among relay cells is quite similar across a wide range of thalamic nuclei and species, and because passive, intrinsic electrical properties are also thought to be quite similar for neurons throughout the brain, it seems a good guess that the passive cable properties seen for geniculate relay cells generally apply to other thalamic relay cells. This is, nonetheless, a point that has not been experimentally verified.

When modeled as passive cables, relay cells of the lateral geniculate nucleus in cats seem to be electrotonically compact, suggesting that steady-state (i.e., DC) voltage changes applied even at the most distally located synaptic sites will attenuate by less than half *en route* to the soma. In the water analogy developed earlier, this would be like a hose with relatively little leakage because the holes are either few or small. Also, current injected into one point in the dendritic arbor by an active synapse will have significant effects throughout the dendritic arbor as well as at the soma. This is illustrated in Figures 1A and 1B for two typical relay cells. Here, we have modeled the cable parameters based on morphological features; electrophysiological measures of input resistance and membrane time constant; and assumed values for membrane resistance and capacitance, cytoplasmic resistance, etc. We then inject steady-state current into a distal dendritic locus of each model, which roughly mimics a synaptic activation there, and we determine how this affects the membrane voltage at various dendritic loci and at the soma (and axon hillock).

It should be noted that this example and further considerations of Figure 1 show the effect of the spread of *steady-state* current injections or those that vary slowly with time. This is because the resistive-

A. Cable Properties

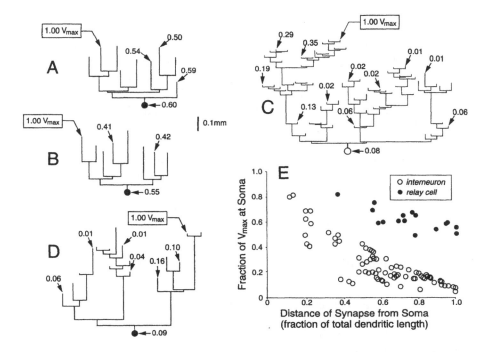

Figure 1
(A–D) Cable modeling of the voltage attenuation that occurs within the dendritic arbors of two relay cells (A and B) and two interneurons (C and D) after the activation of a single synapse (i.e, of a single voltage injection). The cells from the cat's lateral geniculate nucleus were labeled by intracellular injection of a dye *in vivo*, and the stick figures represent a schematic view of one primary dendrite from each cell with all of its progeny branches. Each branch length is proportional to its calculated electrotonic length. The site of voltage injection is indicated by the boxed value labeled 1.00 V_{max} (maximum voltage). Voltage attenuation at various terminal endings within the arbor and soma is indicated by arrows and given as fractions of V_{max}. (E) Attenuation at soma of single voltage injection placed at different dendritic endings as function of anatomical distance of the voltage injection from the soma. Each voltage injection mimics the activation of a single synapse. The abscissa represents relative anatomical distances normalized to the greatest extent of each arbor, and the plotted points represent values from the four cells shown in A to D. Redrawn from Bloomfield and Sherman (1989).

capacitative properties of the membrane act like a low-pass temporal filter with a time constant of roughly 10 to 50 msec, sometimes longer. In practice, this frequency-dependent attenuation along a cable means that faster postsynaptic potentials will attenuate more during conduction to the soma than will slower ones. As noted in Chapter V, postsynaptic potentials result from activation of two different classes of postsynaptic receptor, *ionotropic* or *metabotropic*. Activation of the former leads to postsynaptic potentials sufficiently fast that they will be significantly attenuated by the resistive-capacitative properties of the membrane, whereas activation of the latter produces such slow postsynaptic potentials that they will be minimally attenuated by these membrane properties.

A major reason that these relay cells are electrotonically compact is that their dendrites are relatively thick through most of their extent. The cross-sectional area of a dendrite is proportional to the square of its diameter, but the surface membrane area is only linearly proportional to the diameter. Thus the ratio of its cross-sectional area to its membrane area for a dendrite increases with increasing thickness. Because the relative resistance of the cytoplasm or membrane is inversely related to its area, the larger the ratio of cytoplasmic area to membrane area, the lower the resistance of the cytoplasm relative to that of the membrane. As a result, a thicker dendrite allows more current from synaptic activation to flow down the path of least resistance through the cytoplasm to the soma, and less of this current will leak out through the membrane.

2. CABLE PROPERTIES OF INTERNEURONS AND RETICULAR CELLS

Interneurons of the cat's lateral geniculate nucleus appear to be much larger electrotonically than are relay cells, meaning that, if the dendrites act as passive cables, voltage changes created at distal dendrites would be likely to be much more attenuated at the soma. Two examples of interneurons based on modeling from morphological and electrophysiological data are shown in Figure 1C,D, with the same method as applied to the relay cells to compute measures of membrane voltage expected at the soma and various dendritic loci when a current is injected at one distal dendritic locus (Bloomfield and Sherman, 1989). A major reason for the electrotonically extensive dendritic arbor of interneurons seems again to be related to dendritic diameter: although dendrites of interneurons are roughly as long as those of relay cells, those of interneurons are much thinner, and this would result in more current from synaptic activation leaking across the membrane *en route* to the soma.

As is the case for relay cells, to date the only formal cable modeling of interneurons recorded *in vivo* has been limited to the lateral geniculate nucleus of the cat (Bloomfield and Sherman, 1989). It is not at all clear whether all thalamic interneurons are similar (see Chapter II). More information is needed about the general properties of interneurons and their variation across thalamic nuclei and species.

Cable modeling of neurons of the thalamic reticular nucleus in rats suggests that they are electrotonically extensive (Destexhe *et al.*, 1996). If the only output of a reticular cell is the axon, as is the case with relay cells, one might conclude from teleological arguments either that such cells must be relatively compact electrotonically or that their dendrites are not passive and may actively conduct postsynaptic potentials toward the soma. Otherwise, synaptic inputs located peripherally in the dendritic arbor would be ineffective in influencing the neuronal output, and the creation of such inputs would seem pointless. If, however, a neuron has dendritic outputs (with or without axonal ones), it might make sense for its dendritic arbor to be relatively extensive because inputs that can

A. Cable Properties

influence dendritic outputs nearby need not influence the soma or axon hillock to affect the cell's output. Indeed, as we shall argue in the following paragraphs for interneurons, it makes sense for cells with dendritic outputs to be electrotonically extensive. Thus one might predict that reticular cells in the rat would have dendritic outputs, with reticular cells forming dendrodendritic contacts with each other, and evidence for this exists (Pinault et al., 1997), although it has also been questioned (Ohara and Lieberman, 1985). Furthermore, evidence for such dendrodendritic contacts has been found among reticular cells of the cat (Ide, 1982; Deschênes et al., 1985), but not the monkey (Williamson et al., 1994). Unfortunately, no published evidence of cable properties of reticular cells for species other than the rat exist, and given claims of interspecies differences in dendrodendritic connections among these cells, their cable properties might prove particularly interesting.

3. Implications of Cable Properties for the Function of Relay Cells and Interneurons

In view of their cable properties, it is interesting to compare and contrast how relay cells and interneurons integrate synaptic inputs. Because relay cells have a single axonal output, it is sufficient to consider synaptic integration in terms of how inputs affect membrane voltage at the soma or axon hillock. The dendritic architecture and branching pattern of these cells suggest rather efficient current flow throughout the dendritic arbor, meaning that postsynaptic potentials will attenuate relatively little from the dendritic site of origin to the soma. As indicated previously, and shown in Figure 1, the maximum voltage attenuation for steady-state voltage changes from the most distally located synapse to the soma is estimated to be less than one-half (Bloomfield et al., 1987; Bloomfield and Sherman, 1989). In relay cells, it thus seems likely that all active synapses, regardless of their location in the dendritic arbor, may significantly influence the axon hillock and that synaptic integration in relay cells involves large-scale summation of all such active synapses.

Interneurons appear to function quite differently. Many and perhaps all interneurons have axons (Hamos et al., 1985; Montero, 1987; see Chapter II for more detailed discussion). However, in addition to the synaptic outputs of these axons, interneurons also have synaptic terminals emanating from peripheral dendrites. These are the F2 terminals described in Chapter III, and they form inhibitory synapses onto relay cells. Interestingly, most and perhaps all are *postsynaptic* as well as *presynaptic*. In the lateral geniculate nucleus, they receive synaptic input from various terminals, mostly retinal, but also some from brain stem or from GABAergic axons forming RS or F1 terminals, respectively. Their synaptic relationships are described in Chater III. Typically, they engage in the triadic synaptic arrangements in which the F2 terminal is postsynaptic to a retinal terminal, and both the F2 and retinal terminals are presynaptic to the same relay cell dendrite.

The typical interneuron thus has two distinct types of output pathway, which are summarized schematically in Figure 2. The axonal

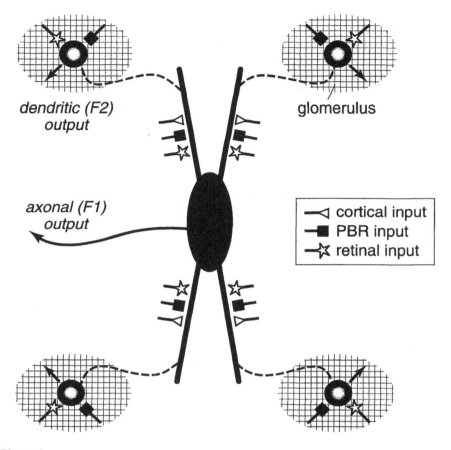

Figure 2
Schematic view of hypothesis for functioning of interneurons in the cat's lateral geniculate nucleus. Retinal and nonretinal inputs are shown both to the glomeruli and to the proximal dendrites. The glomerular inputs are onto F2 terminals, leading to F2 outputs from the dendrites, whereas the inputs to the proximal dendrites and soma lead to F1 outputs from the axon. The dashed lines indicate the electrotonic isolation between glomeruli and the proximal dendrites plus soma. This isolation suggests that the two sets of synaptic computations, peripheral for the glomerular F2 outputs and proximal for the axonal F1 outputs, transpire in parallel and independently of one another. Most glomeruli are also functionally isolated from one other.

terminal outputs are controlled effectively only by inputs to the soma or proximal dendrites because the inputs to more distal dendrites and especially onto the F2 terminals are electronically too distant to have much effect on membrane potential at the axon hillock. Evidence to support this comes from pharmacological studies (summarized in the next chapter) in which drugs thought to depolarize the F2 terminals directly have no recordable effects in the soma (Cox *et al.*, 1998). In turn, events at or close to the soma are too distant electrotonically to have much effect on the F2 terminals. The F2 terminal outputs are controlled by their local inputs, so we have multiplexing in the interneuron: this cell can permit both synaptic input/output routes—via the axonal F1 terminals and via the dendritic F2 terminals—to operate simultaneously. It also seems likely that clusters of F2 terminals are themselves effectively isolated from each other so that the dendritic arbor of the interneuron contains many local circuits performing independent input/output operations. As indicated in Chapter II, all geniculate interneurons so far studied with intracellular dye injection after electrophysiological recording have axons, although the number studied this way is few (Friedlander *et al.*, 1981; Hamos *et al.*, 1985; Sherman and Friedlander, 1988). The methods used to sample these interneurons before filling require that they fire action potentials, otherwise they would be missed, and thus any existing axonless interneurons, which possibly would not generate action potentials, might not have been detected in these experiments (see also Chapter II). An interneuron without an axon would still provide outputs by way of the dendritic F2 terminals and thus may not require action potentials to function.

B. Membrane Conductances

As noted previously, it used to be thought that cable modeling alone provided a reasonably complete and accurate picture of how neurons integrate their synaptic inputs to produce specific firing patterns. However, starting about two decades ago, we began to realize that neurons and their dendritic arbors do not act like simple cables. Instead, they display numerous nonlinear membrane properties that further complicate the job of understanding neuronal input/output relationships.

These membrane nonlinearities are due to various membrane conductances that can be switched on or off, thereby affecting the flow of electric current in the form of charged ions into or out of the cell. In addition to affecting membrane potential, these conductance changes also affect membrane resistance and thus the cell's passive cable properties described previously. Take, for example, the interneuron. Cable modeling based on one specific membrane resistance leads to one value of electrotonic length, and as noted earlier, typical values for this suggest that these cells are electrotonically extensive. If membrane resistance can vary on the basis of different conductances described more fully later, then so can electrotonic extent. Perhaps in some functional modes when the membrane is relatively leaky because a conductance is active (e.g., a K^+ conductance; see later), the interneuron is so extensive electrotonically that effects at dendritic F2 terminals are isolated from the soma, but if the conductance can then be switched off, the membrane resistance increases, leading to a more electrotonically compact cell with a possible breakdown of the isolation between soma and F2 terminals.

Synaptic inputs create changes in ionic conductances (see Chapter V), which underlie changes in membrane potential or postsynaptic potential. Many of the variable membrane conductances seen in cells can also be affected by membrane voltage, being regulated simply by changing membrane potential, so synaptic inputs can have other effects on certain conductances through postsynaptic potentials. Still other conductances are controlled by the changing intracellular concentration of specific ions (e.g., Ca^{2+}), which as shown by examples provided in the following, can be controlled by certain synaptic events. Thalamic cells are fairly typical of neurons throughout the brain in having a rich array of these membrane conductances. Because the discovery of these conductances is still ongoing, it is likely that their full array has yet to be defined. Because these conductances ultimately change membrane potential, and many also have complicated temporal properties, they represent a complex influence on the axon hillock that is not a part of the cable model, but that must be taken into account. Thus cable modeling, although useful, is severely limited. We must now look at these nonlinear membrane conductances and at the quite dramatic effects they have on how a cell responds to its synaptic inputs, and in Chapter V we shall add to this account the synaptic properties of thalamic neurons.

1. Voltage-Independent Membrane Conductances in Relay Cells

The transmembrane voltages that are the signature of neurons and other electroresponsive cells (e.g., muscle cells) are created initially by specific ionic pumps that create differential concentrations of various ions across the membranes. However, the membranes are not completely impermeable to ions, and they constantly leak across the membrane. The ionic conductances underlying these leaks generally do not depend on membrane voltage. The result of the leaks and pumps is a stable, equilibrium concentration gradient for each ion, and this leads to the typical resting potential. This for a typical neuron is usually between -65 and -75 mV. Of the ions that dominate this process, namely K^+, Na^+, and Cl^-, the passive leakage is greatest for K^+. This is the so-called K^+ "leak" conductance. However, if only K^+ leaked across the membrane, the resting potential would be at the reversal potential for K^+, or about -100 mV. The smaller leakage of Na^+ and Cl^-, which are driven by more positive reversal potentials, combines with the K^+ "leak" conductance to create the resting membrane potentials observed, typically between -65 and -75 mV. As we shall see in Chapter V, one important effect of certain synaptic actions is a change in this K^+ "leak" conductance.

2. Voltage-Dependent Membrane Conductances in Relay Cells

a. Action Potentials

Voltage-dependent conductances should not seem mysterious, because the action potential as classically defined for the squid giant axon by Hodgkin and Huxley (1952) is itself the result of voltage-dependent Na^+ and K^+ conductances and is thus an excellent example of these

B. Membrane Conductances

nonlinear membrane properties. For this reason, it is useful to review these properties for the action potential of the squid giant axon, which are summarized in Figure 3. Many readers are likely to be familiar with this material, and they are encouraged to move ahead to the next section. Others, however, will find it a useful introduction to the sections that follow. Although the details of kinetics and voltage dependency and also the nature of the various K^+ conductances differ somewhat across cell types, and the actual conductances for thalamic cells probably differ quantitatively from this illustration, this nonetheless serves as an excellent example for many other voltage-dependent conductances described for thalamic neurons. As shown in Figure 3A, the action potential has a very rapid rise or upstroke from rest and a somewhat slower fall or downstroke that slightly overshoots rest to create a period of relative hyperpolarization, known as the afterhyperpolarization.

First, we can consider the Na^+ channels. Figure 3B shows the time course of conductance changes for Na^+ channels during a typical action potential. At rest, the voltage-dependent Na^+ channels are closed. However, they open rapidly when the membrane is sufficiently depolarized

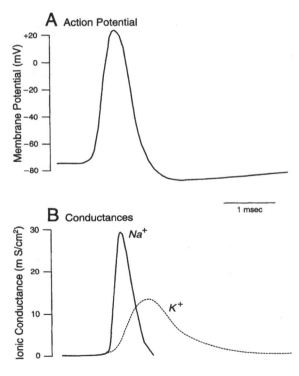

Figure 3
Typical action potential. (A) Voltage trace. Note that just after the action potential is a period of modest hyperpolarization, known as the *afterhyperpolarization*. (B) Voltage-dependent Na^+ and K^+ conductances underlying the voltage changes of the action potential.

from rest. Typically the voltage threshold for this effect occurs at around −50 mV when the membrane is depolarized from more hyperpolarized levels, such as a resting potential of about −70 mV. The rapid opening of these channels, which creates a rapid increase in membrane conductance to Na^+, allows Na^+ to flow into the cell, thereby further depolarizing it. This is the upswing of the action potential. The resulting depolarization serves to *inactivate* the Na^+ channels rapidly. Such inactivation means that, no matter what the membrane voltage, the channels will not open. With the Na^+ conductance now shut down, the cell begins to repolarize to its previously hyperpolarized level, and as we see later, the onset of a voltage-dependent K^+ conductance speeds this process of repolarization. This repolarization to the original, relatively hyperpolarized, level removes the inactivation of (or *de-inactivates*) the Na^+ channels. Note here that the Na^+ channels actually exist in three quite different states: (1) at rest, when *de-inactivated*, the channels are closed, but they can be rapidly opened by activation by means of a suprathreshold depolarization; (2) when *activated* by a suprathreshold depolarization, they are open; and (3) further depolarization *inactivates* the channels and closes them, leaving them temporarily unopenable. There are thus two closed-channel states—inactivated and de-inactivated—and one open state—activated. Moving from the inactivated state to the de-inactivated state not only requires hyperpolarization but also requires a bit of time (1 msec or so). During this period, the channels cannot be opened, and this determines the refractory period, which in turn sets a limit to the maximum firing frequency of cells of about 1000 spikes/sec, which is about the most that a thalamic neuron can do during a burst. There is a similar time dependency (~1 msec) required following depolarization to move from the activated to the inactivated state. Note also that, although we normally think of depolarization as an excitatory event because it promotes cell firing, depolarization that is sufficiently strong and prolonged will actually prevent action potentials because it inactivates the Na^+ channels.

Whereas we have seen that the Na^+ channel has two closed states—inactivated and de-inactivated—and one open state, some voltage-dependent channels do not have an inactivated state, and the K^+ channel considered next is an example. Channels such as this have only two main states—activated and deactivated.

We have seen in Figure 3B that the voltage-dependent K^+ channels participate in the action potential of the squid giant axon. These channels, like Na^+ channels, are also opened by depolarization, and opening K^+ channels serves to hyperpolarize the cell as K^+ flows out of it. However, the kinetics of the K^+ channels are slower than those of the Na^+ channels (Figure 3B), so the increased K^+ conductance is seen somewhat delayed with respect to the Na^+ conductance. Just about when the Na^+ channels become inactivated because of the depolarization, the K^+ channels open, and this serves to repolarize the cell more rapidly than would occur by merely passive means. It also creates the afterhyperpolarization (shown in Figure 3A). Another important difference between these Na^+ and K^+ channels other than their kinetics is that the K^+ channels show no

B. Membrane Conductances

appreciable inactivation during depolarization similar to that of the Na^+ channels. Instead, the K^+ channels close simply because the resulting hyperpolarization brings the membrane potential down below the threshold needed to open these channels, and this *deactivates* (as opposed to inactivates) the K^+ conductance. This is why the K^+ channel has only two states—activated and deactivated.

This picture, of a single, voltage-sensitive K^+ conductance involved in repolarizing the cell after the action potential in the squid giant axon is too simple for many mammalian neurons, for which a series of different K^+ conductances may be triggered as part of the action potential. For instance, in rat hippocampal neurons, four distinct K^+ conductances are associated with the action potential: two are purely voltage dependent and two are Ca^{2+} dependent, meaning that an increase of internal Ca^{2+}, caused perhaps by activation of voltage-dependent Ca^{2+} channels (see later), will activate these K^+ conductances (Storm, 1987; Storm, 1990). However, the specific nature of all of the K^+ conductances associated with action potentials has not yet been shown for thalamic neurons.

The voltage-sensitive ion channels involved in the Na^+ and K^+ conductances of the action potential are located in the soma and axon hillock (and also often in dendrites; see later), where they are concerned with the initiation of action potentials, and all along the axon, where they produce the conduction of the nerve impulse. As a result, a sufficient depolarization in the soma or axon hillock that triggers the Na^+ and K^+ conductances will propagate down the axon as a wave because the depolarization caused by Na^+ entry depolarizes nearby voltage-sensitive Na^+ channels. The refractory period dominates the wake of the action potential as it sweeps down the axon and prevents a backward propagating spike. A special case occurs for myelinated axons, where the voltage-sensitive Na^+ and K^+ channels are concentrated at the nodes of Ranvier, so instead of a smooth, wavelike propagation, the action potential jumps from node of Ranvier to node of Ranvier down the axon in a saltatory fashion. In general, myelinated or not, thicker axons conduct action potentials to their targets with a faster conduction velocity. However, this saltatory conduction in myelinated axons produces a much faster conduction velocity than occurs in unmyelinated axons of the same diameter. All axons of thalamic relay cells so far studied appear to be myelinated. However, the phrase "so far studied" represents a serious proviso because myelination of these relay cell axons has only been studied for a small handful of thalamic nuclei in a few species. In any case, for myelinated axons the myelination allows axons to conduct their signals more rapidly.

Although the presence and role of the Na^+ and K^+ channels along the axon are quite clear, what is less certain is their possible dendritic location and role. In fact, this has not yet been adequately addressed for thalamic cells. However, the possibility is interesting because studies of hippocampal and neocortical pyramidal cells indicate the presence of Na^+, K^+, and Ca^{2+} channels in the dendrites (Kim and Connors, 1993; Johnston *et al.*, 1996; Magee *et al.*, 1998; Hoffman and Johnston, 1998) that may affect the transmission of EPSPs generated in dendrites. They may serve to promote transmission of EPSPs toward the soma and will also play a role in "back propagation" of the action potential from the soma (Schiller *et al.*, 1997; Stuart *et al.*, 1997a,b), which means that an action potential generated in the soma or initial segment may propagate throughout the dendritic arbor, or perhaps just those dendrites expressing these channels, even when the somatic action potential was evoked by EPSP(s) generated in one or a few dendrites. Both of these factors can have dramatic effects on synaptic integration, and it is thus of considerable interest to determine whether thalamic neurons also have these channels along their dendrites.

More recently discovered conductances, now recognized as playing a major role in controlling the firing properties of relay cells and thus modifying the relay of information through the thalamus will be described next. Most of these conductances operate through voltage-dependent ion channels in the membrane of the soma and/or dendrites, although processes other than membrane potential control some of them. Some of these, such as the high threshold Ca^{2+} conductance and Ca^{2+}-dependent K^+ conductance, described later, can be initiated by the large depolarization of the action potential itself. Because most of these other conductances are voltage-dependent, much like the Na^+ and K^+ conductances of the classic action potential, they will be affected by postsynaptic potentials. That is, postsynaptic potentials will often turn other membrane conductances on or off, resulting in additional membrane currents that will ultimately affect the membrane potential at the axon hillock. This, in turn, means that these other conductances can have a quite dramatic effect on the pattern of action potentials fired by the postsynaptic cell. It is important to keep in mind that of all of these, only the voltage-dependent Na^+ and K^+ ion channels underlying the action potential are located along the axon in appreciable numbers. The only way that a thalamic cell can send a message to cortex is thus by the action potential, but the determinants of whether and how the cell fires depend heavily on the conductances that are considered next.

b. Low-Threshold Ca^{2+} Conductance

Apart from the conductances underlying action potentials, this is the most important voltage-dependent conductance for relay cells. It is ubiquitous in relay cells in all dorsal thalamic nuclei of all mammals studied to date (Deschênes *et al.*, 1984; Jahnsen and Llinás, 1984a,b; Hernández-Cruz and Pape, 1989; McCormick and Feeser, 1990; Scharfman *et al.*, 1990; Bal *et al.*, 1995), so it is clearly a key to understanding the function of the thalamic relay. This conductance controls whichever of two distinct response modes, *tonic* or *burst*,[2] is operative when a thalamic relay cell responds to afferent input. The tonic mode of firing occurs when the

---

[2] "Tonic" used in this sense refers to a response mode of a thalamic relay cell, and here it is paired with "burst." X and Y cells, two relay cell types found in the A-laminae of the cat's lateral geniculate nucleus, display both response modes. This use of "tonic" should not be confused with another use of "tonic" when paired with "phasic" to refer to a cell type: "tonic" for X and "phasic" for Y. Throughout this account, we shall use "tonic" only to refer to response mode and not to cell type.

B. Membrane Conductances

low-threshold Ca^{2+} and certain other associated conductances are inactivated. The relay cell then responds more like the linear neuronal integrator described by cable modeling: its response to input is characterized by a steady stream of action potentials of a frequency and duration that correspond fairly linearly to input strength and duration. (A potentially important contributor to the linear relationship between stimulus intensity and response during tonic firing may be a voltage-dependent K^+ conductance underlying a current known as I_A, and this is described more fully in Section B.2.d.) During the burst mode, which occurs when the low-threshold Ca^{2+} conductance is de-inactivated and thus able to be activated, the neuronal response to a depolarizing input consists of brief bursts of action potentials separated by silent periods. As we shall see later, there is a less linear relationship during burst firing between input and the action potential response relayed to cortex in terms of the amplitude or temporal properties of the input. It is worth emphasizing that this response during burst mode bears no resemblance to the known firing patterns of afferent inputs because, for instance, retinogeniculate axons show no evidence of burst firing. The further significance of these firing modes for thalamic relays is the subject of Chapter VI, and here we shall confine ourselves to the underlying membrane properties related to the production of tonic and burst firing.

As noted, the inactivation state of the low-threshold Ca^{2+} conductance is responsible for whether the cell responds in tonic or burst mode because the relay cell responds in tonic mode when this conductance is inactive and burst mode when it is de-inactivated. The inactivation state itself depends on membrane voltage, much like the Na^+ conductance of the conventional action potential. However, the threshold for activating this conductance is at a lower, that is, more hyperpolarized, level than that for the action potential. When activated, it produces a spikelike, triangular depolarization of roughly 20–30 mV and lasting for roughly 50 msec: this is the *low-threshold spike*.

Figures 4A through 4C illustrate aspects of its voltage dependency. The conductance underlying this low-threshold spike is inactivated when the membrane is depolarized more positive than about −60 mV (Figures 4A and 4B), but it is de-inactivated at more hyperpolarized levels from which it can be activated by a suitably large depolarization (Figure 4C), such as an EPSP. Note that the channels underlying this Ca^{2+} conductance, which are known as *T channels* ("T" for transient), have the same three states—activation, inactivation, and de-inactivation—as do the Na^+ channels associated with action potentials. Because of the channel name, the membrane current resulting from the low-threshold Ca^{2+} conductance is known as the *T current*, or I_T. It is this influx of Ca^{2+} ions through the T channels that produces the low-threshold spike. Thus the low-threshold Ca^{2+} conductance, low-threshold spike, and I_T all refer to closely related phenomena.

The large depolarization associated with the low-threshold spike produces, at its peak, a high frequency burst of 2–10 conventional (i.e., Na^+/K^+) action potentials. In this way, I_T and the resultant low-threshold

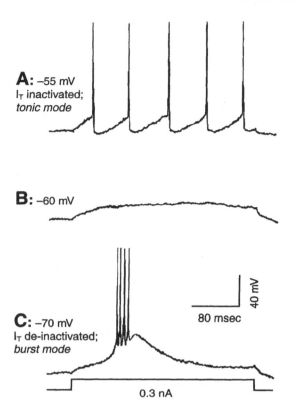

Figure 4
Tonic and burst firing modes for a relay cell from the cat's lateral geniculate nucleus recorded intracellularly in an *in vitro* slice preparation. The cell was held at different initial membrane voltages as indicated by adjusting the current injected into the cell through the recording electrode. Each of the three recording traces shows the response to the same depolarizing 0.3 nA current pulse (shown under the bottom recording trace) injected into the cell, also by means of the recording electrode. (A) Tonic response (I_T inactivated). When the cell is relatively depolarized ($-55\ mV$), the current injection creates an initially passive membrane depolarization sufficient to evoke a stream of conventional action potentials. (B) No response. At this level of polarization ($-60\ mV$), I_T is still mostly inactivated, and the injected current pulse depolarizes the cell insufficiently to evoke action potentials, resulting in a purely resistive-capacitative response. (C) Burst response (I_T de-inactivated). When the cell is relatively hyperpolarized ($-70\ mV$), the same depolarizing pulse now triggers a low-threshold Ca^{2+} spike. Riding the crest of this low-threshold Ca^{2+} spike is a burst of four conventional action potentials (the number of action potentials in the burst can vary up to about 10).

spike provide an amplification that permits a hyperpolarized cell to generate action potentials in response to a moderate EPSP, and the resulting firing pattern is bursty. This contrasts with the maintained discharge of single action potentials during tonic firing.

The voltage dependency of I_T is qualitatively like that for the Na^+ channel involved in generation of the action potential, although the relevant voltage ranges and kinetics are quite different. Figure 5 repre-

B. Membrane Conductances

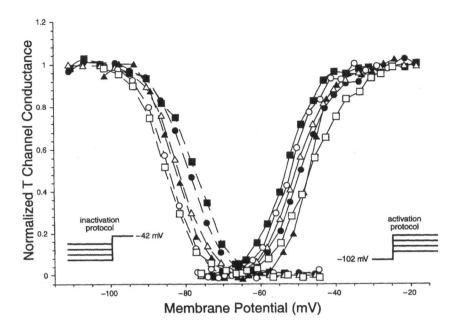

Figure 5
Voltage dependency of activation and inactivation of I_T for six cells of the ventral posterior nucleus in rats. The cells were recorded *in vitro* and in isolation after acute dissociation, which usually leaves the soma and stumps of primary dendrites. Recordings were made in voltage clamp mode to measure I_T, and from these measures, the underlying conductance could be computed. The ordinate shows the conductance measures after each curve was normalized with respect to the largest conductance evoked. On the left are shown the *inactivation* curves (dashed lines). The points for these curves were obtained by first holding the membrane at various hyperpolarized levels (which are plotted on the abscissa) for 1 sec and then stepping up to −42 mV. The voltage protocol for this is seen in the inset at left. Note that the more hyperpolarized the initial holding potential, the more I_T is evoked, because more initial hyperpolarization produces more de-inactivation of I_T. On the right are shown the *activation* curves (solid lines). Here, the points were obtained by initially holding at −102 mV, which would completely de-inactivate I_T, and then stepping up to the various membrane voltages plotted on the abscissa. The voltage protocol for this is seen in the inset at right. Note that larger depolarizing steps activate more I_T. Data kindly supplied by J. R. Huguenard for replotting from Figure 1A of Huguenard and McCormick (1992).

sents a more complete and quantitative description of the voltage dependency of I_T measured from six different relay cells in the ventral posterior nucleus of the rat (Huguenard and McCormick, 1992). Similar data have also been found for relay cells in other nuclei and in other species (Deschênes *et al.*, 1984; Jahnsen and Llinás, 1984a,b; Hernández-Cruz and Pape, 1989; McCormick and Feeser, 1990; Scharfman *et al.*, 1990; Bal *et al.*, 1995), including the ventral anterior nucleus, medial geniculate nucleus, lateral geniculate nucleus, pulvinar, and other, unspecified thalamic regions, from ferrets, guinea pigs, hamsters, or cats. These data are based on measuring the actual I_T during voltage clamp recording.[3] This technique pumps just the right amount of current into or out of the cell during recording to balance any activated currents (like I_T), thereby keeping the membrane voltage constant, or "clamped," at a predetermined voltage. The amount of current pumped in or out provides a measure of the actual membrane current activated by I_T, and this value can be converted to a measure of the related membrane conductance, which is actually plotted.[4] In Figure 5, I_T is activated in two analogous experiments, one to measure the "inactivation" of I_T, and the other to measure the "activation" of I_T.

The insets in Figure 5 illustrate the voltage regimens used to determine the points for the activation and inactivation curves. To obtain each inactivation curve (*dashed lines*), the cell is first clamped at different levels aimed to partly inactivate I_T, which is then assessed by measuring the membrane current evoked by a sudden clamping of the cell to a new depolarized level (−42 mV) sufficient to activate any de-inactivated I_T. To measure activation (*solid lines*), the cell is first clamped at a very hyperpolarized level (−110 mV) to completely de-inactivate I_T, and the current evoked by suddenly clamping the cell to various depolarized levels is determined. Notice that these curves of relative conductance lie between 0 and 1. The intermediate levels reflect partial activation or inactivation of I_T, and this reflects the probabilistic percentage of T channels being

---

[3] Voltage clamp is a useful technique, but there are problems associated with it that should be kept in mind. The main one is that it is difficult to maintain equipotential conditions throughout the cell at all times. Even when the electronics are sufficiently fast to keep pace with fast transient conductance changes, there remains the problem that neurons with extensive dendritic arbors act like cables. Thus if current is injected into the cell at the soma, say, to depolarize and clamp it to a specific membrane potential, some of this current will leak across the membrane as it travels out through the dendritic arbor (see above). There is then less current available to depolarize the membrane as one proceeds further out along the arbor, and the result is a gradient in clamped membrane voltage rather than true isopotentiality throughout the neuron. The same problem exists with attempts to hyperpolarize the cell. If T-type Ca^{2+} channels exist throughout the soma and dendrites, as seems to be the case (Destexhe *et al.*, 1996, 1998b; Zhan *et al.*, 2000), they will experience different membrane voltages at different locations. This, in turn, means that they will contribute differently at different locations during voltage clamp because the differing voltages produce different levels of inactivation. There are several ways of dealing with this problem, and the most common is to record from acutely dissociated thalamic relay cells, because the dissociation removes most or all of the dendrites. Maintaining isopotential recording in such cells during voltage clamp is much more likely without the extensive dendritic arbor, but the problem here is the assumption that a cell with its dendrites ripped away will reveal normal physiological properties. Another implicit assumption is that the behavior of channels in the soma, which in theory can be studied more rigorously in dissociated cells, is identical to that of channels in the missing dendrites. This assumption seems reasonable, but it would still be necessary to know the distribution of channels in the dendrites to reconstruct completely how the cell would behave with respect to the channels under investigation (see Destexhe *et al.*, 1996, 1998b for examples of how this may be done). It is thus important to understand the limitations of the voltage clamp method, although it is frequently used because it is still judged to be the best technique available to address these issues. The data from Figure 5 are in fact taken from acutely dissociated cells.

[4] Some authors simply plot I_T directly for these curves. However, the actual I_T flowing into the cell through Ca^{2+} is a function of both the number of open T channels and the driving force, which is the difference between the reversal potential for Ca^{2+} (roughly +150 mV) and the membrane potential at which I_T is evoked. Thus the more positive the potential at which I_T is evoked, the lower the driving force. Some authors thus convert the current recorded during current clamp to an estimate of membrane conductance, which is not influenced by the driving force and does reflect the number of channels opened.

B. Membrane Conductances

in a given state during these experiments. During the inactivation experiments, the more hyperpolarized the initial holding potential, the more T channels switch from the inactive state to become de-inactivated. Likewise, during the activation experiments, the larger the depolarizing pulse, the more T channels become activated and opened.

There are two points to note about these curves. First, there is very little overlap between activation and inactivation curves. An overlap is a sort of voltage "window," or a membrane voltage range in which I_T is constantly present because it is both partially de-inactivated and partially activated, but this actually seems not to occur to any significant extent in these relay cells of the ventral posterior nucleus in the rat. However, the extent to which this is also true for other relay cells in other species or thalamic nuclei is an open question. This question is an important one to address because the presence and extent of such a window is important in understanding how a relay cell fires (Tóth et al., 1998; Hughes et al., 1999). That is, if a window exists and the membrane voltage lies within it, the relay cell may begin to spontaneously and rhythmically discharge; furthermore, if a more depolarized relay cell is hyperpolarized (e.g., by means of an inhibitory synaptic input) into the voltage window, it may begin to burst rhythmically, which might seem to be a counterintuitive response to a hyperpolarizing input. However, it seems that for relay cells to do this, they must have an abnormally high input resistance brought on by blocking much of the "leak" K^+ conductance (Tóth et al., 1998; Hughes et al., 1999), and it seems that relay cells in a more physiological state do not show bursting or other responsiveness on entering the voltage window (Guitierrez et al., 1999). Second, although the activation curve appears to be fairly steep, there is nonetheless an *apparent* dynamic voltage range of about 20 mV in which partial depolarizing steps will evoke a partial I_T. This, however, is largely due to the conditions of voltage clamp recording. Thalamic relay neurons in the working brain are not voltage clamped. In such a cell, a depolarizing EPSP might activate only a partial I_T, but without the voltage clamped, this partial I_T would further depolarize the cell, activating more I_T, and so on in a sort of autocatalytic process (Zhan et al., 1999). There is thus a form of positive feedback in the generation of the low-threshold spike.

The result is a much narrower dynamic range than might be deduced from the activation curve of Figure 5. By "dynamic range," we mean the range of depolarizations over which a graded low-threshold spike can be evoked. The dynamic range is effectively zero, or, more accurately, so narrow that it usually cannot be measured in practice because the positive feedback means that most depolarizations are either too small to activate any low-threshold spike (i.e., they are below threshold) or they are so large that they evoke the maximum low-threshold spike (i.e., they are above threshold). This is effectively an "all-or-none" response, like that of the action potential, and it is illustrated in Figure 6, which shows data from a typical relay cell in the cat's lateral geniculate nucleus recorded in the "current clamp" mode without any attempt to clamp the voltage.

Figure 6A shows the result of injecting small, incremental steps of current into the cell from an initial membrane potential sufficiently

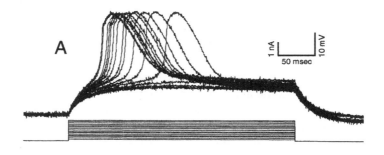

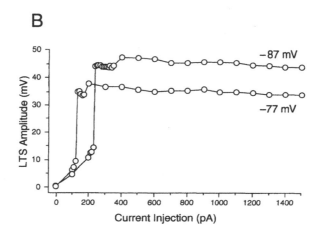

Figure 6
Intracellular recording from a cat's lateral geniculate relay cell showing the all-or-none nature of the low-threshold Ca^{2+} spike. Recordings were made *in vitro* in current clamp mode with application of tetrodotoxin to eliminate action potentials. (A) Fifteen voltage traces superimposed to show responses to current injections (shown under the voltage traces). From an initial holding potential of −87 mV, the current pulses of incremental 10-pA amplitude are injected into the cell. The cell's input resistance is roughly 100 MΩ, and thus each 10-pA current injection depolarizes the cell by roughly 1 mV. As the injected current is increased incrementally, only resistive-capacitative responses are obtained until the next incremental step activates a low-threshold Ca^{2+} spike. Increasing the current steps has no substantive effect on the amplitude of the Ca^{2+} spike, although its latency decreases with stronger injections of current. (B) Plots of injected current versus evoked voltage for same cell as in A. From initial holding potentials of −77 mV or −87 mV, which would significantly but not completely deinactivate I_T, current pulses are injected into the cell as indicated on the abscissa. For small pulses, the threshold for I_T activation is not reached and a resistive-capacitative response results. As the pulses increase in amplitude, a sudden threshold is reached, leading to a Ca^{2+} spike, which does not increase further in amplitude with more injected current.

hyperpolarized to de-inactivate a significant number of the T channels. For this experiment, tetrodotoxin has been added to block the Na^+ channels underlying conventional action potentials, allowing better visualization

B. Membrane Conductances

of the low-threshold Ca^{2+} spike (Zhan et al., 1999). Smaller current injections produce a simple resistive-capacitative response, but as the threshold voltage for the Ca^{2+} spike is reached, the first suprathreshold current injection activates a Ca^{2+} spike. Large suprathreshold injections activate Ca^{2+} spikes that are not significantly larger, although they do occur with shorter latency. Thus, in terms of amplitude, the Ca^{2+} spike behaves like an "all-or-none" spike with a voltage threshold, again much like the action potential. This is illustrated for the same cell by the curves in Figure 6B, which show what happens when small current steps are injected into the cell from each of two initial membrane potentials negative enough to start with considerable de-inactivation of the T channels. An incremental current step of 10 pA, which produces a depolarization of the cell of only roughly 1 mV, is sufficient to activate the near-maximum low-threshold Ca^{2+} spike. Although this experiment was carried out with current injections to depolarize the cell enough to evoke Ca^{2+} spikes, this would also happen with any synaptic input (e.g., from retina) that produced sufficiently large EPSPs.

Because the action potentials represent the only signal sent to cortex by thalamic relay cells, they are thus the only output of the relay cell seen by cortex. We can begin to ask how burst and tonic firing affect the input/output relationships of the relay cells. Because the size of the low-threshold Ca^{2+} spike is monotonically related to the number of action potentials it evokes (Zhan et al., 2000), there is a fairly simple relationship between these measures. Thus the ordinate for the curves of Figure 6B would be monotonically related to firing frequency had tetrodotoxin not been used to prevent action potentials. This is illustrated in Figure 7 for the same cell as in Figure 6 but before application of tetrodotoxin so that action potentials could be recorded. During burst firing, after activation from initial membrane potentials of -77 or -87 mV, which are levels of significant de-inactivation of the T channels, a sudden jump in firing frequency is seen, which corresponds to the threshold activation of Ca^{2+} spiking (Zhan et al., 1999). Thereafter, larger current injections have only a modest effect on the initial firing frequency. This can be contrasted to tonic firing after activation from initial membrane potentials of -67 or -57 mV, which are levels that inactivate most of the T channels. Now the relationship between input (i.e., the current steps) and output (i.e., the firing frequency) is much more linear without a sudden jump or discontinuity as seen with burst firing.

Figures 6B and 7 thus indicate that the input/output relationships are much more linear during tonic firing than during burst firing. Another type of nonlinearity is shown in Figure 4. During tonic firing (Figure 4A), the response lasts as long as the injected current, but during burst firing (Figure 4C), the response does not faithfully represent the duration of the injected current. Fourier analysis of results of injecting sinusoidal currents into geniculate relay cells of the cat confirm that burst firing provides a significantly more nonlinear input/output relationship than does tonic firing (Smith et al., 2000).

In addition to the voltage dependency of I_T, there is also a time

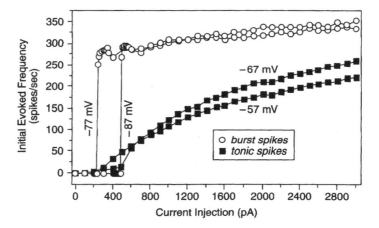

Figure 7
Plots of injected current versus initial firing frequency for same cell as in Figure 6, recorded before tetrodotoxin was applied. Initial firing frequency was calculated from the first six action potentials evoked. With beginning holding potentials of −77 mV or −87 mV, burst firing was evoked, and the relationship between current injection and firing frequency shows a sudden, nonlinear jump. With beginning holding potentials of −67 mV or −57 mV, tonic firing was evoked, and the relationship between current injection and firing frequency is much more linear.

dependency, and the inactivation state of I_T is actually a complex function of membrane voltage and time. Activation, inactivation, and de-inactivation occur with variable time constants, but those for activation tend to be much shorter than the other two. For instance, full inactivation following depolarization or full de-inactivation following hyperpolarization usually takes on the order of 100 msec or longer. In principle, the latter is qualitatively like the refractory period for action potentials, except it lasts much longer. Thus the low-threshold Ca^{2+} spike is followed by a refractory period that limits the rate of such Ca^{2+} spiking to 10 Hz or less. Also, because low-threshold Ca^{2+} spikes are evoked from relatively hyperpolarized levels, because the depolarization caused by the low-threshold spike is itself sufficient to inactivate much of I_T, and because the depolarization and/or Ca^{2+} entry into the cell associated with I_T leads to voltage- and/or Ca^{2+}-dependent K^+ conductances that act to hyperpolarize the cell, the cell rapidly returns to its hyperpolarized state after each low-threshold spike. This tends to prohibit tonic responses, and, given the relative refractory period of I_T, the result is relative silence between low-threshold spikes. This silent period between low-threshold Ca^{2+} spikes enhances the burstiness of the response.

Finally, although I_T activates very quickly, recent evidence in cats indicates that activation requires a moderate rate of depolarization (Gutierrez et al., 1999). If the rate of depolarization is too slow, a thalamic relay cell can be taken from a hyperpolarized state in which I_T is fully de-inactivated to a depolarized state in which I_T is fully inactivated and tonic

B. Membrane Conductances

firing ensues without ever firing a low-threshold Ca^{2+} spike. The rate of rise of most EPSPs (e.g., from retina) is fast enough to activate I_T, but some EPSPs by way of metabotropic receptors (described in Chapter VI) may be slow enough to convert the firing mode of the relay cell from burst to tonic without evoking a burst.

c. Conductances Associated with I_T

As just noted, activation of the low-threshold Ca^{2+} spike itself leads to voltage- and/or Ca^{2+}-dependent K^+ conductances, which repolarize the cell to its former hyperpolarized level, thereby initiating the process of I_T de-inactivation. Another conductance, often associated with the low-threshold Ca^{2+} conductance, is activated by membrane hyperpolarization and de-activated by depolarization. This *hyperpolarization-activated cation conductance* leads, by means of influx of cations, to a depolarizing current, which is called I_h (McCormick and Pape, 1990a,b). It is sometimes called the "sag current" because it is activated by hyperpolarization and causes the membrane potential to drift back, or sag, toward the initially more depolarized level. Activation of I_h is slow, with a time constant of >200 msec. The combination of I_T, the above-mentioned K^+ conductances, and I_h can lead to rhythmic bursting, which is often seen in recordings from *in vitro* slice preparations of thalamus.

Figure 8 illustrates the series of conductances leading to transmembrane currents associated with I_T (McCormick and Pape, 1990a,b; Huguenard and McCormick, 1992; McCormick and Huguenard, 1992). This is almost certainly an oversimplification of the actual conductances involved. The sequence in this schematic diagram starts with a hyperpolarization sufficient in amplitude and duration to de-inactivate I_T (1). (The actual starting point in the cycle is arbitrary.) This will later activate I_h (2), providing a depolarization that activates I_T (3) and the associated burst of conventional action potentials (4). The low-threshold Ca^{2+} spike and action potentials initiate the above-mentioned K^+ conductances that hyperpolarize the cell (5). However, until the cell has been hyperpolarized for some time (~100 msec for I_T and ~200 msec for I_h), I_T is inactivated and I_h is de-activated (6,7). The prolonged hyperpolarization will eventually activate I_h, but this activation is so slow that, before it begins, I_T becomes fully de-inactivated (1), and the process is repeated. This leads to prolonged rhythmic activation of low-threshold Ca^{2+} spikes and bursts of action potentials. The Ca^{2+} spiking can occur at a variety of frequencies, typically at 1 to 3 Hz, the actual value depending on other parameters, including the presence of local feedback inhibitory circuits that may become involved during rhythmic bursting. This bursting can be interrupted only by a sufficiently strong and prolonged depolarization to inactivate I_T and prevent I_h from activating. Appropriate membrane voltage shifts can thus effectively switch the cell between rhythmic bursting and tonic firing. Such rhythmic bursting may be seen during *in vitro* recording or during sleep and may require local circuitry, especially connections between thalamic reticular cells and relay cells, in addition to the conductances shown in Figure 8. The rhythmic bursting seen during sleep is

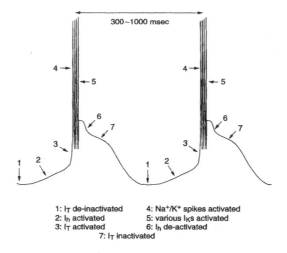

Figure 8
Schematic illustration of various voltage-dependent conductances thought to contribute to cyclic burst firing of thalamic relay cells. Starting at a hyperpolarized membrane potential, I_T is de-inactivated (1), then I_h becomes activated (2), depolarizing the cell to activate I_T (3), which in turn activates action potentials (4). Several voltage- and Ca^{2+}-dependent K^+ conductances then become activated (5), and after a variable period of time, I_T becomes inactivated (6) and I_h de-activates (7). Passive repolarization enhanced by the active K^+ conductances repolarize the cell, which starts a new cycle. Redrawn from Figure 14B of McCormick and Pape (1990b).

largely synchronized among relay neurons, a process that requires local circuitry, particularly the involvement of the thalamic reticular nucleus (Steriade *et al.*, 1993), but the processes suggested by Figure 8 do not imply any such synchrony among cells.

These associated conductances have been most thoroughly studied *in vitro*, and the following relationships have been proposed on the basis of these studies. When the cell starts off relatively depolarized, it fires in tonic mode because I_T is inactivated. When initially hyperpolarized, either by increased activity in inhibitory inputs, decreased activity in excitatory inputs, or both, the cell fires in burst mode or not at all because any sufficiently strong activating (i.e., depolarizing) input will activate I_T. This burst firing would always be *rhythmic*. According to this proposal, random or arrhythmic bursting does not occur. Switching between these two modes is effected by changing membrane potential: depolarized cells will respond in tonic mode with a stream of unitary action potentials, and hyperpolarized cells will respond in burst mode with rhythmic clusters of 2 to 10 action potentials. This proposal has also been extended to the more physiological *in vivo* condition, implying that, depending on membrane potential, relay cells will respond either tonically or with rhythmic bursting, the former being the condition during awake, alert behavior, and the latter occurring during various phases of sleep or drowsiness. The idea here is that the sequence of conductances activated during burst firing (see Figure 8) are much more powerful than those produced by synaptic activation by means of the driving inputs. In this sense, tonic firing would be the *only* relay mode for a thalamic nucleus, and burst firing would represent a functional disconnection of the relay cell from its driving inputs. As we show in Chapter V, this is not strictly correct because arrhythmic bursting appears to be a response mode of relay cells during waking behavior and can be a genuine relay mode. We shall be discussing three forms of bursting: (1) arrhythmic; (2) rhythmic and synchronized, meaning that large populations of thalamic cells fire with the same rhythm and in synchrony, which occurs during slow wave sleep and certain forms of epilepsy; (3) rhythmic and nonsynchronous. Obviously, to distinguish between the last two requires recording activity in populations of thalamic cells.

One final and particularly interesting feature of I_h is that it can be modulated by serotonin, noradrenalin, and histamine (McCormick and Pape, 1990a,b; McCormick and Williamson, 1991). Application of any of these neuromodulators increases the amplitude of the evoked I_h, and it does this by altering the voltage dependency of the underlying ion channels. The detailed mechanisms of this effect have not yet been explained.

d. I_A

I_A is generated by a voltage-dependent K^+ conductance (or several related K^+ conductances) found in most cells throughout the central nervous system (Adams, 1982; Rogawski, 1985; Storm, 1990; McCormick, 1991b). It is different from the K^+ conductances described previously in terms of kinetics, and it also has a much lower activation threshold. One of the problems in dealing with intrinsic membrane properties is that there are often several different conductances using different ion channels that can involve the same ion. As we shall see later, K^+ is not the only example of this.

There are certain similarities between I_A and I_T. For instance, I_A has the same three states: activated, inactivated, and de-inactivated. In thalamic relay cells, the voltage dependence of I_A is rather similar in shape to that of I_T but shifted in the depolarized direction (see Figure 9A; Pape et al., 1994). Thus, I_A is inactivated at depolarized membrane potentials and activated by a depolarization from a hyperpolarized membrane potential. Like I_T, the activation of I_A occurs with a much faster time course than does the inactivation or de-inactivation. Note an important distinction between I_T and I_A: the I_T is carried by Ca^{2+}, which flows *into* the cell and *depolarizes* it, whereas the I_A is carried by K^+, which flows *out of* the cell and thereby *hyperpolarizes* it. When I_T is activated by a small depolarization, it produces a large, depolarizing Ca^{2+} spike, which, as noted previously, can be viewed as a nonlinear amplification of the activating depolarization. When the I_A is activated, it hyperpolarizes the cell, which tends to offset the original, activating depolarization. The result is a slowing down and reduction of the initial depolarization. If the cell responds in tonic mode, the I_A will delay and reduce the frequency of action potentials. However, the effects of I_A during burst mode are more complex and

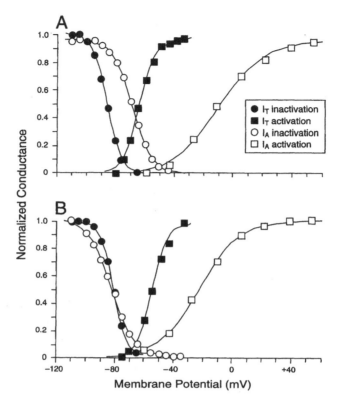

Figure 9
Activation and inactivation curves for both I_T and I_A in cells of the rat's lateral geniculate nucleus, generated in the same fashion as described in Figure 5, with voltage clamp recording after acute dissociation. (A) Curves for a relay cell. Note that the curves for I_T are shifted in a hyperpolarized direction with respect to those for I_A. (B) Curves for an interneuron. Note that the inactivation curves for I_T and I_A are largely overlapped. See text for significance of these differences between relay cells and interneurons.

less well understood. If I_T is activated by a strong depolarization, it will activate very quickly and from a more hyperpolarized level than needed to activate I_A; it thus appears that a low-threshold Ca^{2+} spike will be fully activated before I_A has much chance to affect it. However, as noted earlier (see Figure 6), I_T activated by just suprathreshold stimuli has a longer latency, slower component before the all-or-none, autocatalytic, low-threshold Ca^{2+} spike fires. We would expect that I_A would be activated during this longer latency, which would further delay and perhaps reduce the size of the low-threshold Ca^{2+} spike. Although this has to be experimentally verified, it remains the case that I_A affects burst firing only for a narrow range of activating stimuli for I_T that is barely suprathreshold.

The similarity in the voltage dependencies of I_T and I_A suggest that these currents may frequently interact. Because I_T depolarizes the cell

B. Membrane Conductances

whereas I_A hyperpolarizes, they would tend to offset one another if both were activated. Whether a postsynaptic potential activates one or both depends critically on the initial membrane voltage and temporal properties of voltage changes. If the membrane starts off sufficiently hyperpolarized to completely de-inactivate both conductances, then a depolarization would activate I_T before activating I_A, and the result would be a low-threshold spike with a burst of action potentials riding its crest, and this pattern would be only subtly affected by the later activation of I_A. Notice also from Figure 9A that there is a limited membrane potential near -70 mV for which I_T is thoroughly inactivated but I_A is still slightly de-inactivated. An EPSP (or other depolarizing event) elicited from this range of membrane potentials will activate a small I_A without any I_T, so only tonic firing is possible. However, as noted previously, the I_A will serve to slow down the buildup of the EPSP, which will delay and reduce the level of tonic firing. Finally, because the de-inactivation kinetics are much faster for I_A than for I_T, brief hyperpolarizations followed by depolarizations will selectively activate I_A and not I_T.

It is not clear what function I_A serves. One suggestion is that it extends the dynamic range of input/output relationships for neurons by reducing its slope (Connor and Stevens, 1971). That is, I_A, when activated by an EPSP, would sum with that EPSP to reduce the overall level of depolarization, and this enhances the range of EPSPs that evoke firing before saturation, or the maximal firing frequency, is reached. By opposing the buildup of an EPSP, activation of I_A would ensure that low to moderate stimuli would not maximally depolarize a cell, thereby preventing such stimuli from driving the cell to response saturation. The cell is thus able to signal the presence of stronger stimuli, and its dynamic range is enhanced. For relay cells, such a proposed function only makes sense for tonic mode because the response of low-threshold spikes in burst mode already has a limited dynamic range, and the low-threshold spike will be evoked before I_A can influence it much.

e. High-Threshold Ca^{2+} Conductances

In addition to the Ca^{2+} conductance underlying I_T, there are two or more much higher threshold Ca^{2+} conductances that are located in the dendrites and synaptic terminals (Llinás, 1988; Johnston et al., 1996; Zhou et al., 1997). One involves L-type Ca^{2+} channels ("L" for "long-lasting" because it slowly inactivates) and the other, N-type channels ("N," wryly, for "neither," being neither T nor L type; it inactivates more rapidly than the L-type channel). Other types of high-threshold Ca^{2+} channels also exist (Wu et al., 1998). The threshold higher than for T channels means that a much larger depolarization (to about -20 mV) is needed to activate these Ca^{2+} conductances, and they are the reason that Ca^{2+} enters the cell as a result of an action potential. In synaptic terminals, these conductances represent a key link between the action potential and transmitter release because the action potential will activate these channels, and the resultant Ca^{2+} entry is needed for transmitter release. Less is known about these Ca^{2+} conductances in dendrites, but by providing a regenerative spike that can travel between the site of an activating EPSP and the soma, they may help ensure that distal dendritic inputs that are strong enough to activate this conductance will significantly influence the soma and axon hillock. These channels might also play a role in back propagation of action potentials that depend on Na^+ channels through the dendritic arbor.

f. Ca^{2+}-Dependent K^+ Conductances

Several K^+ conductances can be activated by increased Ca^{2+} concentrations. These commonly occur as the result of an action potential, which activates high-threshold Ca^{2+} channels described in the preceding paragraph, and the increased Ca^{2+} concentration that results can activate Ca^{2+}-dependent K^+ conductances. These will produce a hyperpolarization. Some of these are fast to activate and will help to repolarize the cell and produce an afterhyperpolarization. Others that are slower to activate will build up with more and more action potentials, leading to spike frequency adaptation (Adams, 1982; Powers *et al.*, 1999). This phenomenon results in the slow reduction of the firing frequency of a neuron to a constant stimulus, and the higher the frequency of initial firing, the more adaptation that occurs. Spike frequency adaptation has been demonstrated for thalamic cells (Smith *et al.*, 1999).

g. Persistent Na^+ Conductance

Finally, a persistent and noninactivating Na^+ conductance, which is activated by a strong depolarization, exists in thalamic relay cells creating a plateau depolarization (Jahnsen and Llinás, 1984a,b). When activated, it promotes sustained, tonic firing. This plateau potential involves different ion channels from those subserving the conventional action potential because those related to action potentials are blocked by tetrodotoxin, whereas those related to the plateau potential are not.

3. INTERNEURONS

Interneurons, possibly because of their relatively small size, are much more difficult to record than are relay cells. Until recently, functional criteria to distinguish interneurons from relay cells during recording were lacking. Several such criteria now exist, especially for the *in vitro* preparation. However, except for these very recent experiments, few physiological data have been published regarding interneurons, and much less is known about their membrane properties.

a. Action Potentials

The important issue as to whether all thalamic interneurons exhibit action potentials was discussed earlier in Section C.2. As we have noted there, it is possible that axonless interneurons exist and that these do not fire action potentials, but clearly some (if not all) possess axonal outputs and fire action potentials. In any case, recent evidence indicates that the

B. Membrane Conductances

output from presynaptic dendritic terminals can be activated without an action potential (Cox *et al.*, 1998).

b. I_T and I_A

It had been thought that, unlike relay cells, most interneurons do not possess any measurable I_T, although evidence for some I_T was found in a few interneurons (McCormick and Pape, 1988). A recent analysis suggests that interneurons do indeed actually exhibit I_T and can discharge low-threshold Ca^{2+} spikes, but that the interaction of I_T and I_A in interneurons frequently obscures the former. The reason this happens in interneurons but not in relay cells seems to relate to subtle differences between cell types in the voltage dependencies and peak amplitudes of these conductances (Pape *et al.*, 1994). Voltage dependencies are summarized in Figure 9B. For relay cells, inactivation and activation of I_T is shifted to more hyperpolarized membrane potentials relative to I_A, and this is also true for activation of these currents in interneurons; however, inactivation of both conductances largely overlap in interneurons. Nonetheless, this implies that in both cell types, I_T will be activated before I_A by a depolarizing input. However, in relative amplitude terms, the I_A to I_T ratio is much larger in interneurons than in relay cells (not shown), so I_A can swamp I_T in interneurons, preventing a low-threshold Ca^{2+} spike, if the activating depolarizing input is large enough. This might explain why I_T is generally more difficult to observe in interneurons than in relay cells. However, an even more recent study suggests that bursting may be as common in interneurons as in relay cells (Zhu *et al.*, 1999), so the issue of how readily interneurons may express burst mode firing in behaving animals remains unresolved.

4. Reticular Cells

Reticular cells, in addition to conventional action potentials, also display a collection of conductances related to a low-threshold Ca^{2+} conductance (Huguenard and Prince, 1992). The low-threshold Ca^{2+} conductance is qualitatively similar to that seen in relay cells with a similar voltage dependency, and thus reticular cells also display burst and tonic response modes. However, there are some interesting quantitative differences between relay and reticular cells regarding this firing property.

a. I_T

One difference is in the temporal domain (Huguenard and Prince, 1992). Reticular cells show much slower activation and inactivation kinetics than do relay cells.[5] Furthermore, the inactivation of I_T in reticular cells

---

[5] This current for reticular cells is sometimes written with an added "s" subscripted to "I_T", where the subscripted "s" stands for "slow."

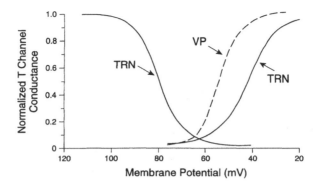

Figure 10
Activation and inactivation curves for typical thalamic reticular nucleus cell (solid curves) of the rat. For comparison is an activation curve for a typical relay cell from the ventral posterior nucleus of the rat (dashed curve). Redrawn from Figure 5B of Huguenard and Prince (1992). Abbreviations: TRN, thalamic reticular nucleus; VP, ventral posterior nucleus.

is voltage independent, so that it remains slow to inactivate even during the low-threshold spike. This contrasts with relay cells, which inactivate more quickly as the low-threshold spike develops. The result of the slow kinetics and voltage independent inactivation is much more prolonged low-threshold spikes in the reticular cells giving rise to many more action potentials.

This pattern of burst firing of reticular cells has a dramatic effect on their target relay cells. Reticular cells are GABAergic and thus inhibit relay cells, and the slow kinetics of I_T in reticular cells means that de-inactivation and thus the refractory period after a low-threshold Ca^{2+} spike is prolonged. It follows that the long bursts in reticular cells will produce powerful and prolonged inhibition of relay cells, and this, in turn, serves well to de-inactivate I_T in relay cells. After the burst of the reticular cell, the inhibition in the relay cell ceases because of the silence of the reticular cell, so passive repolarization of the relay cell will ensue, and this is further promoted by activation of I_h (see earlier). This relative depolarization can trigger I_T in the relay cells. In Chapter VI, we shall come back to this property of how bursting in reticular cells can affect relay cells.

Another feature distinguishing I_T in reticular cells is the depolarization needed for activation. Figure 10 compares I_T activation curves for relay and reticular cells (Coulter *et al.*, 1989; Huguenard and Prince, 1992; Huguenard and McCormick, 1992). These examples are from the rat; the relay cells are from ventral posterior nucleus, and the reticular cells are from the adjacent thalamic reticular nucleus. Note that not only is the activation curve for the reticular cells shifted toward more depolarized values than that for the relay cells, but it is also shallower. This means that larger depolarizing potentials are needed to activate I_T in reticular cells than in relay cells. There may be many reasons for this difference,

B. Membrane Conductances

but one suggested by modeling relates to the finding that many of the actual T-type Ca^{2+} channels are located on peripheral dendrites of both cell types (Destexhe et al., 1996, 1998b). The larger electrotonic structure of reticular cells in the rat (noted previously) compared with relay cells suggests that attempts to activate I_T from current injected in the soma, as in voltage clamp experiments such as shown in Figure 10, would require more current and more depolarization for reticular cells if the channels to be activated are electrotonically more distant. This may thus be an artifact of the voltage clamp method, especially because the more physiological way of activating dendritically located T channels may involve EPSPs generated from synapses on the dendrites close to the channels.

There is another implication of having the T-type Ca^{2+} channels located peripherally in reticular cells. This could permit peripheral synaptic inputs to control I_T locally, perhaps to amplify an EPSP from a specific synapse by generating a low-threshold Ca^{2+} spike there, and this synapse could then have a more powerful effect at the axon hillock. This would also allow synaptic inputs to control I_T locally in different parts of the dendritic arbor. Thereby, a complex pattern of activated, inactivated, and de-inactivated patches of I_T could result.

One important proviso to this discussion of I_T and burst firing in reticular cells: the activity of these cells has not been recorded in awake animals. Although we know from such studies of relay cells that burst firing is seen during the waking state (see preceding and Chapter V), it is not certain whether reticular cells exhibit both response modes during normal, alert behavior. This needs to be resolved.

b. Other Conductances

Earlier we described how, for relay cells, the combination of I_T, I_h, and probably one or more K^+ conductances combine to create rhythmic bursting in relay cells. There are at least two analogous but different conductances in reticular cells that combine with I_T to produce rhythmic bursting (Bal and McCormick, 1993). Interestingly, these are triggered not by altered membrane voltage but by increased intracellular Ca^{2+} caused by the activation of I_T. One is a Ca^{2+}-dependent K^+ conductance that produces a current known as $I_{K[Ca]}$. By opening K^+ channels, $I_{K[Ca]}$ allows K^+ to exit and thereby hyperpolarize the cell. The kinetics of $I_{K[Ca]}$ are quite slow, but it is sufficiently powerful to produce a marked afterhyperpolarization in the reticular cell after each low-threshold spike. This hyperpolarization de-inactivates I_T, and the slow inactivation of $I_{K[Ca]}$ allows the cell to repolarize, triggering another low-threshold spike, and so on. In many reticular cells, only a few burst cycles are seen because the cell slowly depolarizes to inactivate I_T and produce tonic firing. This is because the low-threshold spikes produce another Ca^{2+}-dependent conductance, called I_{CAN}, that seems to be nonspecific for cations. These cations enter the reticular cell, thereby depolarizing it. I_{CAN} has much slower kinetics than does $I_{K[Ca]}$, so several burst cycles are expressed before I_{CAN} slowly depolarizes the cell sufficiently to switch firing mode from burst to tonic.

Few published studies have attempted to define voltage or ion-sensitive membrane conductances for reticular cells, so it remains unclear how many others may exist. For instance I_A seems to be a ubiquitous property of neurons throughout the brain, but there are no published reports, positive or negative, regarding the presence of I_A in reticular cells. Obviously the firing patterns of reticular cells are important to thalamic relays, because these cells provide powerful inhibition to relay cells, so understanding their intrinsic properties is an important goal. We need more study of this problem.

C. *Summary and Conclusions*

If we want to understand how nerve cells communicate with each other, an essential question that needs to be addressed is how a neuron's cellular properties—or more specifically, the passive and active properties of its dendritic (and somatic) membranes—affect its responses to synaptic input. This obviously also applies for thalamic neurons. Indeed, the interplay between a relay cell's cellular properties and the nature of its synaptic inputs (described in Chapter V) is at the heart of understanding the functioning of thalamic relays.

Cable properties of relay cells, as suggested by their dendritic arbors, indicate relatively little electrotonic spread along the dendrites, meaning that synaptic inputs even on the most distal dendritic locations will have some impact at the soma and axon hillock. However, it must be remembered that attenuation along a cable is frequency dependent, meaning that faster events, or faster postsynaptic potentials, will attenuate more during conduction to the soma than will slower ones. The cable properties of interneurons, in contrast, may be quite different. These cells have two outputs: a conventional one by way of the axon and another by way of terminals from peripheral dendrites. Cable modeling suggests that the latter are electrotonically isolated from the soma and axon, suggesting that interneurons might "multiplex," having two routes for input/output computations. Reticular cells seem to be electrotonically extensive, and they, like interneurons, may have dendritic and axonal outputs.

All of these thalamic neurons have numerous voltage-dependent conductances in their dendritic membranes, and such active conductances can override issues related to cable properties. For instance, if interneurons have enough voltage-dependent Na^+ channels in their dendrites, they can allow peripherally located synapses to have strong effects at the soma. Most is known of the voltage-dependent channels in relay cells. These include channels that underlie a variety of membrane currents, including I_T, I_A, and I_h, among others, and interneurons and reticular cells seem to have a similar complement of such channels. The state of these channels, especially their inactivation or activation, can strongly affect how the relay cell responds to driver input and thus how it relays information to cortex. Understanding how these voltage-dependent channels are controlled by various synaptic inputs and how they might interact with one another is an ongoing challenge for students of the thalamus.

D. Some Unresolved Questions

1. To what extent do the properties described in this chapter apply to thalamic cells in general? Are there important differences between species, thalamic nuclei, or relay cell types?

2. How do voltage-dependent conductances interact with one another and with the cell's cable properties to affect how the cell responds to various inputs?

3. Do some or all dendrites of thalamic relay cells, interneurons, or reticular cells have Na^+, Ca^{2+}, and/or K^+ channels? If so, what effect does this have on synaptic integration, particularly with respect to back propagation of action potentials?

4. How functionally isolated are dendritic F2 outputs of interneurons from each other and from the soma? Are there any interneurons that lack action potentials? Do the action potentials of interneurons invade the dendrites, affecting the F2 terminals? If they do, what is the functional effect?

5. Given the apparent importance of voltage-dependent conductances seen with *in vitro* methods, what is the range of membrane potentials typically seen in thalamic cells of awake, behaving animals?

6. How general are presynaptic dendrites in reticular neurons? Do they function like those of interneurons, or differently? For any one species, do they exist, what is their distribution, and how does this relate to the definable cable properties of these cells in different species?

CHAPTER V

Synaptic Properties

In Chapter III, we described the main synaptic inputs to thalamic cells. The axon terminals of these inputs release neurotransmitters that bind to postsynaptic receptors, and thereby affect their postsynaptic thalamic targets. Exactly how these synaptic inputs affect the postsynaptic cell is crucial to understanding the functional circuitry of the thalamus. Not long ago, this seemed a relatively simple task: an action potential in an afferent axon leads to release of neurotransmitter from its synaptic terminals; this affects the postsynaptic cell by producing a stereotypical, fast EPSP or IPSP, the sign of the postsynaptic potential depending on the neurotransmitter delivered; and the postsynaptic cell then linearly sums all of these postsynaptic potentials, much like a passive cable would, and the resulting membrane potential determines the postsynaptic firing frequency.

We now appreciate that the functioning of synapses is much more complicated. For one thing, a neurotransmitter can have very different postsynaptic effects on any one cell, depending on the specific postsynaptic receptors activated. For another, as described in Chapter IV, the postsynaptic cell does not act like a linear cable but instead exhibits a variety of voltage-dependent, transmembrane ionic conductances. Thus in addition to or even instead of seeing synaptic inputs as having a fairly simple and direct effect on the postsynaptic cell's firing rate, we must think in terms of the effects such inputs have on this cell's voltage-dependent conductances, among which the Na^+ and K^+ conductances underlying the action potential are special cases. The most important of the other conductances for thalamic relay cells, emphasized in Chapter IV, is the voltage-dependent Ca^{2+} conductance underlying I_T, because this determines whether the cell fires in burst or tonic mode, and the final response state plays a crucial role in the nature of the thalamic relay. In this chapter, we shall try to summarize many of the key functional features of the several different types of afferent in the thalamus with a focus on how synaptic activation affects both the firing rate and response mode of relay cells.

A. Ionotropic and Metabotropic Receptors

One property common to many pathways in the brain and to most of the major inputs to thalamus, regardless of the neurotransmitter they use, is that they can activate two very different kinds of postsynaptic receptor. These are *ionotropic* and *metabotropic* receptors, and both are found on relay cells, interneurons, and reticular cells. These receptor types, when activated, produce profoundly different actions on the postsynaptic cell.

1. Different Types of Metabotropic and Ionotropic Receptors

Among the major neurotransmitters released by afferents to thalamus are glutamate, GABA, and acetylcholine, and each of these can bind to ionotropic and metabotropic receptors. Other transmitters involved in thalamic circuitry include noradrenalin, serotonin, and histamine. The nature of the receptor types activated in the thalamus by these other neurotransmitters is not yet completely known, but preliminary evidence indicates that these activate mostly and in some cases only metabotropic receptors. Figure 1 schematically shows the main differences between an ionotropic and a metabotropic receptor (Nicoll *et al.*, 1990; Mott and Lewis, 1994; Recasens and Vignes, 1995; Pin and Duvoisin, 1995; Pin and Bockaert, 1995; Conn and Pin, 1997; Brown *et al.*, 1997).

It should be noted that most transmitters can activate several types of receptor within each of the ionotropic or metabotropic categories. These different actions are often first appreciated on the basis of differential sensitivity to specific agonists or antagonists, and once the different receptor types are recognized, further study usually reveals subtle differences in other properties. Receptors sensitive to glutamate serve as an example of this heterogeneity. Three major ionotropic glutamate receptor types are recognizable on the basis of their sensitivity to different agonists, and these are AMPA, kainate, and NMDA (Mayer and Westbrook, 1987; Nakanishi *et al.*, 1998; Ozawa *et al.*, 1998). A similar variety exists for the metabotropic glutamate receptors. In addition to differing with regard to agonist and/or antagonist sensitivity, some of these metabotropic receptors are associated with different second messenger pathways. To date, eight different types that are classified in three groups have been described in brain tissue, and the number recognized may well continue to grow (Recasens and Vignes, 1995; Pin and Duvoisin, 1995; Pin and Bockaert, 1995; Conn and Pin, 1997). However, in the thalamus, there appear to be only two main groups, called *group I*, which includes *types 1* and *5*, and *group II*, which includes types 2 and 3 (Godwin *et al.*, 1996a), and these are considered further in Section B.1. The preceding examples can,

A. Ionotropic and Metabotropic Receptors

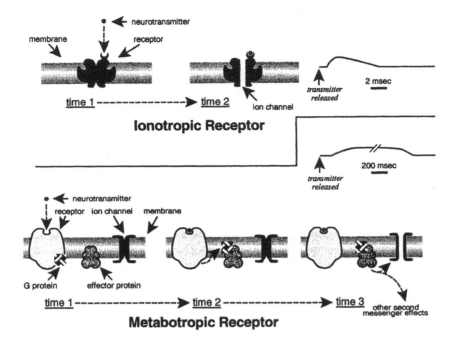

Figure 1
Schematic depiction of an ionotropic and a metabotropic receptor, each shown repeatedly at different times. For the ionotropic example, time 1 represents the period before binding to the neurotransmitter, and time 2 is the period after binding. The binding causes a conformational change that opens the ion channel, which forms the central core of the receptor complex. For the metabotropic receptor, time 1 is the period before neurotransmitter binding, and just after binding (time 2) a G-protein is released, which reacts with an effector protein to produce a cascade of biochemical reactions eventually resulting in opening or closing of an ion channel (time 3). Not shown is the possibility that for some receptors the G-protein can directly affect the ion channel.

in general, be extended to receptors activated by other transmitters used by afferents to thalamus, including GABA, acetylcholine, noradrenaline, serotonin, and histamine.

2. FUNCTIONAL DIFFERENCES BETWEEN IONOTROPIC AND METABOTROPIC RECEPTORS

As shown in Figure 1, when the neurotransmitter binds to an ionotropic receptor, it acts in a fairly direct fashion through a conformational change in the receptor to open a specific ion channel, which is actually embedded in and thus part of the receptor. Flow of ions into or out of the cell through these channels leads to the evoked postsynaptic potential. Because of the direct linkage between receptor activation and opening of the ion channel, the evoked postsynaptic potentials are fast: they have a short latency, a fast rise to peak, and are over generally within 10 or so milliseconds. The same is true for IPSPs evoked after activation of ionotropic receptors: a fast IPSP results.

When the neurotransmitter binds to a metabotropic receptor, a much more complicated series of events is triggered. The conformational change in the receptor ends in the activation of a G-protein, which in turn leads to a cascade of biochemical reactions in the membrane and/or cytoplasm of the relay cell (see Figure 1). This process is known as a *second messenger pathway* because the postsynaptic effects of the neurotransmitter are carried indirectly through second messengers by these reactions. Several chains of reaction result, and one of these eventually causes specific ion channels to open or close. In the thalamus, mostly only K^+ channels are affected in this way by metabotropic receptors. Note that either effect on ion channels, opening or closing of a K^+ channel, may occur with metabotropic receptor activation, whereas ionotropic receptor activation produces only ion channel opening. Postsynaptic potentials produced in this way by metabotropic receptors are quite slow. They begin with a long and somewhat variable latency, usually >10 msec, take tens of milliseconds to reach their peak, and they remain present for a long time, typically hundreds of milliseconds to several seconds or even longer.

Metabotropic receptor activation in other parts of the brain can often produce additional cellular effects through the second messenger systems, and some of these effects can act on the cell nucleus to change gene expression. For instance, metabotropic receptors have been implicated in long-term plastic changes in neocortex and hippocampus that may underlie such phenomena as learning and memory and various developmental processes (Gereau and Conn, 1994; Reid *et al.*, 1997; Huber *et al.*, 1998). However, although both immunocytochemical and pharmacological data indicate that metabotropic receptors are richly distributed in the thalamus, we have as yet no idea what effects other than postsynaptic potentials result from their activation. This is clearly an issue that requires attention. In contrast, after ionotropic receptor activation, the postsynaptic potential is usually the only effect seen postsynaptically. However, in some cases an entering ion can trigger secondary effects. The NMDA receptor is a good example. Activation of this ionotropic glutamate receptor generates an EPSP that depends in part on Ca^+ entry into the cell, and this change in the internal Ca^+ concentration itself can often produce other long-term changes such as long-term potentiation in neocortex and hippocampus (Collingridge and Bliss, 1987; Cotman *et al.*, 1988). However, the NMDA receptor is ionotropic and produces a relatively rapid EPSP in terms of duration and latency that does not itself require a second messenger link. This EPSP, although longer in duration and latency than the non-NMDA ionotropic glutamate receptors, is still much faster than that produced by activation of metabotropic glutamate receptors. If it were proven that changes in internal Ca^{2+} concentration in thalamic relay cells can lead to long-term effects on their responsiveness, then it might become necessary to ask whether the burst firing considered in the previous chapter, which involves activation of I_T and a consequential Ca^{2+} entry, can also trigger second messenger pathways and produce long-term effects.

Finally, there is good evidence from other brain regions and limited evidence in thalamus that the patterns of presynaptic stimulation needed to activate ionotropic versus metabotropic receptors are quite different (e.g., McCormick and Von Krosigk, 1992). Single action potentials or a few closely spaced in time that invade the presynaptic terminal are generally able to activate ionotropic receptors. However, activation of metabotropic receptors typically requires long trains of action potentials. One explanation for this is that metabotropic receptors tend to be further away from the synaptic junction, forming an annulus around them, as has been shown in the hippocampus (Lujan et al., 1996), than are ionotropic receptors, and thus activation of metabotropic receptors requires more transmitter released because of the extra diffusion (and dilution) involved. Whatever the explanation, this is particularly interesting for the pathways that can activate both types of receptor in the thalamus, and several examples are provided later. Perhaps these circuits function in a way that depends critically on activity patterns of the inputs: low levels of brief activity may activate only ionotropic receptors, and as the activity levels increase in frequency and/or duration, metabotropic activation may be added. Unfortunately, this property has not yet been systematically studied in thalamus.

B. Synaptic Inputs to Relay Cells

Figure 2 illustrates the neurotransmitters and postsynaptic receptors for the best understood inputs to relay cells of the lateral geniculate nucleus and ventral posterior nucleus. This is further summarized in Table 1, which shows inputs that are common to many thalamic nuclei and thus exclude ones such as the inputs to the lateral geniculate nucleus from the pretectum and from the superior colliculus.

1. Driving Inputs to Relay Cells

a. Basic Features of Driving Synapses

As noted previously and considered in more detail later in Chapter IX, by "driving inputs," we mean the inputs containing the main information to be relayed to cortex. For first order relays, this is carried, for example, by retinal axons to the lateral geniculate nucleus and by axons of the medial lemniscus to the ventral posterior nucleus; in higher order relays, this is carried by cortical layer 5 cell inputs to various thalamic nuclei, such as parts of pulvinar.

Studies of transmitter actions of these driving inputs have largely been limited to certain rodents (rats, guinea pigs, and hamsters) and to cats and ferrets, and mostly to the lateral geniculate nucleus and ventral posterior nucleus. The detailed summary in the following is based on these observations. These driving inputs all use an excitatory amino acid

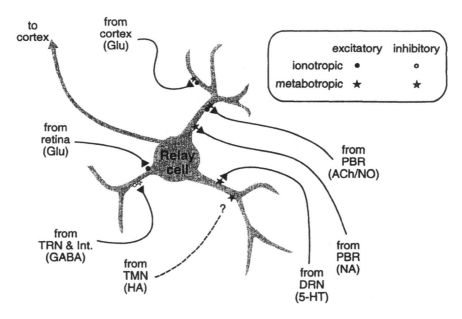

Figure 2
Inputs to thalamic relay cell in lateral geniculate nucleus showing neurotransmitters (in parentheses) and ionotropic or metabotropic types of postsynaptic receptors. The question mark indicates uncertainty about the synaptic relationships of the histaminergic input (see text for details). Abbreviations: 5-HT, serotonin; ACh, acetylcholine; DRN, dorsal raphé nucleus; Glu, glutamate; HA, histamine; Int., interneuron; NO, nitric oxide; NA, noradrenaline; PBR, parabrachial region; TMN, tuberomamillary nucleus; TRN, thalamic reticular nucleus.

(most probably glutamate) as a neurotransmitter, and this activates ionotropic glutamate receptors, which, as we have seen previously, can be divided into two main types: NMDA and non-NMDA. Available evidence from the lateral geniculate nucleus suggests that *every* retinal axon can activate both NMDA and non-NMDA receptors on relay cells (Scharfman *et al.*, 1990). However, the non-NMDA receptors in the thalamus have not been further subdivided into the AMPA and kainate receptors

TABLE 1 INPUTS TO RELAY CELLS OF LATERAL GENICULATE NUCLEUS AND VENTRAL POSTERIOR NUCLEI, INCLUDING TRANSMITTERS AND RECEPTOR TYPES

Source	Transmitter	Receptor type
Retina	Glutamate	Ionotropic
Cortex (layer 6)	Glutamate	Ionotropic and metabotropic
Parabrachial region	Acetylcholine	Ionotropic and metabotropic
Parabrachial region	Noradrenaline	Metabotropic
Dorsal raphé nucleus	Serotonin	Metabotropic
Tuberomamillary nucleus	Histamine	Metabotropic
Thalamic reticular nucleus	GABA	Ionotropic and metabotropic

B. Synaptic Inputs to Relay Cells

described for some other parts of the brain (see preceding and Mayer and Westbrook, 1987; Nakanishi et al., 1998; Ozawa et al., 1998), so it is not clear which of these subtypes is present and what specific role in synaptic function each might play. Consequently, we shall continue to use the term "non-NMDA" to refer to these. Interestingly, retinal inputs to relay cells of the lateral geniculate nucleus do *not* activate metabotropic glutamate receptors (see Figure 2 and McCormick and Von Krosigk, 1992). It would be of great interest to know how general this property is for driver inputs in other thalamic nuclei because, as we shall see, all modulatory inputs to the lateral geniculate nucleus can activate metabotropic receptors, and most also activate ionotropic receptors, and the difference may be crucial to the distinction between drivers and modulators (see Chapter IX).

Activation of the non-NMDA receptors produces a prototypical, fast excitatory postsynaptic potential caused by entry of Na^+ and perhaps other cations. The response associated with NMDA receptor activation is unusual for three reasons. First, it has a voltage dependency so the more hyperpolarized the cell, the less that receptor activation produces an excitatory postsynaptic potential. This is because the ion channel portion of the NMDA receptor becomes clogged with Mg^{2+} ions, preventing influx of cations to depolarize the cell. Prior depolarization of the cell prevents this Mg^{2+} block, and then activation of the NMDA receptor produces an excitatory postsynaptic potential. This excitatory postsynaptic potential is slower than that produced by activation of the non-NMDA receptor but much faster than metabotropic receptor action. Second and as noted previously, although NMDA receptors are ionotropic, their activation involves considerable influx of Ca^{2+}, and this can, in turn, activate certain second messenger pathways providing other postsynaptic effects. Third, activation of NMDA receptors requires the presence of glycine as a cofactor in addition to glutamate. There is a glycine site generally associated with the NMDA receptor that must also be activated along with the glutamate site. This is an unusual glycine site because the more common glycinergic receptor associated with inhibition elsewhere in the brain is blocked by strychnine, but the NMDA glycine site is unaffected by strychnine. What is particularly odd about this glycine requirement for NMDA activation is that there is no known source of glycinergic input to thalamic relay cells. One could argue that metabolic levels of glycine present in the neuropil are sufficient to activate the glycine site in the NMDA receptor. However, how is the level of glycine modulated to affect the NMDA response, and if the level it is not modulated, why would a receptor to an invariant substance ever evolve? There is much yet to be learned about the role of NMDA receptors in synaptic function for the driving inputs.

b. Effect of Driving Afferents on the Response Mode of Relay Cells

We normally think of synaptic inputs as excitatory or inhibitory on the basis of whether the evoked postsynaptic potential is more or less likely to activate action potentials or, more specifically, to activate the voltage-dependent Na^+ and K^+ conductances that produce action potentials. We also must consider the effects of synaptic inputs on activating other conductances, particularly the low-threshold Ca^{2+} conductance. Clearly, the response mode of relay cells, burst or tonic, strongly affects their responses to inputs and thus will affect their relay of driving inputs to cortex. The question here is: Can the pattern of activity in driving inputs by themselves change the response mode of their postsynaptic relay cell targets? Unfortunately, we do not yet have a clear, empirical answer to that question but theory and indirect experimental evidence suggest that driving inputs do not play a major direct role in controlling response mode.

The theoretical reason is that driving inputs to relay cells activate only ionotropic and not metabotropic glutamate receptors. These inputs clearly produce EPSPs, and thus the issue becomes whether this depolarization can convert a relay cell mode from burst to tonic, which would require the EPSP to inactivate I_T. The problem lies in the kinetics. EPSPs from ionotropic glutamate receptors are fast, typically decaying to low levels after 10–20 msec. As noted previously, the activation of I_T is fast and requires the rate of depolarization (dV/dt) to be above a certain minimum, so such an EPSP could (and in fact does) readily activate I_T. However, inactivation is much slower, and the EPSP would be over and done before inactivation proceeded very far, and the cell would repolarize to its former value. The balance of T channels between inactivation and de-inactivation would thus be little affected, and the cell would again be primed to respond to inputs in burst mode. To switch to tonic firing by inactivating I_T would require a much longer depolarization than would be expected by means of a single EPSP from activation of ionotropic glutamate receptors. However, a sufficiently prolonged depolarization might be possible even with ionotropic glutamate receptor activation if a long, high-frequency train of EPSPs were evoked, permitting temporal summation that would sustain the depolarization for roughly 100 msec or more.

One might then ask whether a train of EPSPs by means of ionotropic glutamate receptors[1] (e.g., via retinal synapses) could produce temporal summation that would sustain the depolarization sufficiently to inactivate I_T. The answer is that this is possible with two provisos. One is that driving inputs from retina may lead to the sort of temporal summation of postsynaptic potentials that can inactivate I_T. If the EPSP lasts for, say 25 msec, then summation will begin with firing rates exceeding 40 spikes/sec, certainly a reasonable rate for retinal afferents (Bullier and Norton, 1979). The potential problem here is that electrical activation of retinal afferents produces multisynaptic IPSPs that significantly shorten the monosynaptic EPSP to <10 msec (Suzuki and Kato, 1966). If this happens with natural visual stimulation, it is not clear that even very high rates of retinal firing (e.g., >100 spikes/sec, a rate that may be seldom

---

[1] For this consideration, we assume only AMPA receptors being involved. A consideration of any role of NMDA receptors in this process is complicated by the fact that they are voltage-dependent and will not contribute to an EPSP unless the cell is already fairly depolarized. However, what role NMDA receptors might play in this process is not clear.

B. Synaptic Inputs to Relay Cells

sustained by retinal afferents) may not elicit a sufficient depolarization to inactivate I_T, because the evoked IPSPs might counteract this.

The second and more important proviso is that, for a cell starting off with I_T de-inactivated, any EPSP with a sufficiently fast rise time will activate I_T before inactivating it, so a burst will be evoked; if the EPSP is relatively transient and weak in amplitude, only the burst will be evoked, but if it is sustained and strong enough, the burst may be followed, after a pause, by tonic firing. This is in fact seen when geniculate cells of the cat are activated either *in vivo* by visual stimuli (Lu *et al.*, 1992) or *in vitro* with current injection (Smith *et al.*, 2000). The point here is that EPSPs with a very fast rise time may, with temporal summation, inactivate I_T, but not before activating I_T and an associated burst of action potentials relayed to cortex. Only a prolonged EPSP with a *slow* rise can inactivate I_T without activating a burst (see Chapter IV, Section B.2.b). The obvious way to do this is by activating metabotropic receptors (see also following). Another way might be to activate many EPSPs unsynchronized temporally from many, convergent sources to produce a composite depolarization with a slow rise time. This could be possible through modulatory inputs because there seems to be a large convergence of these inputs to relay cells, but driving inputs from the retina mostly derive from one or a very small number with little convergence (Sherman and Guillery, 1998).

The preceding paragraphs suggest that driving inputs by ionotropic glutamate receptors cannot readily inactivate I_T. However, it is important to realize that only the lateral geniculate nucleus in guinea pigs and cats and the medial geniculate of rats has to date been tested for the type of receptor (ionotropic or metabotropic glutamate receptor) activated by the driving (retinal or collicular) input (McCormick and Von Krosigk, 1992; Godwin *et al.*, 1996b; E. L. Bartlett and P. H. Smith, unpublished observations). It is tempting to generalize this to other thalamic nuclei in other species, but more empirical evidence here is important.

Better than theoretical constructs, especially with provisos, would be actual evidence regarding any role of driving inputs in the control of response mode of relay cells. The evidence available to date is at best circumstantial, but it does suggest that activation of metabotropic receptors is a more efficient way to control response mode than is activation of ionotropic receptors (Godwin *et al.*, 1996b). On teleological grounds, one might argue that it is better to control response mode by means of the modulatory inputs, so response mode is never confused with peripheral stimuli. Unfortunately, teleological grounds are often treacherous, and actual data are needed instead.

2. INPUTS TO RELAY CELLS FROM INTERNEURONS AND CELLS OF THE THALAMIC RETICULAR NUCLEUS

a. Basic Properties of Synapses from Interneurons and Reticular Cells

Both interneurons and reticular cells use GABA as a neurotransmitter, and both of these "GABAergic" cells innervate relay cells (Figure 2).

On the basis of the available evidence, GABA seems to act in the thalamus in an inhibitory manner. Relay cells thus exhibit IPSPs when inputs from either interneurons or reticular cells are activated. These potentials are generated through two different receptors, known as $GABA_A$, which is ionotropic, and $GABA_B$, which is metabotropic (Table 1)[2].

Reticular cells, as a population, activate both receptor types on relay cells, although it is not clear if individual reticular axons or axon terminals activate both receptors or are more selective. The situation with interneuronal activation is less clear. There is indirect evidence that at least some if not all of their dendritic terminals activate $GABA_A$ receptors on relay cells (Cox et al., 1998). However, $GABA_B$ responses have yet to be determined for these synapses, and nothing has been reported on the receptors activated by the axons of interneurons.

Activation of the $GABA_A$ receptor opens Cl^- channels. This inhibits the cell, but not very much by the amplitude of the IPSP, because the reversal potential and thus maximum hyperpolarization seen with an increased Cl^- conductance is only to about -70 to -75 mV. Instead, $GABA_A$ activation creates such a large increase in the Cl^- conductance that any tendency to depolarize the cell by opening other conductances, say, by an EPSP, would be offset by moving more Cl^- into the cell. This effectively decreases neuronal input resistance and serves to shunt any EPSPs. In this way, activation of $GABA_A$ tends to clamp the cell near the Cl^- reversal potential, around -70 to -75 mV, far from the activation threshold for action potentials, and thus the cell is effectively inhibited.

Activation of $GABA_B$ receptors is different. This involves an increase in conductance to K^+, probably by increasing the K^+ "leak" conductance described in Chapter IV. Increasing this conductance by $GABA_B$ activation causes K^+ to leave the cell, thereby hyperpolarizing it. Compared with activation of the $GABA_A$ receptor, this more strongly hyperpolarizes the cell toward the K^+ reversal potential of roughly -100 mV but has much less effect on membrane conductance or neuronal input resistance. Thus $GABA_B$ inhibition acts not by shunting the membrane to a subthreshold level, as happens with $GABA_A$ activation, but by strongly hyperpolarizing the cell. Another way of looking at this difference in response is that $GABA_A$ activation actually reduces, or "divides," the EPSP, whereas $GABA_B$ activation has less effect on EPSP generation but provides a larger IPSP

---

[2] Although it happens that GABA in thalamus of adult animals nearly always acts solely in a hyperpolarizing manner, there are exceptions to this action of GABA in other parts of the brain. In the suprachiasmatic nucleus, which is involved in diurnal rhythms, the effects of GABA are themselves diurnal, depolarizing by day and hyperpolarizing by night (Wagner et al., 1997). This neat trick is accomplished by diurnal changes in internal Cl^- concentration of suprachiasmatic cells: this concentration is higher by day, high enough that activation of the GABA receptor causes Cl^- to exit the cell, thereby depolarizing it. Also, GABA can act as a depolarizing transmitter in some developing neural systems and outside of thalamus in adults. Neurotransmitters should no longer be classified as excitatory or inhibitory because this action depends not so much on the neurotransmitter as on the specific postsynaptic receptor and ion channel combination with which it is related. Indeed, most neurotransmitters in thalamus are both excitatory (i.e., depolarizing) and inhibitory (i.e., hyperpolarizing), and examples are given later in this chapter.

B. Synaptic Inputs to Relay Cells

that is summed to any EPSP (Koch *et al.*, 1982). In this sense, $GABA_A$ activation is multiplicative, whereas $GABA_B$ activation is additive. As noted previously, activation of all other metabotropic receptors on relay cells, like activation of $GABA_B$ receptors, also affects the K^+ "leak" conductance, some by increasing it and others by decreasing it.

Another difference between activation of the $GABA_A$ and $GABA_B$ receptors is the time course of the effect. As noted previously, the $GABA_A$ receptor is ionotropic and so the IPSP associated with it is much faster in latency and duration than that associated with the metabotropic $GABA_B$ receptor. The $GABA_A$ IPSP is typically complete after 10 or so milliseconds, whereas the $GABA_B$ IPSP typically lasts at least 10 to 100 times longer.

Finally, because $GABA_A$ is ionotropic, there is no reason to expect its activation to do anything more than transiently increase a Cl^- conductance. Because the $GABA_B$ receptor is metabotropic, its activation turns on second messenger pathways. This, in addition to increasing a K^+ conductance, may also affect other cell properties that can have a prolonged effect on the postsynaptic relay cell. There is as yet no evidence in thalamus for any such effects of $GABA_B$ activation beyond its IPSP, but this is clearly an important possibility that needs to be investigated specifically for thalamus.

b. Effect of Interneuronal and Reticular Inputs on the Response Mode of Relay Cells

As stated previously for driving inputs, we must consider the effects of these GABAergic inputs on the response mode, burst or tonic, of relay cells. It seems unlikely that brief, transient activation of the $GABA_A$ receptor will have much effect on response mode, although under special conditions considered later, $GABA_A$ activation can indeed promote burst firing in relay cells. Normally, $GABA_A$ activation drives the membrane toward about -70 to -75 mV. As shown in Figure 4 of Chapter IV, this will at best partially de-inactivate I_T. Furthermore, de-inactivation of I_T has a slow time course, and fairly complete de-inactivation requires hyperpolarization for at least 100 msec, whereas the $GABA_A$ response to a single input is largely over in 10–20 msec (see preceding). The brief hyperpolarization is also unlikely to activate I_h, so the return to the original resting potential after the $GABA_A$ IPSP is purely passive and not accelerated or amplified by I_h. Thus when the relay cell recovers from the brief $GABA_A$ IPSP and passively depolarizes to rest, that depolarization will not likely activate I_T. By this logic, activation of $GABA_A$ receptors alone by any one input under most normal conditions would seem an unlikely candidate to produce a low-threshold spike. Of course, if multiple active GABAergic inputs converge onto a relay cell and produce extensive temporal summation of their $GABA_A$-activated IPSPs, this could well inactivate I_T.

Indeed, there is evidence that $GABA_A$ activation from reticular cells can produce low-threshold spiking in relay cells. However, this has been seen when large numbers of reticular cells burst synchronously, as happens during various sleep states, or during *in vitro* recording from thalamic slices, and thus it is not clear if this happens during the waking state. The prolonged bursts of action potentials produce a sufficiently prolonged and powerful IPSP to de-inactivate I_T and activate I_h. This happens even when $GABA_B$ receptors are pharmacologically blocked, and thus it must partly represent activation of $GABA_A$ receptors. Presumably, this example of $GABA_A$ inhibition producing burst firing is not likely to occur during periods of wakefulness because it requires massive, prolonged activity of GABAergic afferents from the thalamic reticular nucleus, and such activity seems to occur normally only during sleep.

Activation of $GABA_B$ receptors would seem better suited to produce burst firing in relay cells. This is because it produces a larger hyperpolarization, toward the K^+ reversal potential near -100 mV rather than the Cl^- reversal potential. Also, because the response is metabotropic, it lasts much longer, typically for >100 msec. Thus the amplitude and duration of the $GABA_B$ IPSP are more likely to de-inactivate I_T and activate I_h. Repolarization to rest after the IPSP decays, especially aided by I_h, should then activate I_T. For these reasons, $GABA_B$ inhibition should, in principle, play an important role in controlling firing mode of relay cells. However, the precise roles of $GABA_A$ and $GABA_B$ activation regarding response mode need to be more thoroughly investigated.

3. Inputs from Cortical Layer 6 Axons to Relay Cells

a. Basic Properties of Layer 6 Corticothalamic Synapses

In this section, by "corticothalamic," we refer only to the layer 6 input. These corticothalamic axons, which provide an excitatory input to relay cells, use glutamate as a neurotransmitter, and their synapses activate both ionotropic and metabotropic glutamate receptors on relay cells (Figure 2). In the lateral geniculate nucleus, the ionotropic receptors activated by corticogeniculate axons are the same types that are activated by retinogeniculate synapses. However, because of the very different locations of their synaptic inputs on the dendritic arbor—retinogeniculate synapses are found proximally and corticogeniculate, distally, with effectively no overlap (see Chapter III)—retinogeniculate and corticogeniculate synapses are unlikely to activate the same individual receptors. It is particularly interesting that corticogeniculate synapses also activate metabotropic glutamate receptors on relay cells, whereas retinogeniculate synapses do not (see Figure 2; see also McCormick and Von Krosigk, 1992; Godwin et al., 1996a). We do not yet know if every corticogeniculate axon has access postsynaptically to both ionotropic and metabotropic receptors. That is, it is possible that some activate only ionotropic receptors, some only metabotropic, and some, perhaps, both types. The possibility that different corticogeniculate axons can activate different mixes of receptor type means that different groups of these axons could produce quite different postsynaptic effects on the relay cell and thus subserve different functions. We saw in Chapter III that there is evidence for heterogeneity of corticogeniculate axons from area 17, and the pathway from cortex to the lateral geniculate nucleus involves several different

B. Synaptic Inputs to Relay Cells

cortical areas (Updyke, 1975), so different types of corticogeniculate axon might relate partly to differential postsynaptic activation of ionotropic and metabotropic receptors. Our lack of knowledge here is a severe limitation because understanding corticogeniculate function requires such knowledge.

The available evidence suggests that corticogeniculate synapses activate type 1 metabotropic glutamate receptors on relay cells. The production of inositol-phosphates resulting from this activation leads ultimately, through other biochemical pathways, to reduction of the K^+ "leak" conductance. This depolarizes the cell, creating an EPSP that is quite slow in onset (>10 msec) and lasting longer than hundreds of milliseconds (McCormick and Von Krosigk, 1992).

The implications of the mix of non-NMDA and NMDA ionotropic glutamate receptors activated from cortex are similar to those that apply to the same receptor types activated from the retina, but the metabotropic glutamate receptors also activated from cortex add another dimension to corticogeniculate function. Not only is the EPSP through metabotropic glutamate receptors very slow and prolonged, but also the second messenger cascades and release of intracellular Ca^{2+} pools raise the possibility that activation of this input can have long-term effects on relay cells. However, as noted previously, this is an issue of fundamental importance that has yet to be investigated for thalamus.

Because of the slow and long-lasting response to activation of this metabotropic glutamate receptor, such a response is much better suited than are those associated with ionotropic glutamate receptors to maintain sustained changes in membrane voltage of relay cells. This, in turn, would control general cell excitability—the more depolarized the cell, the more excitable it would be. Also, because the EPSP resulting from activation of metabotropic glutamate receptors involves reduction of the K^+ "leak" conductance, this increases the neuronal input resistance, which in turn results in larger postsynaptic potentials, excitatory and inhibitory, by synaptic activation. Finally, such slow and prolonged membrane potential changes could be quite important in allowing cortex to exert control over voltage-dependent conductances expressed by these relay cells (see later). The slow response, however, would act like a low-pass temporal filter in transferring information across the synapse so that specific firing patterns in the cortical afferents would not be imposed on the relay cells. In contrast, excitatory postsynaptic potentials evoked by ionotropic glutamate receptors, particularly non-NMDA ones, would be faster and perhaps permit better transfer of the firing patterns, but it would be less suitable for sustaining changes in membrane voltage.

This property of corticothalamic axons activating both ionotropic and metabotropic glutamate receptors on relay cells has not yet been extended to other thalamic nuclei.

b. Effect of Corticothalamic Inputs on Response Mode of Relay Cells

Corticothalamic synapses in the lateral geniculate nucleus and ventral posterior nucleus (and presumably elsewhere in thalamus) seem particularly well suited to control the firing mode of the relay cells because they can activate metabotropic glutamate receptors, and these provide a prolonged depolarization that would effectively inactivate I_T. Furthermore, as noted in Chapter IV, to activate I_T requires that the depolarization exceed a minimal rate of rise, or dV/dt, and the rate of rise of the EPSP evoked by metabotropic glutamate receptors seems too slow to activate I_T, but the ensuing depolarization is sufficiently prolonged to inactivate I_T (Gutierrez *et al.*, 1999). Thus evoking a metabotropic EPSP can perform the neat trick of inactivating I_T without ever activating it. The result would be a strong bias toward tonic mode responses, and any bursting activity would be greatly reduced or eliminated. Evidence from *in vitro* studies of thalamus (both the lateral geniculate nucleus and ventrobasal complex) in guinea pigs and *in vivo* studies of the lateral geniculate nucleus in cats indicate that such a process does indeed occur (McCormick and Von Krosigk, 1992; Godwin *et al.*, 1996b).

4. BRAIN STEM MODULATORY INPUTS TO RELAY CELLS

Brain stem modulatory inputs can vary quantitatively among thalamic nuclei (Fitzpatrick *et al.*, 1989), and most of our detailed knowledge of these inputs stems from studies of the lateral geniculate nucleus. Thus the description following is limited to the lateral geniculate nucleus. Whether the same principles apply to other thalamic nuclei remains to be studied as do many unanswered questions for the lateral geniculate nucleus, highlighted later.

a. Parabrachial Inputs

In cats, most of the input to the lateral geniculate nucleus from the brain stem derives from cholinergic neurons in the midbrain and pontine tegmentum surrounding the brachium conjunctivum. We refer to this brain stem area as the parabrachial region.[3] As is summarized by Figure 2, activation of this input produces an EPSP caused primarily by activation of two different receptors (McCormick, 1990, 1992). The first is an ionotropic nicotinic receptor that produces a fast EPSP by permitting influx of cations. The second is a metabotropic muscarinic receptor, known as an M1 type, that triggers a second messenger pathway ultimately leading to a reduction in the K^+ "leak" conductance. This muscarinic response is a very slow, long-lasting EPSP. In this regard, the effect of activating this muscarinic receptor is remarkably similar to the metabotropic glutamate response seen from activation of corticogeniculate

---

[3] Terminology for these brain stem cholinergic afferents to thalamus has been confusing. Among other terms commonly used for these cells are the "pedunculopontine tegmental nucleus" and the "laterodorsal tegmental nucleus." The problem is that, in many species such as the cat and monkey, the cholinergic cells are intermixed with noradrenergic and possibly other afferents to thalamus, whereas in other species, such as rodents, the pedunculopontine tegmental nucleus effectively includes only cholinergic afferents, whereas the noradrenergic afferents are gathered in the "locus ceruleus." For the cat, we prefer the term "parabrachial region" because it does not imply a homogeneous, well-demarcated cell group.

B. Synaptic Inputs to Relay Cells

input (see preceding). The possibility exists that both metabotropic receptors may be linked to the same second messenger pathways and K^+ channels, as seems to be the case for hippocampal neurons, although this has not yet been experimentally tested for thalamic neurons. As would be expected, activation of these parabrachial inputs by *in vivo* or *in vitro* application of acetylcholine effectively converts the firing mode of thalamic relay cells from burst to tonic (McCormick, 1991a; Lu et al., 1993).

In addition to acetylcholine, these axons and their terminals in the lateral geniculate nucleus also contain NO (Figure 2), a neurotransmitter or neuromodulator with a widespread distribution in the brain (Bickford et al., 1993). Relatively little is known concerning the action of NO in the lateral geniculate nucleus, and it is not yet clear what other thalamic nuclei, if any, also receive inputs that might release NO as a neuroactive substance. Two recent studies of the lateral geniculate nucleus suggest different but perhaps complementary roles for the release of NO from parabrachial terminals. In an *in vitro* study, application of NO donors depolarized relay cells and caused them to switch from burst to tonic firing (Pape and Mager, 1992), perhaps complementing the role of acetylcholine in this regard. In an *in vivo* study, similar application of NO donors seemed to promote the activation of NMDA receptors by retinal afferents (Cudeiro et al., 1996; Rivadulla et al., 1996). This, too, could be related to depolarization of the relay cells because such depolarization would relieve the NMDA receptor from the Mg^{2+} block. This study made no comment on the effect of NO on response mode, but we would expect that the depolarization that enables NMDA receptor activation would also promote tonic firing.

5. OTHER INPUTS

Other sparse inputs to the relay cells of the lateral geniculate nucleus are shown in Figure 2 and include noradrenergic axons from cells in the parabrachial region, serotonergic axons from cells in the dorsal raphé nucleus, and histaminergic axons from cells in the tuberomamillary nucleus of the hypothalamus (see Chapter III for details). Interestingly, in most examples in thalamus the dominant receptor type activated by these other inputs is metabotropic and the result is an effect on the K^+ "leak" conductance.

Noradrenalin has two very different effects on relay cells, and these effects act through two different metabotropic receptors, known as α_1 and β adrenoreceptors (see Table 1). The first effect, by activation of the α_1 adrenoreceptors, increases excitability of relay cells in the lateral geniculate nucleus by reducing the K^+ "leak" conductance and thereby producing a long slow EPSP. This has the additional result of promoting tonic firing. This is much like the glutamate activation of metabotropic glutamate receptors or the cholinergic activation of muscarinic receptors just described. The other effect, which operates through the β adrenoreceptors, changes the voltage dependency of I_h in such a way as to increase this depolarizing, voltage-dependent current. This latter effect has been described in Chapter IV.

Effects of serotonin are complex and somewhat controversial. In the *in vivo* preparation, application of serotonin or activation of the dorsal raphé nucleus, which contains the serotonergic cells that innervate the lateral geniculate nucleus, inhibits relay cells (Kayama *et al.*, 1989). However the picture from *in vitro* studies is very different. Here, application of serotonin seems to have no conventional inhibitory or excitatory effect on relay cells (McCormick and Pape, 1990a). It seems plausible that the effects seen *in vivo* result from serotonergic excitation of interneurons or reticular cells, which would indirectly inhibit relay cells (Funke and Eysel, 1995). There is, nonetheless, an unconventional effect of serotonin on relay cells described from *in vitro* studies. By operating through an unknown but probably metabotropic receptor, serotonin has the same effect on I_h as described earlier for noradrenalin (McCormick and Pape, 1990a; see also Chapter IV).

Finally, histamine clearly has effects on functioning of thalamic cells, but the anatomical substrate for this is not entirely clear (Manning *et al.*, 1996). As noted in Chapter III, there is ample innervation of thalamus by histaminergic axons from the tuberomamillary nucleus of the hypothalamus, but electron microscopic surveys of thalamus have revealed very few conventional synapses formed by these axons (Uhlrich *et al.*, 1993). Perhaps this system works entirely without synapses by having the axon terminals release histamine into the extracellular space to diffuse to receptors on thalamic neurons. In any case, application of histamine to relay cells has nearly identical effects to noradrenergic application (McCormick and Williamson, 1991). One effect, operating through one metabotropic receptor (called H_1), reduces the K^+ "leak" conductance. This produces a long slow EPSP that also promotes tonic firing. The other effect, which operates through a different metabotropic receptor (H_2), changes I_h in the same way that noradrenalin and serotonin do (see earlier and Chapter IV).

C. Inputs to Interneurons and Reticular Cells

Relatively few recordings have been made from interneurons and reticular cells. As noted previously, interneurons seem to have two distinct and functionally independent innervation zones that may be electrotonically isolated from each other: one is the cell body and proximal dendrites, where firing of the axon and its F1 terminals is controlled, and the other is the region of (distal) dendritic F2 terminals (see Figure 2 of Chapter IV). Thus in recordings from the cell body of an interneuron, presumably only the former, soma-dendritic inputs are revealed. Although there is evidence for limited dendritic terminal output for reticular cells in some species (see Chapter II), for the most part these cells seem to be organized in a conventional fashion, integrating synaptic inputs to produce an axonal output. Figure 3 summarizes the synaptic inputs to these cells with the types of postsynaptic receptors activated. The many question marks

C. Inputs to Interneurons and Reticular Cells

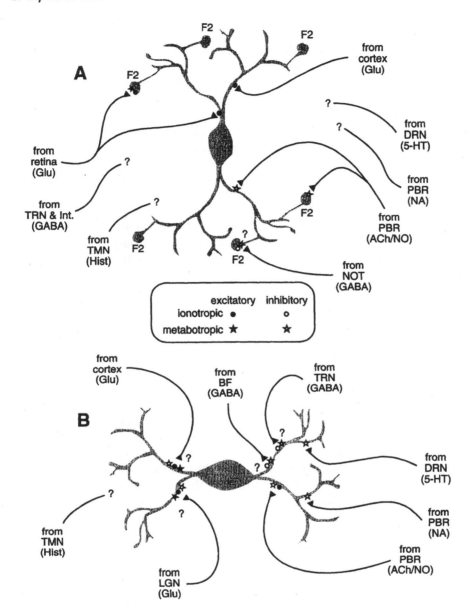

Figure 3

Inputs to interneuron (A) in lateral geniculate nucleus and cell of the thalamic reticular nucleus (B) shown with the same conventions as in Figure 2. The question marks indicate uncertainty about the postsynaptic receptors involved because of lack of information; in some cases we know so little that no synaptic site is drawn, and in others, we have limited information about these sites and associated receptors. Note that the F2 terminals of the interneuron receive synaptic inputs. Abbreviations as in Figure 2 plus BF, basal forebrain; NOT, nucleus of the optic tract.

in Figure 3 reflect the many gaps in our specific knowledge, even for such well-studied examples as the lateral geniculate nucleus and ventral posterior nucleus.

1. GLUTAMATERGIC INPUTS

Retinal axons innervate interneurons but not reticular cells, and relay cell axons innervate reticular cells but generally not interneurons, except perhaps for the interlaminar interneurons thought to be displaced thalamic reticular nucleus cells (Sanchez-Vives et al., 1996; see also Chapter III and Figure 3). In Chapter II, we have summarized evidence that interneurons receive synaptic inputs onto both dendritic F2 terminals as well as proximal dendrites plus the cell body. It appears that the inputs onto F2 terminals control the output of these terminals and the inputs onto proximal dendrites plus cell body control the axonal output. Recordings from cell bodies of interneurons would reveal only the latter function. Such recordings indicate that retinal inputs produce EPSPs based on ionotropic glutamate receptors (Pape and McCormick, 1995; Cox et al., 1998). Indirect evidence, including immunocytochemistry and pharmacological studies of effects on relay cells attributed to interneuronal activation, indicates that retinal input to the dendritic F2 terminals activates metabotropic as well as ionotropic glutamate receptors (Godwin et al., 1996a; Cox et al., 1998; Cox and Sherman, 2000). The metabotropic glutamate receptors on these F2 terminals are type 5, which are subtly different from the type 1 metabotropic glutamate receptors activated by cortical axons on relay cells, but like them, their activation depolarizes the postsynaptic target by decreasing the K^+ "leak" conductance. Exactly how activation of these glutamate receptors controls the F2 terminals is not known, but it seems plausible that this acts by depolarizing these terminals. This, in turn, would lead to an influx of Ca^+ into the cell by means of a voltage-dependent, high-threshold Ca^+ conductance similar to that described in Chapter IV but perhaps using different channel types. The resulting Ca^+ entry leads to transmitter release, GABA in this case. However, there is no direct evidence for voltage changes in the F2 terminal, and one can imagine other scenarios. For instance, activation of the metabotropic receptor on the F2 terminal can affect internal Ca^{2+} by means of second messenger pathways without having any effect of K^+ channels (and thus the membrane voltage), and a change in internal Ca^{2+} could influence transmitter release. Thus the details of precisely how metabotropic activation in the F2 terminal affects transmitter release is unresolved.

The result is that retinal input to the F2 terminal, as part of the triadic synaptic arrangement described in Chapter II, leads to feedforward inhibition of the relay cell through the F2 terminals. This is schematically shown in Figure 4. Thus retinal inputs will evoke a monosynaptic EPSP and often a disynaptic IPSP in the relay cells that receive such triadic inputs and will evoke just a monosynaptic EPSP in those that do not. As noted in Chapter III, X cells have such triadic circuitry involving their retinal input, whereas Y cells do not.

However, because the retinal innervation of the F2 terminal involves a metabotropic receptor, whereas innervation of the relay cell does not, the operation of this circuit might depend on activity patterns in the reti-

C. Inputs to Interneurons and Reticular Cells

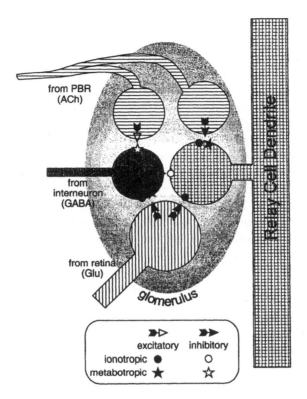

Figure 4
Schematic view of triadic circuitry in the lateral geniculate nucleus showing neurotransmitters and postsynaptic receptors. The triadic circuitry is embedded in a glomerulus into which an appendage from a relay cell intrudes. The arrows indicate the direction of each synapse from presynaptic to postsynaptic. Note that a single retinal terminal forms a triad with an F2 terminal and the appendage of the relay cell dendrite, whereas two terminals from the same parabrachial axon are involved in forming the same functional sort of triad.

nal afferent, because ionotropic and metabotropic receptors seem to operate differently depending on these patterns. That is, as noted earlier, metabotropic glutamate receptors respond relatively poorly to low rates of afferent input. Low or brief rates of retinal firing will activate only ionotropic receptors, leading to brief monosynaptic EPSPs (and brief disynaptic IPSPs in the relay cell due to the ionotropic glutamate receptors present on the F2 terminal); only with increasing rates or duration of firing will the metabotropic receptors on the F2 terminal become active, thereby producing an additional prolonged IPSP in the relay cell. Thus the inhibition relative to the excitation through the triadic circuit actually grows with increasing input strength. This may be important for preventing response saturation in the relay cell for very strong inputs. For retinogeniculate circuitry, this is a form of contrast gain control that would enable the relay cell to continue to inform the cortex about increasing stimulus strength or contrast over a wider dynamic range of contrasts in the visual scene. Presumably, relay cells without triadic circuitry are less able to accomplish this.

Perhaps a similar function for the triad exists in other relays. For example, in the medial geniculate nucleus or ventral posterior nucleus, this could prevent response saturation as the sound or touch increases in amplitude. However, relevant data to address the general function of the triad in thalamus are lacking.

Cortical axons innervate both interneurons and reticular cells. Their innervation of interneurons does not involve F2 terminals. Instead, cortical terminals innervate dendritic shafts of interneurons where they activate only ionotropic receptors that presumably produce only fast EPSPs (Pape and McCormick, 1995). It would thus appear that their main effect is to control axonal outputs of these cells, but there remains the untested possibility that action potentials from the cell body can invade peripheral dendrites to influence F2 terminals.

There is ample evidence that, in general, activation of cortical or relay cell inputs excites reticular cells (but see later). However, the pattern of postsynaptic receptors associated with cortical or relay cell inputs to reticular cells is far from clear, partly because of the lack of data and partly because of complexities unique to these cells. One complexity is that reticular cells receive glutamatergic inputs from these two sources, and it is not yet known which glutamate receptors are associated with which glutamatergic inputs. Nor do we have clear evidence about exactly which types of glutamatergic receptors are found on reticular cells. There is immunocytochemical evidence that these cells contain two types of metabotropic glutamate receptor (Godwin *et al.*, 1996a): type 1 and type 2/3 (types 2 and 3 have currently not been distinguished from each other for thalamus). This leads to another complication. Activation of the metabotropic type 1 receptor *reduces* the K^+ "leak" conductance, producing an *EPSP*, but activation of the metabotropic type 2/3 receptors *increases* the K^+ "leak" conductance, producing an *IPSP* (Cox and Sherman, 1999). The former response usually dominates, and the glutamatergic IPSP can be seen only when the excitatory responses are blocked or minimized. This may reflect the relative numbers of the receptor types, with type 1 outnumbering type 2/3, but no data are available at present. Thus global activation of cortical or relay cell inputs or global application of agonists produces only the EPSP, swamping the IPSP. The distinct possibility remains that, under some conditions, activation of perhaps some cortical or relay cell axons can actually *inhibit* reticular cells: perhaps only a small subset of these axons has terminals associated with type 2/3 receptors or perhaps only certain firing patterns activate these receptors. This clearly needs much more study.

2. CHOLINERGIC INPUTS

On the basis of recordings from cell bodies of interneurons and reticular cells, activation of the cholinergic inputs from the parabrachial region

C. Inputs to Interneurons and Reticular Cells

generally inhibits both cell types (McCormick and Prince, 1986; McCormick and Pape, 1988). This is interesting, because individual parabrachial axons branch to innervate both of these cell types and relay cells, and, as noted previously, their action on relay cells is excitatory. This neat trick— a single axon exciting some targets (relay cells) and inhibiting others (interneurons and reticular cells)—is accomplished by the fact that different muscarinic receptors are activated on these different targets. This is illustrated schematically in Figure 5. We have seen earlier that relay cells have nicotinic and M1 type muscarinic receptors, but another type of muscarinic receptor, a type other than M1,[4] dominates on interneurons and reticular cells. Activation of this "non-M1" type of muscarinic receptor increases the K^+ "leak" conductance, leading to hyperpolarization of interneurons and reticular cells (McCormick and Prince, 1986; McCormick and Pape, 1988). However, cells of the thalamic reticular nucleus also respond to this cholinergic input by means of another, nicotinic receptor that leads to fast depolarization. Nonetheless, the main effect of cholinergic stimulation of these cells as seen at their cell bodies seems dominated by the muscarinic, inhibitory response. Because axonal outputs from these interneurons and reticular cells inhibit relay cells, activation of this cholinergic pathway thus disinhibits relay cells (see inset to Figure 5).

An interesting aspect of cholinergic effects on interneurons is that parabrachial terminals also innervate many dendritic F2 terminals of these cells (Erişir et al., 1997), and the effects of cholinergic activation of these terminals are likely to be invisible to somatic recording of the interneuron. Typically, a retinal terminal and parabrachial terminal do not innervate the same F2 terminal. Recent indirect evidence based on recording from the relay cells postsynaptic to F2 terminals is that cholinergic afferents inhibit the F2 terminals and thereby disinhibit the relay cells (Cox and Sherman, 2000). This is illustrated in Figures 3 through 5. It appears that the cholinergic parabrachial terminals contacting the F2 terminals activate only M2 muscarinic receptors, and this presumably increases the K^+ "leak" conductance, thereby hyperpolarizing the F2 terminal and reducing GABA release, although as noted earlier, there is no direct evidence regarding how activation of these muscarinic receptors reduces transmitter release in the F2 terminal. Notice that a sort of functional triad is formed by the parabrachial axons innervating relay cells and interneuronal F2 terminals, because the same axon contacts both, but usually with different presynaptic terminals (Figure 4). In this regard it differs from the triad formed from retinal terminals, where a single terminal contacts both an F2 terminal and an appendage of a relay cell dendrite (Figure 4), and it differs functionally, because the retinal input increases transmitter release from the F2 terminal, whereas the parabrachial input reduces this release. Thus the effect of firing in these cholinergic afferents from the parabrachial region is reduced inhibition in relay cells. This results because the cholinergic input affects the axonal F1

---

[4] This probably includes M2 types, but M3 and M4 types may also be involved. We simply refer to these as "non-M1" types.

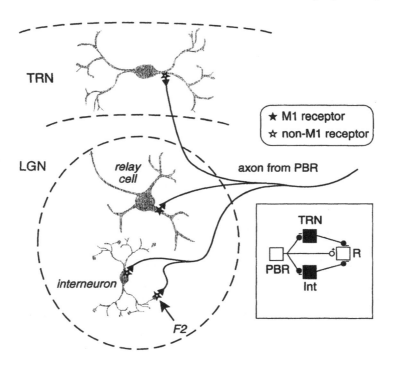

Figure 5
Pattern of innervation of thalamic cells by single parabrachial axon. All three major cell types—relay cells, interneurons, and reticular cells—are innervated by branches of the same parabrachial axon. However, the metabotropic (muscarinic) receptors activated are not the same on these cells. For relay cells, an M1 type of receptor is found, activation of which produces a slow EPSP. For interneurons and reticular cells, a non-M1 type of receptor is found, activation of which produces a slow IPSP. Not shown are nicotinic receptors, which appear to be found on relay and reticular cells but not interneurons and which produce fast EPSPs. (Inset) Schematic diagram of circuit properties. Inputs from parabrachial region (PBR) inhibit (−) interneurons (Int) and cells of the thalamic reticular nucleus (TRN), and these postsynaptic cells, in turn, inhibit relay cells (R). The parabrachial axons excite (+) relay cells. Thus the effect on relay cells of increased activity in parabrachial afferents is direct excitation and indirect disinhibition, the latter by means of inhibiting inhibitory inputs to the relay cells. Abbreviations as in Figure 2, plus LGN, lateral geniculate nucleus.

outputs by hyperpolarizing the cell body and/or it affects the dendritic F2 outputs as described earlier.

3. GABAergic Inputs

The GABAergic pathway from the basal forebrain to the thalamic reticular nucleus (see Chapter II) likely inhibits the reticular cells, although there has been no published physiological or pharmacological study of

C. Inputs to Interneurons and Reticular Cells

this pathway. We noted in Chapter II that this pathway does not directly innervate the lateral geniculate nucleus and thus does not innervate interneurons.

Reticular cells innervate each other by axon collaterals and, possibly in some cases, by dendrodendritic synapses (Pinault et al., 1997). Again, we can assume that these local interactions are inhibitory, but there is as yet no physiological confirmation of this assumption. It is thought that these local interconnections are important in synchronizing cells of the thalamic reticular nucleus, particularly during sleep when they fire in rhythmic and synchronous bursts (see Chapter IV and Steriade et al., 1993).

Electron microscopic studies indicate that interneurons receive F1 terminals on their dendritic stems and branches and occasionally on their dendritic (F2) terminals (Guillery, 1969a,b; Hamos et al., 1985; Erişir et al., 1998). These F1 terminals are thought to be inhibitory and GABAergic, but to date we can account for the sources of only a few of these F1 inputs, and in no case has the physiology or pharmacology of these inputs been determined. In one limited electron microscopic study of the axonal output of thalamic reticular cells to the lateral geniculate nucleus (Cucchiaro et al., 1991), these outputs seemed to be directed at relay cells, but a small minority innervated interneurons. There is also a GABAergic input from the pretectal region to the lateral geniculate nucleus that seems to target F2 terminals of interneurons (Cucchiaro et al., 1993). Presumably, this input inhibits F2 terminals, and rather indirect *in vivo* data are consistent with this possibility (Fischer et al., 1998). Such a pathway may be unique to the lateral geniculate nucleus. However, GABAergic inputs from areas other than the thalamic reticular nucleus have been described to other dorsal thalamic nuclei. In the ventral anterior nucleus of the monkey, a GABAergic projection from the globus pallidus and substantia nigra contacts both relay cells and interneurons (Ilinsky and Kultas-Ilinsky, 1997). In the rat's medial geniculate nucleus GABAergic input from the inferior colliculus contact relays neurons (Peruzzi et al., 1997), but because there are virtually no interneurons in the rat's medial geniculate nucleus (Arcelli et al., 1997), the possibility that interneurons in this nucleus might receive such input in other species remains open.

4. NORADRENERGIC INPUTS

Anatomical studies in the cat indicate that noradrenergic axons from the brain stem innervate the dorsal thalamus (see Chapter III), and here they may possibly innervate interneurons and the thalamic reticular nucleus (Morrison and Foote, 1986; de Lima and Singer, 1987; Fitzpatrick et al., 1989). Noradrenalin depolarizes reticular cells by reducing the K^+ "leak" conductance, presumably acting through a metabotropic receptor (McCormick and Prince, 1988; McCormick and Wang, 1991). However, noradrenalin has no clear effect on interneurons (Pape and McCormick, 1995).

5. Serotonergic Inputs

Anatomical data indicate that serotonergic axons from the dorsal raphé nucleus innervate the thalamic reticular nucleus and dorsal thalamus. However, the effects of serotonin are poorly understood. Serotonin depolarizes reticular cells by blocking the K^+ "leak" conductance, again presumably by means of metabotropic receptors (McCormick and Wang, 1991). Effects of serotonin on interneurons are confusing because its application produces a slight depolarization of some interneurons while not clearly affecting others (Pape and McCormick, 1995).

6. Histaminergic Inputs

Very little published work is as yet available on the effects of histamine application to interneurons or reticular cells. Such application *in vitro* does depolarize interneurons, but the receptors involved are unknown. There is some evidence that this depolarization is an indirect result of activating excitatory inputs to interneurons and not a direct effect on interneurons (Pape and McCormick, 1995). Preliminary results from reticular cells indicate that histamine applied *in vitro* inhibits them by increasing a Cl^- conductance (Lee and McCormick, 1998).

D. Summary

In this chapter we have stressed the importance of the postsynaptic receptor in shaping the responses of neurons, and particularly of thalamic relay cells, to their afferent inputs. Two major classes of receptor are found on relay cells, and these are ionotropic and metabotropic receptors. One clear difference in their behavior is that activation of ionotropic receptors results in relatively fast postsynaptic potentials with a more rapid onset and briefer duration, whereas activation of metabotropic receptors produces postsynaptic potentials with a slower onset and longer duration. In addition, second messenger systems turned on when metabotropic receptors are activated could have other effects on relay cells beyond creation of postsynaptic potentials, including such long-terms effects as regulation of gene expression, but this has not yet been studied for thalamic neurons.

The differential timing of activation of these receptor types has implication for functioning of relay cells. Obviously, the sustained membrane potential changes associated with activation of metabotropic receptors will affect overall excitability of the cell. Furthermore, activation of these receptors seems particularly well suited to controlling the inactivation states of many of the voltage-dependent conductances of relay cells, such as I_T, I_h, and I_A, because these have relatively long time constants for inactivation and de-inactivation. With I_T, for example, activation of excitatory metabotropic excitatory receptors (e.g., metabotropic glutamate, M1 muscarinic, and noradrenergic) will produce a long, slow EPSP ideally suited to inactivating I_T, and activation of metabotropic $GABA_B$ receptors will produce a long, slow IPSP that will effectively de-inactivate I_T. On the other hand, EPSPs and IPSPs activated by means of ionotropic receptors are too transient, without extensive temporal summation, to have much effect on the inactivation state of I_T. Thus activation of metabotropic but not of ionotropic receptors seems particularly appropriate for controlling I_T and thus the response mode, burst or tonic, of relay cells. The importance of this response is emphasized in Chapter VI.

In this context it is especially interesting that, for the lateral geniculate nucleus of the cat, the driver (retinal) input to relay cells activates only ionotropic glutamate receptors, whereas all other (modulator) inputs activate metabotropic receptors, although many also activate ionotropic receptors. (It is of critical importance to determine whether this distinction for driver and modulator inputs holds for the rest of thalamus.) This suggests that one of the important roles for modulatory inputs is to control response mode of relay cells, and evidence for this exists in studies related to corticogeniculate and cholinergic parabrachial inputs. This also suggests that retinal input is less effective in controlling response mode. By activating only ionotropic glutamate receptors, retinal axons would produce faster EPSPs, which avoids loss of high frequency visual information that would be compromised by slower EPSPs by means of metabotropic receptors.

E. Some Unresolved Questions

1. Does the distinction between driver and modulator inputs as regards the type of receptor activated—ionotropic or metabotropic—that is seen in the lateral geniculate nucleus apply to the rest of thalamus? That is, is it generally true that driver inputs activate only ionotropic (glutamate) receptors, whereas all modulatory inputs activate metabotropic receptors (and often ionotropic receptors as well)?

2. Do the metabotropic receptors in the thalamus have any long-term actions comparable to those seen in hippocampus or neocortex?

3. Do low-threshold spikes from activation of I_T produce enough Ca^{2+} entry to generate similar second messenger cascades and perhaps other long-term effects?

4. What is the functional significance of the mix of receptor types, especially ionotropic and metabotropic, related to most afferent pathways? Do certain patterns of activity activate different receptors relatively selectively? Do individual axons in these pathways activate different combinations of receptors associated with the entire pathway?

5. Can driver inputs to thalamic relay cells effectively control switching between tonic and burst response modes or can this effectively be done only by modulator inputs?

6. Can it be demonstrated that interneurons have two functionally independent synaptic input zones: one controlling F2 terminals and the other, the axon?

7. Can back propagation of the action potential invade dendrites of thalamic neurons to affect synaptic integration there? Can this occur in interneurons as a means of affecting the F2 terminals? Does normal synaptic activation of dendrites of thalamic cells produce action potential initiation in dendritic locations or only in the cell body or axon hillock?

8. Do all corticogeniculate axons from layer 6 activate metabotropic and ionotropic receptors, or do some activate just one or the other?

CHAPTER VI

Function of Burst and Tonic Response Modes in the Thalamocortical Relay

As noted in Chapter IV, thalamic relay cells display a variety of voltage-dependent membrane properties. The best understood and undoubtedly the most important are those underlying conventional action potentials, because these represent the only way for thalamic relay cells to transmit information to cortex. Perhaps the next in importance are those underlying the low-threshold Ca^{2+} spike, because the activation state of the conductance underlying this spike determines which of two response modes, burst or tonic, the relay cell will display. Because in each of these response modes the thalamic cells send a different pattern of action potentials to cortex, the obvious question is: What is the significance of the distinct burst and tonic response modes for thalamic relay functioning? Several different, although not mutually exclusive, answers have been proposed to this question, and these are considered in this chapter.

A. Rhythmic Bursting

Initial studies from both *in vitro* slice and *in vivo* preparations have emphasized two features often associated with the burst response mode: (1) when relay cells exhibit burst firing, they frequently (but, as we shall see, not always) show *rhythmic* bursting, and (2) relay cells through large regions of thalamus become synchronized in their rhythmic bursting (Steriade and Deschênes, 1984; Steriade and Llinás, 1988; Steriade *et al.*, 1993; McCormick and Bal, 1997). Interactions between relay cells and cells of the thalamic reticular nucleus are critical for synchronizing such rhythmic bursting (Steriade *et al.*, 1985). It is worth noting that detection of synchrony usually requires simultaneous recording from two or more cells, a technically difficult feat not often accomplished, so most studies of rhythmic bursting have not directly assessed any synchrony among cells. The frequency of this rhythm (i.e., the inverse of the fairly constant time interval between bursts) can vary, depending on a variety of other factors, but typically is in the range of 0.3 to 10 Hz. Although circuitry probably plays a role in rhythmic bursting, isolated relay cells also have some capacity to burst rhythmically. As noted in Chapter IV, the combination of various membrane currents, including I_T, I_h, and various K$^+$ currents, can lead to rhythmic bursting in an isolated cell, which has been suggested as a mechanism for this property in some studies (McCormick and Huguenard, 1992).

The first *in vivo* studies of the response modes in cats demonstrated that, when an animal entered slow wave sleep,[1] individual thalamic relay cells began to burst rhythmically, and that such rhythmic bursting was not seen during awake, alert states (Livingstone and Hubel, 1981; Steriade and McCarley, 1990; Steriade et al., 1993). The details of what underlies the pattern of synchronized, rhythmic bursting are not entirely known, but it seems to involve circuit features with the thalamic reticular nucleus playing a key role (Steriade et al., 1985). Apparently, when large ensembles of reticular cells burst rhythmically, their connections with each other help them maintain synchronous firing. The result is a strong reticular volley that will hyperpolarize interconnected groups of reticular cells and relay cells, removing inactivation of their T channels. In relay cells, this will activate I_h (see Chapter IV), which helps to depolarize the relay cell and fire a low-threshold Ca^{2+} spike. All relay cells do so in unison, and this will depolarize reticular cells, firing a burst in them. After the burst in reticular cells, various K$^+$ conductances come into play (see Chapter IV, Section C.1.b), which de-inactivates I_T in these cells, and the cycle is repeated. Thus rhythmic bursting ensues and remains synchronized.

This led to the hypothesis that awake animals have depolarized thalamocortical cells that operate strictly in tonic mode and thus reliably relay information to cortex; during slow wave sleep, the cells become hyperpolarized and thus burst rhythmically, which reduces relay of information to cortex. In this regard, rhythmic bursting is thought to represent a state during which normal driving inputs (e.g., retinal or auditory inputs for geniculate relay cells) have less impact on the firing patterns of their postsynaptic relay cells, and the relay of information to cortex is reduced or interrupted. The reasons for this are discussed more fully in

---

[1] This is also called "synchronized" sleep and is characterized by rhythmic, low-frequency, high-voltage waveforms in the electroencephalogram (EEG). The other main stage of sleep is called "desynchronized" and is characterized by a high-frequency but low-voltage EEG without much rhythmicity. Thalamic cells appear to fire mostly in tonic mode during desynchronized sleep. Rapid eye movements also occur during this latter sleep phase, so it is also known as "REM" sleep. Relatively little is known specifically about thalamic relay properties during these sleep phases, although transmission through the thalamus seems somewhat depressed during synchronized sleep but not during desynchronized sleep. For further details about these sleep states and thalamic functioning, the reader is referred elsewhere (e.g., Favale et al., 1964; Dagnino et al., 1965; Dagnino et al., 1966; Ghelarducci et al., 1970; Dagnino et al., 1971; Marks et al., 1981; Llinás and Pare, 1991).

A. Rhythmic Bursting

Chapter IX. According to this view, information would be effectively relayed to cortex only during tonic firing. During rhythmic bursting, the firing pattern of the relay cells would be largely controlled by their intrinsic membrane properties plus the activity of reticular cells. However, several studies of thalamic cell properties during slow wave sleep have emphasized that rhythmic bursting is not an invariant property of relay cells because some show very little rhythmic bursting and others show tonic firing interspersed with rhythmic bursting (McCarley et al., 1983; Ramcharan et al., 2000). We still have much to learn about thalamic relay properties during sleep.

More importantly, we have much to learn about relay properties and the role bursting may play during the vigilant, wakeful state. Studies of visual response properties of relay cells in the lateral geniculate nucleus of lightly anesthetized or awake, behaving animals suggest that bursting is not limited to sleep. These data, reviewed in the following paragraphs, indicate that both tonic and burst response modes are normally used by thalamic relay cells to transmit visual information to cortex.

1. VISUAL RESPONSES OF GENICULATE RELAY CELLS

If the burst mode indeed represented a complete failure of the relay through thalamus, as suggested by earlier studies, it would follow that, *in vivo*, a geniculate relay cell sufficiently hyperpolarized to de-inactivate the Ca^{2+} conductance underlying its low-threshold spike should either remain silent or begin bursting rhythmically, regardless of which visual stimuli, if any, are presented. Recording from lightly anesthetized cats *in vivo* shows that cells in burst mode in the absence of any visual inputs commonly fire arrhythmically with randomly occurring bursts (Lu et al., 1992; Sherman, 1996). Such burst firing has also been reported in awake, alert animals (McCarley et al., 1983; Nicolelis et al., 1995; Guido and Weyand, 1995; Ramcharan et al., 2000), including humans (Lenz et al., 1998; Radhakrishnan et al., 1999; see also later). This general lack of rhythmicity seen in the awake state may explain why bursting was missed in earlier descriptions of thalamic cell responses during wakefulness. That is, the rhythmicity seen during sleep is nearly impossible to miss because any regularities in responsiveness are relatively easy to detect, but the occasional appearance of bursts at unpredicted intervals might well be missed unless one were looking for them.

Not only is arrhythmic bursting seen during spontaneous activity (i.e., the firing of the cell when the visual stimulus is absent) in geniculate relay cells *in vivo*, but these cells, while still in the burst mode, also respond quite reliably to visual stimuli (Sherman, 1996). The bursts then follow the temporal properties of the visual stimulation rather than any intrinsic pacemaker frequency (see Figures 1 and 2). During burst firing, the response is in the form of bursts riding the crests of low-threshold spikes rather than the streams of unitary action potentials that occur when the same cell responds in tonic mode (Guido et al., 1992; Mukherjee and Kaplan, 1995). Recent but preliminary evidence suggests that the

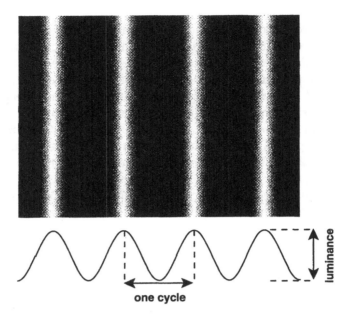

Figure 1
Sine wave grating used as a visual stimulus. The grating is shown at the top and is vertically oriented in this example. Below it is its sinusoidal luminance profile. The grating may repeat over many cycles. Typically, such a visual stimulus would be temporally modulated in various ways, most simply by drifting the grating in a direction orthogonal to its orientation (horizontally in this example). When such a grating drifts, each point along its path is stimulated by light that sinusoidally varies in intensity.

same response properties can be seen in the lateral geniculate nuclei of awake, behaving cats and monkeys (Guido and Weyand, 1995; Ramcharan *et al.*, 2000). Indeed, analysis of spontaneous activity in widespread areas of the thalamus, including somatosensory and motor relays, shows that arrhythmic bursting is present in relay cells of awake, behaving rats (Nicolelis *et al.*, 1995), cats (Guido and Weyand, 1995), monkeys (Ramcharan *et al.*, 2000), and humans (Lenz *et al.*, 1998; Radhakrishnan *et al.*, 1999).

Thus geniculate cells clearly respond to visual stimuli quite vigorously in either tonic or burst mode; the pattern of the response depends on the activation state of the underlying low-threshold Ca^{2+} conductance at the time the visual stimulus is presented. This state, in turn, depends on the membrane potential at that time (see Chapter IV). Recent analysis of the pattern of burst and tonic firing to various visual stimuli indicates that both firing modes can convey roughly equal amounts of information (Reinagel *et al.*, 1999). Burst firing conveys information more efficiently (with less noise), but other experiments show that tonic firing conveys information more linearly (Guido *et al.*, 1992). That is, although the total amount of information relayed is roughly similar during both response modes, it should be clear from Chapter V (Figure 4 in Chapter V) and

A. Rhythmic Bursting

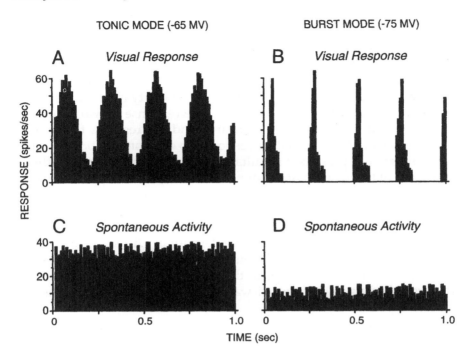

Figure 2
Responses of a single representative relay cell in the lateral geniculate nucleus of a cat to a drifting sinusoidal grating as in Figure 1. The cell was recorded intracellularly in a lightly anesthetized cat, and the grating was drifted across the cell's receptive field. Current was injected into the cell through the recording electrode to alter the membrane potential. Shown are average response histograms, plotting the mean firing rate as a function of time, averaged over many epochs of that time. Thus in (A) and (C), the current injection was adjusted so that the membrane potential without visual stimulation averaged −65 mV. This represents tonic firing, because I_T is mostly inactivated at this membrane potential. In (B) and (D), the current injection was adjusted to a more hyperpolarized level, permitting burst firing. The top histograms (A and B) reflect firing in response to four cycles of the drifting grating, and the bottom histograms (C and D) reflect spontaneous activity when the gratings have been removed. Note that the response profile during the visual response in tonic mode (A) looks like a sine wave, but the companion response during burst mode (B) does not. Note also that the spontaneous activity is higher during tonic than during burst firing (C, D).

Figures 2 and 3 that burst and tonic response modes represent different types of stimulus/response transformation, and that these response modes almost certainly represent different forms of relay of information to cortex.

The obvious question then is: What is the functional significance of the two modes for transmission of retinal information by geniculate relay cells? This key question can be broken down into two different, but ultimately related, questions. The first, considered in this section, asks, what is the possible significance of burst versus tonic firing in terms of the type of information each mode relays to cortex? The second, considered next in Section B, asks, what is the possible significance of the two firing modes in terms of the transmission of this information from thalamocortical axons to cortical cells?

One of the differences between burst and tonic firing involves linear summation of visual stimuli. Strictly speaking, linearity in the context of visual receptive field properties means that the response to two or more individual stimuli can be summed linearly, and the result is the same response that would be evoked in response to the combination of the two or more stimuli presented simultaneously; any departure from this indicates lack of linear summation, or nonlinearity, in the response.[2] Such linearity is frequently tested with a sinusoidal stimulus, such as a sinusoidal grating as illustrated in Figure 1. This stimulus could be presented in several ways, but a common way discussed later is to drift it through the receptive field of the cell at a constant velocity in a direction orthogonal to the orientation of the grating. Such a sinusoidal stimulus makes it easier to apply Fourier techniques to the study of responses, and this in turn offers a fairly straightforward way to assess the linearity of these responses.[3]

Figure 2 shows examples of burst and tonic responses in the same cell to a drifting grating like that shown in Figure 1, and this illustration suggests two prominent differences between tonic and burst mode. One is that the tonic mode displays much greater linear summation than does the burst mode. In this example, because the visual stimulus changes contrast in time with a sinusoidal waveform, a cell that responds linearly should show a response profile that is correspondingly sinusoidal. Thus the sinusoidal response profile during tonic mode firing (Figure 2A) reflects a linear transformation between the visual stimulus and the response. In contrast, the response profile during burst mode firing (Figure 2B) reflects a nonlinear distortion because the response here is nonsinusoidal to the sinusoidal stimulus. This very likely results from the nonlinear amplification of the low-threshold spike described in Chapter IV (see Figures 6 and 7 in Chapter IV), which provides a similar response regardless of the amplitude or duration of any suprathreshold stimulus. This would have the effect of creating a response to a sinusoidal stimulus that was dominated by an initial burst near the beginning of each cycle,

---

[2] This definition of linearity can be extended to any neuron as follows: the response to a complex combination of inputs equals the sum of responses of the neuron to each component stimulus.

[3] Fourier techniques refer to the analytical methods developed by the French mathematician, J. Fourier (1768–1830). He showed that any complex waveform could be synthesized or analyzed by the linear addition of pure sine waves appropriately chosen for amplitude, frequency, and phase. Determining the component sine waves of a complex waveform is *Fourier analysis*, and creating a complex waveform from sine waves is *Fourier synthesis*. A neuron can be tested for linearity by stimulating it with a sine wave input (e.g., a sinusoidally varying visual stimulus such as shown in Figure 1) and Fourier analyzing the response profile. The Fourier sine wave component that matches the input sine wave in frequency is the linear response component, and all other sine wave components of the response comprise nonlinear response components. For further details on this subject, see Shapley and Lennie (1985).

and thus instead of the response during burst firing being graded in a sinusoidal fashion (as during tonic firing), it has a pronounced peak near the beginning of each cycle. These impressions of more linear summation for tonic than burst firing have been confirmed by Fourier analysis of the responses of the cells during the two response modes as summarized in Figure 3A (Guido et al., 1992; Guido et al., 1995) and also by analysis of responses to flashing spots (Mukherjee and Kaplan, 1995). It should be noted that the difference in spontaneous activity contributes to the difference of the responses because the higher level during tonic mode (Figures 2A and 2C) helps to prevent nonlinearities caused by half-wave rectification[4] in the response.

The other prominent difference between response modes illustrated by Figure 2 is the difference between spontaneous and visually driven activity. That is, the lower spontaneous activity during burst mode coupled with vigorous visual responsiveness during either mode (Guido et al., 1995) suggests the possibility that the ratio between signal (visual response) and noise (spontaneous activity) is actually improved during burst mode. This, in turn, suggests that cells in burst mode might be more capable of detecting a stimulus than when in tonic mode. This possibility has been tested formally by using techniques of signal detection theory to create receiver operating characteristic curves for responses during tonic and burst mode; these curves test the ability of the cell to detect a visual stimulus against background noise (Green and Swets, 1966; Macmillan and Creelman, 1991). Figure 4 shows how this technique is used to assess stimulus detectability for burst and tonic firing. As shown in Figure 3B, every geniculate cell so tested displays considerably better detection of the visual stimuli when in burst mode than when in tonic mode (Guido et al., 1995). Furthermore, the more difficult a stimulus is to detect (e.g., stimuli of lower contrast), the greater the detection advantage of the burst over the tonic mode. This is because a less salient stimulus would produce a smaller response during tonic mode, but such a stimulus, if detectable, would provide nearly the same response during burst mode as would a more salient stimulus.

Besides the differences in linear summation and signal detection, the response mode has an effect on temporal tuning (Mukherjee and Kaplan, 1995; Smith et al., 2000). Geniculate neurons firing in tonic mode respond to a broader range of temporal frequencies than do the same cells firing in

---

[4] A visual stimulus can be excitatory and/or inhibitory. For instance, the sine wave grating in Figure 1 would excite an on-center geniculate cell when the bright half of each cycle falls across the receptive field center, and it would inhibit the same cell when the dark half of each cycle falls across the receptive field center. Passage of a full cycle of the sine wave across the receptive field thus would produce both an increase and decrease in firing of the geniculate cell. To encode the full response to a complete cycle of the stimulus linearly would require that the spontaneous activity be high enough to permit a sculpting out of the inhibitory phase of the response. Put another way, negative firing rates (i.e., below zero) cannot be relayed to cortex, and if there is too little spontaneous activity, the inhibitory responses will be clipped and not relayed. In the extreme case of no spontaneous activity, only the responses to the excitatory half of the stimulus are relayed to cortex, and this is half-wave rectification, which is a significant nonlinearity.

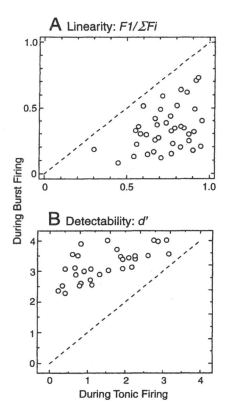

Figure 3
Measurements of linearity and detectability for population of relay cells recorded from the lateral geniculate nucleus of lightly anesthetized cats during both tonic and burst firing modes. (A) Linearity. This was assessed by Fourier analyzing the response of each cell to a drifting grating and dividing the fundamental (F1) component, which is linear, by the sum of the first 10 nonlinear components. The denominator thus gives an estimate of the extent of nonlinearity in the response, and the larger this fraction, the more linear the response. Each point represents a single cell, and the abscissa reflects linearity during tonic firing, whereas the ordinate reflects the same measure during burst firing. The line of slope one is also shown, and every cell falls below this line, indicating that tonic firing is always more linear. (B) Detectability. This was assessed by receiver operating characteristic (ROC) analysis (see text and Figure 4 for details). As in A, each point represents a single cell, reflecting its ability to detect the same stimulus during tonic and burst firing. The abscissa shows detectability during tonic firing, and the ordinate, during burst firing. The line of slope one is also shown, and every cell falls above this line, indicating that burst firing is always better for stimulus detection.

burst mode. That is, in tonic mode, geniculate cells respond well to the lowest temporal frequencies of visual stimulation and continue to respond as temporal frequency increases until the resolution limit is approached; in burst mode the cells respond poorly if at all to very low temporal frequencies (<1 Hz), respond best to middle frequencies (~4 Hz), and respond poorly if at all to higher frequencies (>10 Hz). In engineering

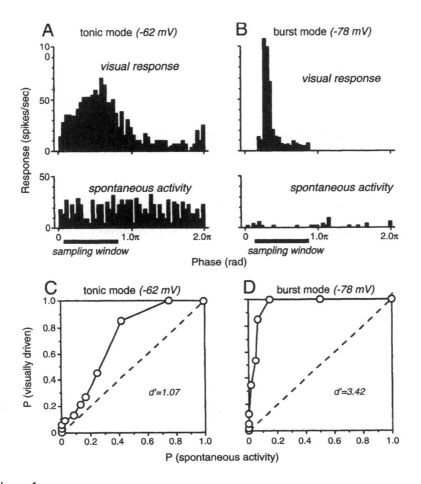

Figure 4
Example of signal detection estimate based on derivation of receiver operating characteristic (ROC) curves (Green and Swets, 1966; Macmillan and Creelman, 1991). This is another relay cell recorded intracellularly from the lateral geniculate nucleus of a lightly anesthetized cat. As in Figure 2, current injection was applied to create either tonic firing (A, C) or burst firing (B, D). The histograms in A and B represent responses to a drifting grating and spontaneous activity based on averaging responses to many trials as in Figure 2. Under the histograms and roughly centered on the peak of the visual response are shown the *sampling windows*. These represent the time periods during each stimulus trial (i.e., one cycle of the grating) during which spikes are counted, and an equivalent period is also used to count spikes during spontaneous activity. From these counts, the ROC curves in C and D are constructed. These curves represent the cumulative probabilities that at least a criterion number of spikes, from zero to the largest number found in any sampling window for any single trial, will be found in each window, and this is calculated for each of the criterion numbers of spikes. Thus for a low criterion number of spikes (e.g., 0) the probabilities of finding at least that number are high (e.g., 1), and vice versa for a high criterion number. The probabilities seen during spontaneous activity are plotted against those seen during visual responses. The line of slope one is also shown in each ROC curve, and this represents the locus of probabilities for which the cell's responses to a visual stimulus are no better at signaling the presence of that stimulus than are the cell's responses during spontaneous activity, when no stimulus is present. The extent to which the curves lie above the line of slope one provides a reliable estimate of the ability of the cell to detect the specific visual stimulus used. From the area under the curve, a value (d') can be computed that reflects detectability (Green and Swets, 1966; Macmillan and Creelman, 1991).

terms, this means that cells in tonic mode behave as low-pass temporal filters, whereas those in burst mode act as band-pass filters. This difference suggests that, in burst mode, geniculate cells respond more selectively to sudden changes in the visual world and will not respond well to static images or gradual changes.

It thus seems clear that both response modes efficiently relay visual information to cortex. Burst mode appears better for initial detection of stimuli. This may be effective during visual search when the less accurate analysis permitted by the nonlinear responses is nonetheless sufficient for target acquisition. It may also be useful when attention is directed elsewhere (e.g., to another part of visual space or to another sensory modality) as a sort of "wakeup" call for novel and potentially interesting or dangerous stimuli. This notion is in many ways similar to the "searchlight" hypothesis for burst firing first advanced by Crick (1984) in relation to the possible function of the thalamic reticular nucleus in attention. The nonlinear distortion associated with burst mode suggests that, although the stimulus can be readily detected, it will not be as accurately analyzed while the relay is in this mode. Tonic mode, with its more linear relay of visual information, would permit more faithful signal analysis. The differences between these response modes in temporal tuning (Mukherjee and Kaplan, 1995; Smith *et al.*, 2000) are also consistent with this hypothesis because the low-pass tuning during tonic mode would effectively relay information contained in lower temporal frequencies that result from stimuli being fixated or tracked and thus imaged fairly stably on the retina, which would be expected for stimuli analyzed in detail. Sudden changes in the visual world (e.g., the appearance of a novel stimulus) or visual search would not have much representation of low temporal frequencies, so if cells in burst mode are concerned with such tasks, they need not be sensitive to these lower frequencies.

Such analyses of and speculations about firing modes for thalamic nuclei other than the lateral geniculate nucleus have not yet appeared in the literature. For other sensory relay nuclei, it is a fairly simple matter to extrapolate these ideas from vision to audition or somesthesis. For example, a change or sudden appearance in a somatosensory stimulus would be better detected during burst firing, but the detailed and accurate analysis of such a stimulus would be better accomplished when the relay cells fired in tonic mode. This notion could even be extrapolated to thalamic relay nuclei in general. To the extent that all thalamic nuclei relay driving input to cortex, the hypothesis would suggest that any change in the pattern or sudden appearance of a novel or unexpected driving input would be detected better when the postsynaptic relay cells responded in burst mode and that tonic firing provides more accurate representation of the driving patterns of input in information relayed to cortex. What is vitally needed is experimental testing of this idea to support or reject this notion of the possible role of response mode in relaying information to cortex. Indeed, experimental data on burst and tonic firing of thalamic relay cells during normal behavior are likely to prove crucial to our understanding of thalamic functions.

This idea can naturally be extended to higher order thalamic relays, which receive their driving afferents from layer 5 of cortex itself (see Figure 1 of Chapter VIII for further details). The higher order thalamic relays can be expected to use the burst and tonic modes in the same way as the first order relays, but now in the transfer of information passing from one cortical area to a second. A burst mode in a higher order relay might then represent a state that is ready for sudden shifts in the pattern of outputs coming from the relevant, driver afferents that arise from layer 5 pyramidal cells in the first cortical area. These shifts would serve as a "wake-up call" so the second cortical area, the target cortex of the higher order thalamic relay, could then, by way of its layer 6 efferents, switch the mode of the relay to tonic firing so that a more faithful record of the message sent from the first cortical area by means of its layer 5 efferents can be transferred to the second cortical area.

B. Effect of Response Mode on Transmission from Relay Cells to Cortical Cells

1. Paired-Pulse Facilitation and Depression

The previous section has emphasized the possible effects of response mode on the different attributes of information that might be relayed to cortex: better linearity during tonic firing and better signal detection during burst firing. An additional and potentially important issue relates to the possible consequence of these two firing modes for transmission across thalamocortical synapses. This issue arises because interspike intervals differ in burst and tonic mode and because there is evidence that many synapses in the brain, including thalamic and cortical synapses, behave in a frequency-dependent manner (Thomson and Deuchars, 1994, 1997; Lisman, 1997). This is because the presynaptic interspike interval can strongly influence the size of the evoked postsynaptic potential. This is often explored by comparing the sizes of postsynaptic potentials evoked by a first action potential to one evoked by the next action potential as a function of the intervening time interval. *Paired-pulse facilitation* (see Figure 5A) describes the situation of the second postsynaptic potential being larger than the first for interspike intervals smaller than a certain value, usually several milliseconds to a few tens of milliseconds. *Paired-pulse depression* (see Figure 5B) is the opposite: the second evoked postsynaptic potential is smaller for a range of interspike intervals, typically tens to hundreds of milliseconds. There may be many different cellular mechanisms for these phenomena, and there is still considerable debate about them. One idea is that these phenomena are largely presynaptic and that they relate to the probability that a single action potential will cause transmitter release. The essential point here is that not every action potential causes transmitter release in any synapse; instead there is a probability between 0 and 1 associated with this cause-and-effect relationship, and it is possible that the probability for any given synapse can

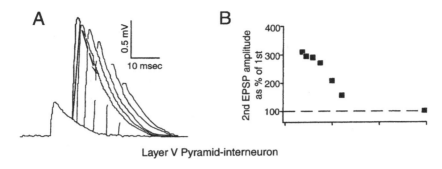

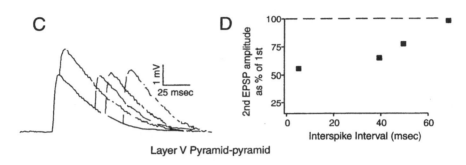

Figure 5
Examples of paired-pulse facilitation and depression. These are taken from simultaneous intracellular recordings of two cells within layer 5 of neocortex, one being postsynaptic to the other. Pairs of spikes with variable interspike intervals were elicited by means of intracellular current injection in the presynaptic cell, leading to pairs of EPSPs in the postsynaptic cell. All data generously provided by Alex Thomson. (A) Paired-pulse facilitation shown in EPSPs elicited in postsynaptic cell. This is a composite of several superimposed pairs of EPSPs, with the second EPSP being evoked at variable latencies after the first. The amplitude of the EPSPs is measured from the baseline, which, because of temporal summation, is often more depolarized in the second EPSP. Note that the second EPSP is larger in amplitude than the first for a range of intervening intervals. (B) Plot of EPSP amplitudes in A as function of interspike interval. (C) Paired-pulse depression shown in EPSPs elicited in postsynaptic cell; conventions as in A. Note that the second EPSP is smaller in amplitude than the first for a range of intervening intervals. (D) Plot of EPSP amplitudes in C as function of interspike interval. For further details of these paired-pulse effects, see Thomson (2000).

change over time. Note that most axons contribute multiple synapses, sometimes hundreds, to one or more postsynaptic target cells, and that a probability < 1 does not mean that an action potential often has no postsynaptic effect. That is, if the probability for release for each synapse, on average, is 0.5, and an axon with several thousand terminals contacting many cells contacts a given cell with 50 synapses, then an action potential will cause 25 synapses on that target, on average, to release transmitter and produce a postsynaptic potential. The point is that the larger the probability of release, the larger the resultant postsynaptic potential will be.

Depression is least likely to occur in a synapse if the preceding action potential fails to elicit transmitter release because, as noted, this release is always a probabilistic occurrence. Such failure naturally will occur most commonly in synapses with low probability of release. This probability may be closely related to the Ca^{2+} concentration inside the synaptic terminal because the probability of release is monotonically related to internal Ca^{2+} concentration (Dunlap et al., 1995; Matthews, 1996; Reuter, 1996). Synaptic terminals contain high-threshold Ca^{2+} channels (these have been described in Chapter IV, Section B.2.e; see Dunlap et al., 1995; Matthews, 1996; Reuter, 1996) that differ from the T-type channels involved in the low-threshold Ca^{2+} spike. However, like these T channels, those in terminals are voltage-dependent but with a much higher (i.e., depolarized) threshold for activation. An invading action potential depolarizes the terminal sufficiently to activate these channels, leading to Ca^{2+} entry, and, as the internal Ca^{2+} concentration increases, so does the probability of transmitter release. However, when a single action potential and the subsequent Ca^{2+} influx fails to promote transmitter release, the Ca^{2+} concentration will remain elevated for several milliseconds. If a second action potential follows the first while the internal Ca^{2+} concentration remains elevated, it will cause a second wave of Ca^{2+} entry that will sum with what remains from the first, much like temporal summation in postsynaptic potentials. Because transmitter release increases with internal Ca^{2+} concentration nonlinearly as a power function with a power of three to four (Landò and Zucker, 1994), the result will be a higher probability of transmitter release for the second action potential. Because a typical axon innervates any target with many synaptic terminals, as long as the average probability of release for all these synapses is very low (in many connections within cortex this has been computed as <0.1), then most synapses will fail to release transmitter in response to the first action potential. However, the probability of release from each terminal may be significantly enhanced for the second action potential if it follows the first within tens of milliseconds or so.[5] This can result in more transmitter release and thus a larger postsynaptic potential for the second action potential, thus producing paired-pulse facilitation.

Depression occurs because once transmitter is released, it takes time

---

[5] Just how low this probability must be for this to occur is a complex function of many factors. Consider an afferent axon with 20 synapses on one of its target cells. If the average baseline probability of release is 0.1, only 2 synapses will release transmitter, creating a postsynaptic potential of proportional amplitude. The 18 synapses failing to release transmitter will have a greater probability of release to a second action potential if it arrives soon enough, and the 2 synapses that released transmitter will, in turn, have a much lower probability (near zero) of release. For the second action potential to create a larger excitatory postsynaptic potential (EPSP) than the first (i.e., to result in paired-pulse facilitation) would require that > 2 synapses released transmitter, and this would happen if the average probability for the remaining 18 were $\geq 2/18$. Thus the phenomenon of paired-pulse facilitation depends on the initial probability of release, as well as the increase seen in those terminals failing to release initially. The phenomenon of paired-pulse depression described later has a similar dependence on these variables.

(often hundreds of milliseconds) before the probability of release to another action potential returns to baseline levels. This may be partly due to depletion of transmitter stores or to other effects. The result is that for some time after transmitter release the probability of release is reduced. This will result in paired-pulse depression in afferents having an average probability of release for their synapses high enough that many release transmitter to the first action potential.

It follows from the preceding that if an axon contacts a postsynaptic cell with synapses having a low probability of release (i.e., "low p" synapses), these synapses are more likely to show paired-pulse facilitation. This is because most of the synapses will not release transmitter but will instead show an increased probability for some time because of an increased internal Ca^{2+} concentration. Conversely, if the contacts are made with synapses having a high probability of release (i.e., "high p" synapses), they are more likely to show paired-pulse depression because more synapses will release transmitter and show relative refractoriness until their transmitter pools are restored.

Note that this explanation for paired-pulse effects is based on the assumption that they are related to *presynaptic* factors involving probability of release. However, it is possible that postsynaptic factors also play a role in this, perhaps even a dominant role. For example, if the probability of release were unchanged by paired-pulse effects, one could simply consider the probability that an evoked EPSP sufficiently depolarized the postsynaptic cell to fire an action potential. Paired-pulse effects are then related to the nature of temporal summation of the EPSP and activation of NMDA receptors; often a single EPSP does not depolarize the postsynaptic cell sufficiently to overcome the Mg^{2+} block of the NMDA receptor, but two summed EPSPs could do so. Thus the probability of firing an action potential from two summed EPSPs could be more than twice that of a single EPSP, reflecting paired-pulse facilitation, or it could be less than twice the single event probability, thereby reflecting paired-pulse depression.

Whatever the explanation for paired-pulse facilitation or depression, the fact that synapses often show one or the other behavior becomes important in the context of response mode of thalamocortical cells. The issue is the size of the EPSP created by a thalamocortical synapse, and how this size might vary with the response mode of the thalamic cell. This has important implications because the action that thalamic relay cells can have on their postsynaptic cortical cells will depend on the firing mode of thalamic relay cells.

The functional implication that we now develop more fully is that burst firing produces a large EPSP by means of thalamocortical synapses regardless of the paired-pulse effects, whereas tonic firing produces relatively large postsynaptic effects only for synapses showing paired-pulse depression and is relatively ineffective for synapses showing paired-pulse facilitation. We explore the problem from a theoretical point of view here, and stress that at present we have little information on the precise nature of thalamocortical synapses. However, on the basis of the

B. Effect of Response Mode on Transmission

following arguments one might expect that, for tonic firing to be a useful relay mode, at least an important minority of thalamocortical synapses, if not all, must show paired-pulse depression.

A thalamocortical axon contacting a cell with high p synapses (and thus likely to show paired-pulse depression) will likely release transmitter and thus evoke an EPSP for every action potential that follows its predecessor by, say, ≥10 msec. This is the typical pattern for tonic firing. For a high-frequency volley, such as a burst, with 2 to 10 spikes occurring with interspike intervals on the order of 4–6 msec, the first action potential will produce an EPSP, but the following action potentials will be much less likely to because of depression. The result is that both firing modes likely evoke EPSPs in their cortical targets, although for tonic firing, most action potentials produce an EPSP, whereas for burst firing, only the first action potential is likely to do so.

For a thalamocortical axon contacting a cell with low p synapses (and thus likely to show paired-pulse facilitation), the situation is quite different. Tonic firing, with its relatively large interspike intervals, will rarely produce facilitation, and each action potential is likely to produce a relatively small EPSP, the amplitude of which is related to the product of the number of synaptic contacts and the baseline probability of release. If the synapses are very low p and there are not very many, and there seem to be few for each geniculocortical axon contacting a layer 4 cell (Freund *et al.*, 1985), the EPSP might be too small commonly to activate action potentials postsynaptically. Each burst, however, produces a much larger EPSP because, although the first action potential likely will produce a small EPSP comparable to that seen in tonic firing, facilitation produced by subsequent ones will eventually lead to a much larger EPSP. The result for such a synapse is that tonic firing will be relatively ineffective in transmitting its message to cortex, whereas burst firing will be quite effective. A similar notion has been advanced for intracortical circuitry, within which most synapses seem to exhibit paired-pulse facilitation: single spike firing would not be an effective means for intercellular communication, whereas bursting would work quite well (Lisman, 1997). Put another way, synapses showing paired-pulse depression would evoke a large EPSP during both firing modes, but those showing facilitation would evoke large EPSPs only during burst firing.

Although burst firing is effective in producing EPSPs for either synaptic type, for a given synapse, only one of the action potentials in the burst is likely to contribute to the EPSP; the others typically produce no postsynaptic response. That is, a given synapse may release transmitter to the first action potential in a burst and then be depressed, thereby not releasing to later action potentials; alternatively, as each action potential in the burst fails to evoke release, the next has an increasing probability of doing so until release is finally achieved, after which depression sets in. By this scenario, except in the unlikely case in which release probability never exceeds threshold for release for an entire burst, each synapse will give the same contribution to an EPSP during a burst, but the actual transmitter release may be elicited by any one of the action potentials.

This may seem to be a disadvantage for burst firing, but a recent study of information content carried by tonic and burst firing patterns in the lateral geniculate nucleus of cats indicates otherwise (Reinagel et al., 1999). This analysis suggests that, whereas each action potential during tonic firing may encode a specific piece of information, during bursting, each burst encodes information as a single event, and the number or pattern of action potentials within a burst is superfluous to the information content it represents. Thus the idea that a burst produces a single EPSP by means of a thalamocortical synapse, regardless of whether the synapse facilitates or depresses, may not be a limitation in terms of information to be conveyed to cortex.

This scenario implies that there is an important difference between the cortical effect of burst and tonic firing. Regardless of the baseline release probability, for all synapses contacting a postsynaptic cell from one axon, a burst will almost always get through virtually all the synapses. This results in a maximal EPSP in the cortical cell, representing the burst as a unitary event. With tonic firing over the same axon, a given action potential will produce successful release in only a fraction of the synapses, the fraction depending on the baseline release probability. Thus the EPSP will be smaller. For example, imagine an axon contacting a postsynaptic cell with 20 synapses, each having a baseline release probability of 0.5. With a tonic action potential, only 10 of the synapses, on average, will contribute to the EPSP. However, with a burst, each synapse releases transmitter, producing a much larger EPSP. The notion that a burst in a thalamocortical axon will produce a stronger signal in cortex is in keeping with the hypothesis that bursts provide a sort of "wakeup call" to maximize detection of novel stimuli.

Of course, we need to know what type of synaptic properties, especially regarding facilitation and depression, is shown by thalamocortical synapses. In general, we do not have a clear answer to this question. It is indeed possible that there is no single answer because different thalamocortical cell types or even synapses from the same cell but terminating in different cortical layers (see following paragraphs and Chapter III) may exhibit different synaptic properties. Note that the latter possibility, that geniculocortical synapses from a single axon show different paired-pulse effects in different cortical target layers, implies either that the different synapses have different p levels despite originating from the same axon or that this could be a postsynaptic effect, reflecting differences in the postsynaptic cells.

In any case, the tonic firing mode needs to affect cortex strongly for cortex to receive the linear representation of retinal processing needed to reconstruct the visual scene faithfully, and this would occur if a significant fraction of geniculocortical synapses, if not all, showed paired-pulse depression. If instead they showed paired-pulse facilitation, as most synapses in cortex seem to (Lisman, 1997), a strong signal would be transmitted to the postsynaptic cell only if a large number of synapses were formed between each geniculocortical axon and the layer 4 target cell.

B. Effect of Response Mode on Transmission

However, the available evidence suggests that this number is quite small (Freund *et al.*, 1985), and so we tentatively predict that most geniculocortical synapses innervating layer 4 will show paired-pulse depression.

2. POSSIBLE RELATIONSHIP OF FIRING MODE TO THALAMOCORTICAL TERMINAL PATTERNS

From the preceding, it follows that if geniculocortical information is to have a maximum impact in cortex in either firing mode, then geniculocortical synapses should show paired-pulse depression. If they showed paired-pulse facilitation, tonic firing would have relatively little impact. Given the importance of preserving linearity in the messages arriving at geniculocortical synapses, and given that this linearity is preserved to a much greater degree during tonic than during burst firing, we would expect that at least some of the geniculocortical synapses, and perhaps all, show paired-pulse depression. However, if any subset shows strong paired-pulse facilitation, these synapses would activate their targets effectively only during burst firing.

As noted in Chapter III, we can roughly divide thalamocortical terminations into three zones: those concentrating in the middle layers of cortex—mostly layer 4 but also spilling over into layer 3 and the top of layer 5, those terminating in layer 1 with some extension into layer 2, and those terminating in layer 6. Receptive field studies in cats suggest that geniculocortical inputs to layer 4 cells are the driver inputs to these cells (see Chapter X for a more complete discussion of drivers and modulators). That is, receptive field properties in the postsynaptic cell can be largely accounted for on the basis of the receptive fields of the convergent, afferent relay cell axons (Ferster, 1987, 1994; Reid and Alonso, 1996). If geniculocortical inputs to layer 4 represent a major route of geniculocortical information flow, and, it should be stressed that at present we do not actually know that this route is any more important to such flow than inputs to upper layers or to layer 6, then the underlying synapses are likely to show paired-pulse depression, thereby transmitting information during both firing modes. There is limited evidence for this from *in vitro* studies of thalamocortical synapses in striate cortex (Stratford *et al.*, 1996) and in somatosensory barrel cortex of rats (Castro-Alamancos and Connors, 1997). However, a recent study (Usrey *et al.*, 2000) from *in vivo* recording in cats argues that, with short interspike intervals, the second action potential in a pair seen in geniculocortical afferents has a higher probability than does the first in activating a postsynaptic action potential. This is consistent with moderate paired-pulse facilitation, and even if the cellular explanation has a different basis (e.g., temporal summation of the EPSPs, which is a postsynaptic effect as opposed to presynaptic consequences of probability of transmitter release), the result indicates that the short interspike intervals seen in bursts are likely to be more effective in driving cortical cells than are the longer intervals seen in tonic firing. Clearly the somewhat conflicting evidence for efficacy of burst versus tonic firing to drive postsynaptic cortical cells needs to be resolved. Also, to date, nothing has yet been reported regarding paired-pulse effects for thalamocortical synapses in layers other than 4. Clearly this is a question that needs more experimental attention.

Very little is known of the physiology of the thalamocortical projection to the upper layers because we know neither the nature of the synaptic physiology nor the postsynaptic targets. Many afferents, mostly intracortical axons but also thalamocortical axons, terminate in layer 1. The main dendritic targets to be found there are the terminal tufts of those pyramidal cell dendrites that reach up to layer 1; such pyramidal cells can be found in almost any layer below 1. Recent evidence from such pyramidal cells in layer 5 of somatosensory cortex recorded *in vitro* (Larkum et al., 1999; see also Kim and Connors, 1993) suggests that inputs to their layer 1 dendrites by themselves have little effect on the cell body because of their extreme electrotonic distance from the axon hillock and presumed spike initiation zone. However, because of voltage-gated channels in the dendrites of these cells, including Na^+ channels like those underlying the action potential and high-threshold Ca^{2+} channels (i.e., different from the T-channels), an action potential generated in the soma can back propagate into distal dendrites under appropriate conditions.[6] If a synaptic input arrives at the layer 1 dendrites with the right timing after an action potential so that the peripheral dendrites are now depolarized, the EPSP generated in the layer 1 dendrite will activate a high-threshold Ca^{2+} spike, thereby producing a large EPSP at the soma. Thus a peripheral input onto the layer 1 dendrites can produce a large effect on the cell only if it arrives within a few milliseconds of an input to the cell body or more proximal dendrites, including basal dendrites, that fire the cell. Whether this mechanism applies to thalamocortical synapses and their targets is not specifically known, nor has there been any study of paired-pulse effects (i.e., facilitation or depression) of the layer 1 synapses.

The thalamocortical inputs to layer 6 cells are of particular interest because layer 6 cells provide the feedback modulatory input from cortex to thalamus. It is not clear whether all thalamocortical axons contact layer 6 cells. Nor is it known whether the layer 6 cells that receive thalamic afferents are the ones that project to thalamus, because only some of the cells in layer 6 project to thalamus (see Chapter III). These are other issues that need experimental verification. However, if any thalamocortical axons do innervate corticothalamic cells in layer 6, a particularly interesting possibility exists that is most simply described for geniculocortical interactions but would also work in a similar way for other thalamocortical systems. As noted in Section C.2, the layer 6 feedback from cortex seems especially important in the control of response mode. Namely, the direct cortical input to relay cells looking at the same part of visual field activates metabotropic glutamate receptors, which would

---

[6] This may involve neuromodulators, such as acetylcholene (ACh) or noradrenalin (NA), which control the activation states of the various voltage-gated channels in the dendrites (reviewed in Johnston et al., 1996).

B. Effect of Response Mode on Transmission

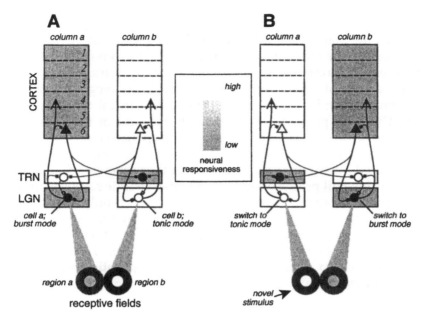

Figure 6
Speculative scenario for possible control of geniculate firing mode by visual cortex. Shown are two nearby columns (columns *a* and *b*) that are mapped to nearby regions (regions *a* and *b*) of visual field. The geniculate relay cells innervating each column thus have receptive fields in the appropriate regions of visual field. See text for further details.

have the effect of inactivating the T channels and promoting tonic firing; and the indirect input by way of cells of the thalamic reticular nucleus could activate strong enough inhibition in relay cells looking at the adjacent, surrounding part of visual field to remove inactivation of the T channels and promote burst firing (see Section C.2).

If even part of the reciprocal connections between thalamus and layer 6 of cortex work this way, it suggests the following, albeit presently speculative scenario (see Figure 6). When a part of the visual field (*region a* in Figure 6A) contains nothing novel or of particular interest, but a neighboring region does (*region b*), the former evokes little activity in its associated cortical column (*column a*), whereas the neighboring region will promote strong firing in its cortical column (*column b*; Figure 6A). Thus the layer 6 feedback pathway is relatively silent in *column a* but active in *column b*. The result is that the geniculate relay cell innervating *column a* (*cell a*) is silent but primed to fire in burst mode, whereas the nearby geniculate cell innervating *column b* (*cell b*) fires in tonic mode. This will have the effect that any significant excitatory stimulus suddenly appearing in the receptive field of *cell a* will evoke a low-threshold Ca^{2+} spike and a burst, which will readily be detected by *column a*. The burst strongly activates the layer 6 cell in *column a*, which in turn inactivates the T current in *cell a*, promoting tonic firing. This will also serve to hyperpolarize and promote burst firing in *cell b* as the activating stimulus for its receptive field diminishes. The result of this "wakeup call" would be to switch the thalamocortical pattern to that illustrated in Figure 6B.

Obviously, if we are to move from speculation to a real understanding of the effect of response mode on thalamocortical transmission, we must have a general understanding of synaptic physiology of thalamocortical synapses and how this varies from one cortical target layer to another. This may prove to be particularly relevant in the context of paired-pulse facilitation or suppression. Our hypothesis is that both response modes will produce efficient drive of cortical cells for synapses showing paired-pulse depression but that for those showing facilitation, only burst firing will produce a significant postsynaptic response in cortex. Perhaps all thalamocortical synapses show paired-pulse depression, and thus both response modes are always fairly effective. Perhaps there is variation in this property related to cell type (e.g., X versus Y or M versus P), to layer of termination (e.g., layer 4 versus 6), or to neuromodulatory effects within cortex (which might affect synaptic transmission) related to behavioral state.

C. *Control of Response Mode*

In Section A, we suggested that the different tonic or burst response modes provide different advantages in the relay of visual information through the lateral geniculate nucleus to cortex. Tonic firing enhances linearity in the relay and is thus better suited for signal analysis; burst firing enhances signal detection. We can thus suggest from the preceding that geniculate relay cells switch between tonic and burst firing modes depending on the state of the visual system or of the animal. That is, the response mode of the relevant geniculate relay cells would depend on whether that part of the animal's visual system is involved in signal analysis or detection, whether the animal is attending to other visual or other sensory stimuli, or whether the animal is more or less alert. This hypothesis can also be readily extended to other thalamic relays. For this to be plausible, there must be a ready means for modulatory inputs to control these response modes. This can be accomplished through effects on the membrane potential of relay cells because the low-threshold Ca^{2+} conductance underlying the burst mode is voltage dependent. Both brain stem and cortical inputs can do this. Other factors, discussed later, could also contribute to control of the response mode, and these may also be influenced by brain stem or cortical inputs.

1. BRAIN STEM CONTROL

The largest brain stem input to the lateral geniculate nucleus derives from cholinergic cells in the parabrachial region, and this is the brain stem input that has been most studied. Electrical activation of the parabrachial region *in vivo* causes dramatic switching of geniculate relay cells from burst to tonic mode (Lu *et al.*, 1993). Likewise, *in vitro* application of ACh

C. Control of Response Mode

eliminates low-threshold spiking, causing bursting cells to fire in tonic mode (McCormick, 1989, 1992). Because activation of parabrachial inputs switches firing mode from burst to tonic, it seems likely that inactivation of this pathway does the opposite, but this remains to be tested empirically.

As noted in Chapter V, parabrachial inputs to the lateral geniculate nucleus activate both ionotropic (nicotinic) and metabotropic (muscarinic) receptors on relay cells. Because membrane voltage changes must be maintained for ≥ 50 to 100 milliseconds to produce a significant change in the inactivation state of the low-threshold Ca^{2+} conductance underlying burst firing, and because EPSPs produced by activation of the metabotropic receptors are sufficiently long to produce this inactivation state, but those activated by ionotropic receptors are not, it seems likely that it is activation of the metabotropic (muscarinic) receptors that is crucial to this control of response mode (see Chapter V).

Parabrachial axons innervating the lateral geniculate nucleus typically branch to innervate interneurons and the adjacent thalamic reticular nucleus as well (Figure 4; Cucchiaro et al., 1988; Uhlrich et al., 1988). Although the precise role of the interneurons and the thalamic reticular nucleus in controlling the switch between tonic and burst modes remains to be defined in the alert animal, it should be clear from their connections that they are likely to be intimately involved in the switch, presumably by means of their activation of metabotropic $GABA_B$ receptors on relay cells.

Other brain stem inputs to thalamic relay cells that were noted in Chapter III included noradrenergic axons from the parabrachial region, serotonergic axons from the dorsal raphé nucleus, and histaminergic axons from the tuberomamillary nucleus. These inputs appear to activate only metabotropic receptors on relay cells, so it seems likely that activation of these other inputs is also well suited to creating the sustained changes in membrane potential needed to control response mode. As noted in Chapter V, activation of noradrenergic or histaminergic inputs evokes a long, slow EPSP in relay cells that promotes tonic firing. Activation of serotonergic inputs may have the opposite effect by directly exciting interneurons and/or reticular cells, thereby producing IPSPs in relay cells that could promote burst firing. However, although it seems likely that these inputs can control response mode, there is as yet no direct experimental evidence that they actually do so.

Precisely how these inputs may act to control response mode is to an important extent related to their innervation patterns. That is, brain stem axons that diffusely innervate the entire thalamus will likely have global effects with little ability to control specific thalamic regions differentially. Unfortunately, the available evidence for the patterning of brain stem inputs, which really must be studied at a single axon level,[7] is rather scarce,

---

[7] Imagine that so many axons are labeled in a pathway from brain stem to thalamus that connections of individual axons cannot be resolved. The overall, global pattern may appear to be diffuse and innervate the entire thalamus. However, it would still be possible that single axons have quite discrete and specific projection patterns, and it is this single axon pattern that would determine whether the pathway can act with local sign. (See Chapter III.)

but there are some hints in the literature. Individual serotonergic axons from the dorsal raphé nucleus seem to be genuinely diffuse in that each seems to innervate most thalamic nuclei, including large regions of the thalamic reticular nucleus, rather indiscriminately (unpublished observations by N. Tamamaki and S. M. Sherman based on labeling individual axons from the brain stem to their thalamic terminal arbors). Thus this pathway is likely to have global effects on response mode throughout much of thalamus. This may be related to global levels of arousal that would have relatively uniform effects on response mode of relay cells throughout thalamus.

Evidence for parabrachial axons indicates that they have much less local sign than do corticogeniculate axons, but they are more specific in innervation patterns than are axons from the dorsal raphé nucleus. In a study of parabrachial axons innervating the lateral geniculate nucleus in cats (Uhlrich et al., 1988), it was found that most of these axons branch to also innervate other visual thalamic structures i.e., parts of pulvinar), but they never innervate nonvisual thalamic nuclei (e.g., the ventral posterior or medial geniculate nuclei). Perhaps others innervate only auditory, somatosensory, or other related thalamic nuclei, but this has not been explicitly demonstrated. The implication is that most parabrachial axons are unimodal in operation, and this is precisely what may be needed to switch firing modes in related thalamic nuclei so, for instance, relay cells of the lateral geniculate nucleus could be maintained largely in tonic mode, whereas those of the ventral posterior or medial geniculate nuclei would fire more in burst mode.

2. CORTICAL CONTROL

The role of corticogeniculate input in the control of the response mode has been more difficult to assess *in vivo* because electrical activation of this pathway also usually activates geniculocortical axons antidromically, obscuring the interpretation of any orthodromic effects. However, because corticogeniculate inputs are the only ones that activate metabotropic glutamate receptors (i.e., retinogeniculate inputs, which are the only other known glutamatergic input to geniculate relay cells, activate only ionotropic glumatate receptors; see Chapter V), it is possible to mimic activation of the corticogeniculate input fairly specifically by applying agonists for this receptor while recording from the relay cells. When this is done either *in vitro* (McCormick, 1989, 1992) or *in vivo* (Godwin et al., 1996b), geniculate cells switch firing mode from burst to tonic.

Like brain stem axons, those from cortex also branch to innervate geniculate cells, as well as reticular cells and interneurons, so understanding the role of the corticogeniculate input in the behaving animal must also take into account neural circuits involving these local GABAergic circuits. By definition, the input from visual cortex to the lateral geniculate nucleus is visual in nature, and as far as we can tell, the layer 6 connections to other thalamic nuclei also represent the same quality of

C. Control of Response Mode

information (i.e., visual, somatosensory, motor) that is relayed through thalamus. This is probably different from the nature of many of the brain stem inputs to thalamus described earlier, which are not known to reflect single sensory modalities and may often relate to several modalities or to no identifiable modality.

However, the situation with corticogeniculate inputs is more complex and less well understood than is that with parabrachial inputs, and there are two related differences to consider. First, while a widespread increase in corticogeniculate activity will directly depolarize relay cells, this will also depolarize reticular cells and interneurons, thereby indirectly hyperpolarizing relay cells. The end result on relay cells is hard to predict from this alone. Second, unlike the parabrachial and dorsal raphé inputs, those from cortex show considerable local sign or topographical fidelity (e.g., retinotopic fidelity in the lateral geniculate nucleus). This makes the local details of the connections with relay cells, reticular cells, and interneurons critical, but at present we do not know these details on which the actual effect on relay cells will depend. Figures 8A and 8B of Chapter III show two plausible circuits (among many possibilities) to illustrate this point and also to show that we do not know enough about the cortical inputs to understand how they are organized to control response mode. For simplicity, the circuits shown do not include interneurons, but one could easily imagine how important it is to know details of their local connectivity as well. With the circuit of Figure 8A of Chapter III, activation of the corticogeniculate axon would both directly depolarize and indirectly hyperpolarize the relay cell, so the final effect on membrane potential would be hard to predict. With the circuit of Figure 8B of Chapter III, activation of any one corticogeniculate axon would have quite distinct effects on neighboring relay cells: for instance, activation of corticothalamic axon *b* would indirectly hyperpolarize relay cells 1 and 3, promoting burst firing, while it simultaneously directly depolarizes relay cell 2, promoting tonic firing.

Data recorded from geniculate relay cells while corticogeniculate axons were activated with glutamate applied very locally to layer 6 favor the circuit drawn in Figure 8B of Chapter III (Tsumoto et al., 1978). If the receptive fields of the geniculate cell and those of cortical cells near the glutamate delivery site largely overlap, the effect of the glutamate is to raise the firing rate of the geniculate cell. If the receptive fields are offset by up to about 2 degrees, the effect is to suppress firing in the geniculate cell. Offsets beyond about 2 degrees result in no discernible effect on the geniculate cell when glutamate is administered in layer 6. Thus, in terms of controlling response mode, this suggests that activation of layer 6 will promote tonic firing in relay cells looking at the same part of visual field while near neighbors of these geniculate cells, looking at adjacent parts of the visual field, will be driven to burst firing, and more distant geniculate cells will be unaffected.

Whatever the details of innervation pattern, the high degree of local sign in the corticogeniculate pathway means that it has the potential to control response mode differently for relay cells looking at different parts of the visual field. The pathway may be even more specific in the sense that individual cortical afferents may relate to unique geniculate cell types, like X versus Y or M versus P, because the evidence presented in Chapter III suggests that different subpopulations of layer 6 cells project to distinct geniculate layers. What is also potentially interesting about the corticogeniculate control of response mode is that, unlike brain stem inputs, the cortical input represents a feedback pathway. This means that cortex can, in principle, control the response mode of its own afferent supply. The feedback pathway emanates from cortical layer 6, and relay cell axons terminate mainly in cortical layer 4 but with a substantial input into layer 6 as well. Thus the layer 6 feedback appears to lie very close in a hierarchical sense to thalamic inputs, and if we could close this loop—that is, follow in detail the processing steps between thalamic input and the layer 6 cells that innervate thalamus—we would likely be able to understand more fully how the cortex controls the firing mode of its thalamic inputs.

D. *Summary*

We can now speculate on the possible role of burst and tonic response modes for vision and by extension for other thalamic relays as well. When the animal is not attending to a particular visual stimulus, either because it is searching for a stimulus, attending visually to a different stimulus, attending by means of another sensory modality, or not attending at all but in a drowsy state, the geniculate relay cells for that particular stimulus will be in burst mode. This suggests that the direct cortical and parabrachial inputs to these geniculate cells are relatively quiescent, although this could also imply strong activity in cortical inputs to reticular cells innervating these geniculate cells. Such firing in burst mode enhances the ability of the geniculate relay cells to detect the presence of a novel stimulus, one that is potentially interesting or threatening. Such enhanced detection, however, is associated with nonlinear distortion of the relayed signal and is thus unsuitable for accurate stimulus processing. Once novel stimuli or major components of a visual scene activate a previously quiescent cortical column, the layer 6 corticogeniculate neurons also become active and effect a switch in firing from burst to tonic in their geniculate targets. The tonic firing now enhances linear processing, permitting the visual system to analyze the scene more faithfully, but this is at the expense of stimulus detectability.

Increases in activation of the cholinergic afferents from the parabrachial region would produce comparable results. However, we know very little of the effects of other afferents, such as serotonergic, noradrenergic, and histaminergic afferents, on response mode. Also, there is very little evidence yet to suggest what controls activity of these afferents, although changes between drowsy and alert states are generally associated with changes in activity of brain stem afferents. Although the details are not defined, the actions of these other afferents may serve to allow affective states to influence perception.

D. Summary

At the cellular level, both cortical and brain stem inputs have similar effects, producing long, slow postsynaptic potentials in relay cells by means of metabotropic receptors. Nonetheless, their roles in controlling response mode are probably quite different and the difference depends on the connectivity each pathway establishes with thalamic circuitry. We can briefly reiterate some of the conclusions stated earlier as follows. The brain stem inputs, which probably lack local sign (see preceding), are likely to have global effects on response mode. These inputs are thought to be involved in arousal and in general levels of attention, and as the level of activity in these pathways diminishes, with a related faltering of attention, thalamic relay cells may tend to respond more in burst mode and thus be ready to provide a "wakeup call" if something changes significantly in the environment requiring attention. Some of the brain stem inputs may also be related to a single sensory modality, although we are basically ignorant of the specific sorts of messages conveyed by these afferents and just how some particular patterns of sensory stimulation activate these brain stem centers, whereas others leave them unaffected. Whatever the mechanisms may prove to be, some brain stem afferents described earlier as having a distribution limited to one sensory modality (Uhlrich et al., 1988) may provide a basis for switching attention between sensory modalities so when we attend, say, to auditory signals, more relay cells in the medial geniculate nucleus are in tonic firing mode, whereas more in the visual or somatosensory relays are in burst mode, primed to submit a "wakeup call" if their sensory domain changes abruptly.

By definition, the cortical input to the lateral geniculate nucleus conveys strictly visual information, and as noted, it is mapped with great retinotopic precision. Thus its control of response mode would be limited to attentional needs within the visual domain. For instance, if the animal were paying attention to one stimulus (e.g., likely with the fovea or area centralis, but conceivably in peripheral retina), geniculate cells mapped to nearby regions (and perhaps to much of the rest) of the visual field would be maintained in burst mode, primed to detect any new stimulus of potential interest or danger. Once a group of these thalamic inputs signals a change in the visual scene with bursts, local processing within cortex might lead to increased firing in the layer 6 feedback pathway, and this, in turn, could lead to a switch in firing from burst to tonic for the specific group of relay cells that had been responding in burst mode. This same increased firing of the layer 6 cells might cause very different changes in response mode in nearby geniculate cells, depending on the details of cortico-reticular-thalamic circuitry as depicted in Figures 8A and 8B of Chapter III. Indeed, more complex corticothalamic circuitry, such as but not limited to that depicted in Figure 8B of Chapter III, permits highly specific and local control of response mode by corticothalamic afferents, biasing some relay cells toward burst mode, and others toward tonic mode. One can even imagine that a highly active focus of layer 6 output representing a small region of visual space would maintain tonic firing in the retinotopically aligned geniculate relay cells while simultaneously maintaining burst firing in relay cells with receptive fields surrounding the target of interest. This could mean that, although tonic firing related to the region of immediate interest optimizes detailed analysis of targets within that area, surrounding areas would be more sensitive to any changes that might require a shift of attention.

Many thalamic relay cells burst rhythmically and in synchrony during certain phases of sleep (Steriade and Llinás, 1988; Steriade et al., 1990, 1993), although the extent to which those of the lateral geniculate nucleus do is a matter currently unresolved (McCarley et al., 1983; Ramcharan et al., 2000). Rhythmic, synchronized bursting seems to imply a functional disconnection of these cells from their main afferents, interrupting the thalamic relay. Our suggested role for the burst mode of firing is not incompatible with this other role. Instead, we are suggesting that, depending on the animal's behavioral state (e.g., alert versus sleeping), burst firing can subserve at least two quite different roles or perhaps two extremes of one role. It can either provide a relay mode in the awake animal for detecting significant but possibly minor changes in specific afferent activity, and after detection, changes in cortical or brain stem inputs can then be used to switch the relay to tonic mode for more accurate analysis. In the sleeping animal, the burst mode may involve more complete functional shutting off of cortical and parabrachial inputs, more analogous to the *in vitro* situation where these inputs are physically removed. Then the switch from burst to tonic mode may occur only in response to major stimuli that are threatening or significant for other reasons, such as an infant's cry. The low levels of activity present in the cortical and parabrachial pathways during the waking state may permit single low-threshold spikes but prevent rhythmic bursting. Bursting, when rhythmic and synchronized, may provide a positive signal to cortex that nothing is being relayed despite the possible presence of sensory stimuli, and this is less ambiguous than no activity, which could either mean no relay or no stimulus. When the bursting is arrhythmic, cortex can interpret this arrhythmicity as representing responses evoked by sensory stimuli.

E. *Some Unresolved Questions*

1. What are the differential roles of tonic and burst firing for thalamic relays?
2. Are both burst and tonic firing seen in relay cells throughout the thalamus during normal behavior?
3. What effects do noradrenergic, serotonergic, and histaminergic inputs to thalamus have on response mode?
4. In which synapses of thalamocortical circuitry does paired-pulse facilitation or depression occur, a question especially important for thalamocortical synapses? As a corollary, what are the response characteristics of thalamocortical synapses in the different cortical layers? That is, can a thalamocortical axon produce paired-pulse facilitation in one layer and depression in another? Finally, do the different classes of thalamocortical cell exhibit different synaptic physiology in cortex?

E. Some Unresolved Questions

5. What is involved in "closing the loop" between thalamic afferents and layer 6 corticogeniculate cells in feedback control of thalamic response mode? How do the details of corticogeniculate circuitry involving the thalamic reticular nucleus contribute to control of response mode?

6. Related to the above, do thalamocortical axons that innervate cells in cortical layer 6 innervate the cells that project back to the thalamus or only a different subpopulation?

CHAPTER **VII**

Maps in the Brain

A. Introduction

1. THE FUNCTIONAL SIGNIFICANCE OF MAPPED PROJECTIONS TO THALAMIC RELAYS

We have seen in earlier chapters that many of the driver afferents to the thalamus are mapped, producing topographically ordered representations of sensory surfaces or of cortical areas within individual thalamic nuclei. It is possible that all driver afferents to the thalamus are mapped in an organized topographical pattern, and it is tempting to propose this as a generalization about the thalamus. For now it is best treated as an open question. From the point of view of the functional organization we are proposing for the thalamus, the extent to which the modulators are mapped is equally important. Some of the modulator afferents are clearly mapped, particularly those coming from layer 6 of the cortex and many of those that come from the thalamic reticular nucleus, but others, like some of those coming from the brain stem, show no evidence of any mapping. That is, the drivers and some of the modulators appear to be organized so that the thalamus can deal with specific and very limited sectors of any one afferent pathway, whereas other modulators have a broader, more global action on the thalamus as a whole or on one or more of the thalamic nuclei. We have argued that modulators can change the nature of thalamic transmission, and the pattern of their projections suggests that they can either do this globally, for instance, as a generalized wakeup call for the whole thalamus, or locally, so that a particular limited part of a driver input, possibly representing a small part of the visual field or the sensory surfaces at the tips of two or three fingers exploring a small object, can bring detailed and accurate information to a small area of the cortex.

In this chapter we look at the way in which earlier workers have thought about maps in the brain, and we look at some of the experimental evidence about maps in the thalamocortical pathways, including the reticular relay. One of the striking features of contemporary evidence about the maps is that there are multiple maps in the cortex. A second feature, which has received less attention but is vital for understanding how thalamus and cortex are interconnected, is that these maps are often mirror reversals of each other and that a single thalamic map can be connected to two mirror-reversed cortical maps or vice versa. These connections require a complex interchange of the axonal pathways between thalamus and cortex, involving divergent, convergent, and crossing pathways. Much of this interchange occurs in the thalamic reticular nucleus, where the relationships of the several maps that relate to any one modality become of crucial importance for understanding how the modulatory activity of the reticulothalamic pathway is organized in relation to the multiplicity of maps.

The multiplicity of maps, their mirror reversals, and complex interconnections represent a key feature of thalamocortical relationships. In this chapter we focus on the maps themselves and on abnormalities of the maps that can be produced genetically or by experimental manipulations. The abnormalities are of interest because they illustrate the importance of the mirror reversals and demonstrate the capacity of the brain to produce new intercrossings in response to abnormally mapped inputs.

Our view of the action of thalamic modulators is one reason for presenting this chapter on maps. There are other reasons. One is that although it seems intuitively obvious that sensory surfaces, such as those dealing with visual, auditory, or somatosensory inputs should be mapped, and although this intuition has been significantly exploited in the past, there are no solid theoretical or empirical grounds to support this intuitive approach, which we explore further in the following. The main sensory pathways are, indeed, mapped within the thalamocortical pathways, but the fact that they are mapped cannot be taken as support for the view that the brain could not function if they were not mapped. For other thalamocortical pathways, such as those from the anterior thalamic nuclei, the medial dorsal nucleus, the lateral dorsal nucleus, the pulvinar, or the intralaminar nuclei, there are no intuitive grounds on the basis of which one can make a very good case as to why they should be mapped, nor do we know of any empirical or theoretical evidence that might be relevant. Yet they are mapped and there appears to be an orderly topographically organized connection for many, possibly all, thalamocortical and corticothalamic pathways (e.g., Cowan and Powell, 1954; Walker, 1938; Updyke, 1977, 1981; Goldman-Rakic and Porrino, 1985; see Chapter III.) The simplest explanation for a universal, ordered topographical mapping of thalamus onto cortex would be a developmental one. One might want to argue that if the thalamocortical axons take a relatively orderly course during development, maintaining neighborhood relationships as they go, then the ordered thalamic radiation that would result could provide a neat explanation for the topographic maps (see Caviness and Frost, 1983;

A. Introduction

Hohl-Abrahao and Creutzfeld, 1991). We have already indicated that the pathways from thalamus to cortex do not represent an orderly radiation, but instead present a very complex network of axonal crossings responsible for a multiplicity of interconnected thalamic and cortical maps. We will explore the nature of this complex network further in this chapter and will show that the simple developmental view outlined above is not viable. This leads us to see the complex network as requiring a set of separate specified pathways for each mapped projection, producing a possibility for interactions among maps, particularly as they pass through the thalamic reticular nucleus, but it leaves a serious question about the functional variable that is mapped in those thalamocortical pathways in which one cannot see a simple representation of a sensory surface.

For sensory maps, at least for the visual ones, which we explore in most detail in what follows, there is evidence from experimental manipulations and from the study of mutants that the maps are modifiable at early stages of development and that in a normal individual they form as development proceeds through these stages. It has been shown that changing the topography of the afferents to the thalamus at early postnatal stages of development produces a change in the complex, mapped pattern of the thalamocortical projections. We look at this evidence in some detail because it demonstrates two important conclusions. One is that the intuitive view of a brain that cannot function without an orderly representation of sensory surfaces does not apply to the thalamus but does appear to apply to primary visual cortex. The second, perhaps more significant conclusion, is that for the visual pathways the nature of the orderly thalamocortical projection is an active response to the nature of the patterned input arriving along the driver pathways at the thalamus. The experimentally modified and the genetically abnormal thalamocortical projections show us that the thalamus is not passive in the formation of the thalamocortical projections. The axons of the relay cells are ordered in accord with demands made by the pattern of the afferents that reach them. For the visual system we can understand the functional demands that underlie this response of the thalamocortical axons. For much of the thalamocortical system we have no idea at all what forces are active as the thalamocortical order that characterizes the normal adult is formed. And yet, unless we want to believe that the thalamocortical pathways dealing with sensory afferents play according to rules that are entirely different from those for the other thalamocortical pathways, we should start to ask about the functional variables that are relevant to the many mapped thalamocortical pathways. We offer no apologies for having no answer to the question we raise; our main aim is to formulate the question so that it can come within the realm of questions that can be attacked experimentally.

2. THE NATURE OF THALAMIC AND CORTICAL MAPS

If we look at the details of maps, we see that in the thalamus different aspects of the sensory inputs may map to different parts of the thalamus, so that, for example, the magnocellular and parvocellular pathways map to separate geniculate layers and the spinothalamic and lemniscal afferents go to separate parts of the ventral posterior nucleus (see Chapter III). In the cortex, maps are split up in different ways; for example, visual afferents concerned with orientation selectivity or ocular dominance are mingled within a single map in a regular, interrupted, and repeating pattern (Hubel and Wiesel, 1977), whereas other afferents are segregated to different cortical areas, a pattern of functional separation that has been strikingly demonstrated in the auditory pathways of bats (Fitzpatrick et al., 1998). It is difficult at present to use this information to formulate a generalization about the basic nature of cortical maps, even though there can be no question that for each major modality, with the exception of olfaction and possibly of taste, the topographic maps of the sensory surfaces play a significant role. That is, although much of what is presented below deals with the thalamic and cortical representation of sensory, particularly retinal, surfaces it has to be recognized that, in terms of the detailed distribution of the functional components within the maps, we are not necessarily addressing single continuous representations either of sensory surfaces or of the nerve cells that give origin to the afferents supplying any one map. In the rest of this chapter we are concerned with the overall layout and orientation of maps and will not deal with the way in which functionally distinct components may be fractionated within a map or distributed over several maps.

B. Early Arguments for Maps

The earliest expressions of the view that the brain must receive orderly mapped representations were based, so far as one can tell, on an intuitive evaluation of the capacities of the brain, not on any experimental evidence. The experimental evidence came later. Thus, when Newton in 1704 (cited by Polyak, 1957) argued that there must be a partial decussation at the optic chiasm, of the sort that is now illustrated in any neuroscience textbook and is shown here in Figure 1C, his argument assumed that the visual pathways had to provide the brain with an orderly representation of the single visual scene transmitted by two pathways, one from each eye, with each image reversed by the lens of the eye. This argument, that the brain needs a single ordered representation, a sort of single internal projection screen of the binocularly viewed visual scene, was not explicitly stated, but Newton's argument could not have been carried to its apparently brilliant and now recognizably correct conclusion if Newton had allowed for the possibility that the brain could have managed with a disrupted, disorderly, or distributed representation of the visual world.

The logic of the argument, however, is not as rigorous as the beautifully correct conclusion might suggest. Descartes had earlier, in 1686 (cited by Polyak, 1957), also recognized the problem presented by a single image transmitted to a single central end-station through two reversing

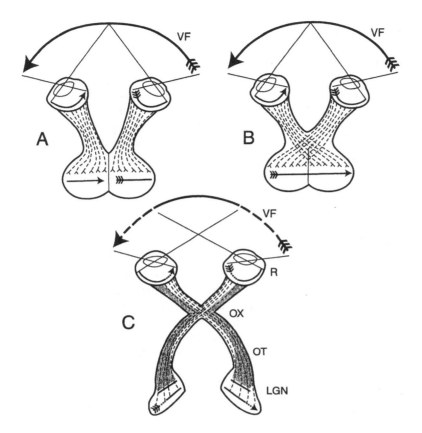

Figure 1
(A) Schematic representation of the visual pathways of an entirely hypothetical "lower" vertebrate, having no binocular vision and no optic crossing. It is important to stress that there is no such creature. However, Cajal uses the figure to show the "discrepancy between the conscious visual image and the object itself" (depicted by the arrows). (B) "The benefits" that a complete chiasmatic crossing would produce in such a creature; the arrow is now continuous in the brain. (C) A heavily modified version of Cajal's scheme showing what happens in an animal with a significant degree of binocular overlap.[1] The uncrossed and the crossed pathways are shown, each going to a different layer of the lateral geniculate nucleus (LGN), where the two representations of the contralateral visual hemifield are in register. The monocular sectors of the visual field and their central representations are shown by the interrupted parts of the arrow and its representation. LGN, lateral geniculate nucleus; OT, optic tract; OX, optic chiasm; R, retina; VF, visual field.

[1] We have shown in Figure 1C, in accord with current evidence (Torrealba *et al.*, 1982; Guillery *et al.*, 1995), the crossed and the uncrossed axons in the optic tract intermingling on their way to the lateral geniculate nucleus and maintaining a rough topographic order. Gross (1998) points out that a schema found in a Newton manuscript implies just such mixing of the nerve fibers in the optic tract, and Gross concludes that Newton thought (wrongly in Gross' view) that the images from the two eyes fused in the optic tract rather than in the visual cortex. Gross bases his view of Newton's supposed error on modern evidence about neuronal receptive fields: that it is not until the visual inputs reach the cortex that single binocularly driven nerve cells can first be recorded in the path from the eye to the cortex. However, the fact that here, too, Newton was serendipitously correct about the mingling of the two sets of fibers in the pathway makes one marvel at his capacity to come to correct conclusions about relationships that almost 300 years later Gross is still not aware of. The functional or developmental significance, if any, of this binocular mingling of axons in the optic tract in rough topographic order, remains unresolved.

lenses. Descartes' scheme showed each optic nerve going to one hemisphere, so that the image from each eye was represented separately at a first central way station in the brain. Descartes then had fibers from that way station going to the pineal gland, which he proposed as the final receiver of the visual image, through a pathway that involved a partial intrahemispheric crossing. Whereas Descartes had placed the partial crossing in the brain, where we now know that it occurs in a bird like the owl with good binocular vision (Karten *et al.*, 1973), Newton had placed it in the chiasm, where we now know it occurs in mammals. Neither could possibly have had any empirical evidence for their conclusion. Although the junction of the optic nerves at the optic chiasm was known, methods for tissue preservation and study of the fiber systems were not adequate for the type of detail needed to trace the crossed or the uncrossed course of fibers through the optic chiasm; and neither author made an empirical claim. The empirical evidence was not obtained for almost 200 years, a period during which assumptions about localization of functions in the brain were first lost and then regained (see Spillane, 1981), all on the basis of arguments that had nothing to do with the rigorous logic applied to the visual system by Descartes and Newton.

Although Gudden (1874) had been able to demonstrate the partial decussation in the optic chiasm of rabbits experimentally by removing one eye very early in development and showing that the surviving optic nerve split into a large crossed and a smaller uncrossed component, it took Cajal's demonstration of individual axons in Golgi preparations (see Cajal, 1911), some crossing in the chiasm, some not crossing, to help to persuade his contemporaries that there is a partial decussation at the optic chiasm of the mammalian brain. Not only did Cajal show the individual nerve fibers, but he produced schematic figures (see Figures 1A and 1B) based on the same implicit logic that Descartes and Newton had used. With these he explained why there had to be a complete chiasmatic crossing in a vertebrate with laterally placed eyes and no binocular visual field (Figures 1A and 1B). The assumption that the brain could not use the broken arrow shown in the brain of Figure 1A is made explicit by Cajal (1995):

"... that correct mental perception of visual space can take place only in a brain in which the center responsible for the perception is bilateral, and both halves act in concert so as to render the two images continuous and in the same direction as projected by the right and left halves of the two retinas."

Then he explained why an animal with forward-looking eyes (see Figure 1C) having a binocular visual field had to have a partial decussation in the chiasm. Finally he used these arguments to explain why in the brains of vertebrates each hemisphere connects to the tactile and other sensory inputs and to motor outputs on the other side of the body. These crossed connections of sensory and motor pathways have not been shown here in Figure 1. The logic that requires them is that, in each hemisphere, visual signals must be related to other sensory and to motor signals relating to the same side of the body. The implication of Cajal's argument is clear: a disrupted representation, illustrated by a broken arrow, could not work to transmit sensory information to the brain.

Although at about the time that Cajal was writing, evidence for orderly, topographic, motor and visual cerebral maps had already been under discussion for some time (Jackson, 1873; Henschen, 1893), he did not refer to this work in his discussion of the chiasm, and for a long time there was no experimental evidence to support the fundamental basis of Cajal's argument that the brain is unable to function with disrupted maps. It is important to recognize that as sensory messages are traced past the first cortical relay, the accuracy of the maps diminishes, and evidence from other parts of the brain, such as the olfactory pathways and the cerebellum, shows that there are regions of the brain that operate with widely distributed or "fractured" representations of the input (e.g., Haberly and Bower, 1989; Welker, 1987). The intuitive approach, which assumes that a central "viewing screen" is needed is not as compelling as perhaps it once seemed. Pattern recognition is possible from distributed, nonmapped systems (Kohonen et al., 1977).

There is a danger, when one is considering the many orderly representations that are found in the thalamocortical pathways, of following the intuitive approach represented in Figure 1C and not asking what the maps are for. We shall be describing the rich and complex interrelations between numerous cerebral maps in the visual and other pathways, and it will be easy for the reader to forget that the fundamental question of why the brain needs orderly representations of sensory surfaces and also of the motor mechanisms, or better, why the brain uses such maps, remains largely unanswered. There are a few examples, such as visual area V1, where one knows enough about local interactions to argue that without mapped projections such interactions might be very costly in axonal connections, but in general the maps are there as a challenge.

C. Clinical and Experimental Evidence for Maps in the Geniculocortical Pathway

1. Establishing That There Are Maps

The empirical evidence in favor of mapped representations in the visual system came first for the cortex and later for the thalamus. Initially it was shown in human patients that localized lesions of the occipital cortex gave rise to localized visual field losses (scotomas) (Figure 2). The history of these studies has a sad link with major international conflicts and with the development of high-speed bullets able to penetrate the skull without killing. These bullets could leave clean entry and exit wounds that allowed quite accurate definition of the cortical damage long before the use of x-rays or modern scanning methods. Studies of visual field losses after damage to the occipital cortex started after the Franco-Prussian war, continued after the Russo-Japanese war, and became highly refined after the

204   CHAPTER VII *Maps in the Brain*

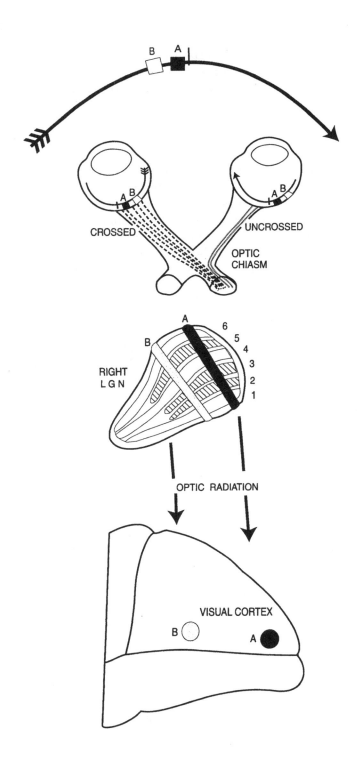

1914–1918, 1939–1945, and subsequent wars (Koerner and Teuber, 1973; Glickstein, 1988).

The fact that local cortical lesions could produce corresponding local visual field losses was evident quite early (Figure 2 and see Figure 1 in Chapter II). The precise extent of the cortex involved in the production of such scotomas and the precise orientation of the visual field map on the cortical surface were the subject of interesting and quite heated debate for some time, but it is now clear that there is a histologically well-defined area in the occipital lobe, the striate cortex, V1, or area 17, which receives a highly organized, binocular projection of the contralateral visual field. The evidence obtained clinically for the human brain was followed much later by experiments that recorded localized cortical activity in response to visual stimuli delivered to limited parts of the visual field (e.g., Daniel and Whitteridge, 1961, for monkey; Tusa *et al.*, 1978, for cat).

Early knowledge of retinal maps in the thalamus depended on the observations of the cortical lesions and on the rapid and severe retrograde cell losses that these lesions produce in the thalamus (see Chapter II, Figure 1). When the thalamus was studied in brains with localized lesions of the visual cortex, it was found that there are correspondingly localized zones of retrograde degeneration (Minkowski, 1914; Garey and Powell, 1967; Kaas *et al.*, 1972, and see Figure 2) in the lateral geniculate nuclei. The fact that each cortical lesion corresponds not only to a particular part of the visual field but also to a particular part of the lateral geniculate nucleus demonstrates that there is a map of the visual field in the nucleus, just as there is in the cortex. We have seen that the lateral geniculate nucleus is laminated and that this lamination has been well studied in carnivores and primates. Figure 2 shows that in a primate each of the six or so laminae receives from either the eye on the same side (shown as hatched layers) or from that on the other side (shown as plain in the figure) but not from both. The relationship of the zones of retrograde degeneration (also labeled A and B in the figure), which pass perpendicular to

---

Figure 2
In the lower part, two localized cortical lesions (A and B) are shown on an outline of a medial view of a right human occipital lobe. The visual field, represented by the arrow at the top of the figure, shows the corresponding visual field losses for the left visual field. There are localized visual losses corresponding to the nasal retina of the left eye and the temporal retina of the right eye. In the lateral geniculate nucleus (shown as a parasagittal section, with posterior to the right) on the side of the lesions in the right hemisphere, each small cortical lesion has produced a localized column of retrograde cell degeneration that goes through all of the geniculate layers "like a toothpick through a club sandwich" (see footnote 2). Within the retina one would expect to see degenerative changes in retinal ganglion cells at the points marked A and B, but it should be stressed that the visual losses are regarded as a direct result of the cortical damage, not of the retinal damage, which is a secondary, transneuronal retrograde degeneration.

the geniculate layers and through all of them, demonstrates an important point about the way in which the visual field maps onto the lateral geniculate nucleus. The figure shows that all of the layers are receiving inputs from the contralateral visual field, but that some layers receive this information through the left eye, whereas others receive it through the right eye. Since a localized lesion in the striate cortex produces a local visual field loss (scotoma) that involves both eyes (a "homonymous visual field loss" involves the same part of the binocular visual field, not of the retina, for each eye), the column of degeneration that passes through the geniculate must correspond to the scotoma for each eye. The astonishing conclusion is that the two maps of the contralateral visual field, one coming from the left eye and the other from the right eye, must be in register through all the laminae of the lateral geniculate nucleus.

2. The Alignment of Maps with Each Other

At first sight this mapping of the visual field in the geniculate layers may not seem as extraordinary as it really is. Figure 2 shows that in order to produce such an alignment of visual field representations, small retinal areas of the left and right eye that are both looking at the same part of visual space must project to geniculate cells that are aligned along a single column (A and B in the figure) running through all of the geniculate layers.[2] From a developmental point of view the two homonymous retinal points labeled A have no shared morphological features, nor do those labeled B. They are not symmetrically placed in the head and do not lie in corresponding parts of the two retinae. So far as our current knowledge goes, they are only related because they look at the same point in the visual field. And yet, the developing nervous system is able to guide to two sets of retinofugal nerve axons to precisely aligned sites in adjacent geniculate layers. This is a remarkable achievement in itself. It is even more notable because the alignment is completed before eye opening (Shatz, 1996; Crair et al., 1998). We do not know at what developmental stage the projections from the visual cortex and from the thalamic reticular nucleus acquire their mapped order. One can regard the formation of the in-register visual field maps as providing a basis on which the

---

[2] Walls (1953) wrote: "The problem of why the lateral geniculate nucleus should ever be stratified has seemed a mystery for half a century. In this time it has been brilliantly shown: first that each LGN embodies an isomorph of one half of the binocular visual field and the adjoining uniocular temporal crescent, with each peripheral or central quadrant of each of two hemiretinas projecting to a specific segment of the nucleus; second, that in the strongly binocular carnivores and primates certain laminae receive only from the contralateral retina and others from the ipsilateral; and third, that physiologically correspondent spots in homonymous hemiretinas are scattered along a single "line" running through all of the gray laminae like one of the toothpicks in a club sandwich. We may marvel at this precision, and we may pardon the old-school localizer for taking refuge in the visual system and crowing a bit. But this knowledge helps us not one jot or tittle to see why the LGN should have lamination per se—not, if no cell in any lamina communicates within the nucleus with any cell in any other lamina. No matter how closely the system may bring binocularly correspondent paths into approximation, if it is not for the purpose of enabling them to "fuse" in the LGN, then the approximation is senseless; and, the lamination that brings it about continues to seem senseless."

corticogeniculate and the reticulogeniculate axons can pass straight through the nucleus, relating to neurons that receive from just one point in visual space but activated from either the left or the right eye. That is, all of these maps, the retinogeniculate, the corticogeniculate, and the reticulogeniculate are in matched topographical order in the adult, but the left and the right eye afferents are kept separate in the lateral geniculate nucleus (see also Chapter III).

We have stressed the apparently uncanny degree of map matching in the lateral geniculate nucleus because it illustrates what appears to be a rather general property of thalamocortical connections, and probably of many other pathways in the brain. It applies to thalamocortical pathways where one thalamic nucleus projects to several cortical areas and also to corticothalamic connections that can bring mapped inputs from several cortical areas to a single thalamic nucleus. We will see that it applies to the connections of the thalamic reticular nucleus as well. One can find many places where two or more maps are brought into precise, or reasonably precise, register with each other. In some instances they are precisely matched within a nucleus or a lamina, in others they form a less rigorous representation, or the matching may be more difficult to demonstrate. For example, in the monkey ventral posterior nucleus, topographically organized projections to different somatosensory cortical areas come from distinct parts of the nucleus, and these parts interdigitate with each other, so that the functionally distinct representations of the same body part come to lie close to each other (Krubitzer and Kaas, 1992). That is, there is a topographical matching but not the precise alignment seen in the geniculate laminae. The visual system brings out problems of mapping in a particularly striking way because it has to deal with the problem of matching two half maps from two almost identical inputs, one from each eye both of them reversed by the lens. However, maps are a common feature of thalamocortical, thalamoreticular, corticoreticular, and corticothalamic pathways, and we can find them in auditory, motor, and somatosensory pathways where we know something about what it is that is being mapped (e.g., Reale and Imig, 1980; Beck *et al.*, 1996; Strick, 1988), as well as in the mamillothalamic pathway (Cowan and Powell, 1954), where the mapped functions are still undefined. There is good evidence that activity patterns can play a significant role in some aspects of the development of thalamocortical projections (Shatz, 1996), but the example of the binocular matching across geniculate layers before eye opening suggests that other mechanisms are likely to be involved as well because the patterns of activity coming from the two eyes are not likely to be matched in terms of the visual field alignment that is matched across laminae.

The essential point for us now is that developmental mechanisms have the capacity to produce finely matched maps. Where we fail to see such accuracy it may be that processing is concerned with features other than the precise topography of the afferents. In relays that are not concerned with location, such as taste or smell, or in relays that are a few steps from a sensory surface that is concerned with location, the maps are more difficult to define, presumably because the simple topography of the peripheral map does not play a central role and other features, often more difficult to isolate for study, begin to dominate.[3] This will prove to be a key to understanding many of the "higher" cortical areas and of the higher order thalamic nuclei that serve these cortical areas. It may also provide a key to understanding how pathways that carry several cortical and thalamic representations of a single modality, some representations more accurately mapped than others, some perhaps not mapped at all, come to relate to each other as they pass through the thalamic reticular nucleus.

Given that the mammalian visual pathways have achieved such a complex developmental feat of binocular matching in the lateral geniculate nucleus, it is reasonable to ask with Walls (see footnote 2) what it can possibly be for. Although there is some evidence for weak binocular interactions in the lateral geniculate nucleus (Sanderson et al., 1971; Schmielau and Singer, 1977), the problem of why the lateral geniculate is laminated and why the visual representation in the layers are aligned is not solved. Whereas the lamination clearly serves to separate left eye from right eye inputs, the mapping brings them into register, as though they are meant to interact or to be subjected to a common input. This dual aspect of the functional separation may provide a clue to thalamic organization in general. On the one hand, the thalamus provides a simple, clear relay for sensory pathways on the way to the cortex. For some of the time that is the main relay function of the thalamus; the cortex receives the messages from each eye that come through the one set of geniculate layers; the two inputs are essentially independent and only start to relate to single cells within the visual cortex (Hubel and Wiesel, 1977). On the other hand, we have argued that there is another function for the thalamic relay and that this serves to *modulate* the activity relayed through the thalamus. This modulation can be global or local, and, for the visual pathways, if it is local it would generally need to relate to the binocular visual field, not just to one or the other retina. The two important modulatory inputs mentioned above, one coming from the cerebral cortex and the other coming from the thalamic reticular nucleus, both have axons that pass through the geniculate laminae like Walls' toothpick (footnote 2). That is, in Figure 2, corticothalamic afferents coming from the small cortical area indicated by (A) distribute to the whole of column A of the lateral geniculate nucleus, and correspondingly for the parts labeled (B). The experimental evidence for this was illustrated in Chapter III (Figure 7), and Figure 11 in Chapter III showed that the same is true for the axons that go from the reticular nucleus to the geniculate layers. It appears that these modulatory connections require the alignment of the two maps, so that enhancement or suppression is not limited to a part of the visual field for just one eye but can act on the same visual field representation for each eye. There are rare occasions when modulatory actions

[3] The olfactory system provides an excellent example of how difficult it is to discover the nature of a map in the central nervous system if one does not know what it is that is mapped. In its early stages the system does not map the spatial distribution of odors in the environment, but instead maps distinct groups of molecular structures (Mombaerts et al., 1996; Belluscio et al., 1999).

may need to be monocular. There is evidence that some modulatory axons from the parabrachial region of the cat innervate a single geniculate layer, thus possibly providing modulation for one eye only (Uhlrich et al., 1988). Possibly some corticogeniculate or reticulogeniculate axons also distribute to left eye or right eye geniculate layers only, but they have not been documented.

Although the arrangement of the binocular inputs to the lateral geniculate nucleus provides a unique relationship that can be found only in the visual pathways, an analysis of how the retinal maps relate to each other helps to illuminate a duality of thalamic functions that is not limited to the visual pathways: transmission of information to cortex and a modulation of that transmission that can be either global or localized. Here it is worth extending this argument beyond the most obvious example of local action, with topographical specificity relating to a part of a map, to specificity based on the functional type of relay cell affected, such as geniculate X cell versus Y cell. Although currently we have limited evidence for such connections, it is likely that the differential distribution of corticogeniculate and reticulogeniculate axons to specific layers of the lateral geniculate nucleus (see Chapter III) relate to such a form of specific functional rather than topographic modulation.

D. Multiple Maps in the Thalamocortical Pathways

1. THE DEMONSTRATION OF MULTIPLE MAPS

Evidence that there might be more than one map of the visual field in the cerebral cortex came from electrophysiological recordings (Allman and Kaas, 1974, 1975; Tusa et al., 1979; Tusa and Palmer, 1980; Van Essen et al., 1992) and from studies of callosal connections between the two hemispheres, which exploited the fact that most of the visually responsive axons in the corpus callosum have receptive fields close to the vertical midline of the visual field, so that a plot of distinct islands of termination of callosal axons could be interpreted as a plot of several representations of the vertical meridian (Cragg, 1969; Zeki, 1969; Olavarria and Montero, 1984, 1989). Studies of cortical responses to other afferents have also shown multiple cortical maps for the somatosensory and auditory pathways (Beck et al., 1996; Reale and Imig, 1980; Kaas et al., 1999) and, similarly, several motor maps have been described (Strick, 1988). Detailed studies of the visually responsive cortex of cats and monkeys now show well over 30 distinct, more or less accurately mapped representations of the visual field (e.g., Van Essen et al., 1992). It is to be stressed that the accuracy of the maps varies from one cortical area to another. Some areas show little or no local sign, others show local sign but appear to represent only a part of the visual field, not all of it.

Many of these cortical areas are thought to represent a different aspect of cortical processing, one, for example, being concerned with color, another with movement (Zeki, 1993; Britten et al., 1993), although for many, no specific functional specialization has yet been recognized. A great deal of attention has been focused in recent years on the way in which these cortical areas are interconnected with each other through corticocortical pathways. Although the functional roles of these corticocortical pathways are almost entirely unexplored in terms of their driver or modulator functions (see Chapter IX; also see Crick and Koch, 1998; Sherman and Guillery, 1998), the interconnections that have been described are complex and have led to complicated schemes of corticocortical communication, the details of which are, fortunately, well beyond the scope of this book.

In addition to this multiplicity of cortical maps, there are also several maps of the visual field in the thalamus itself. These can be displayed by electrophysiological recording (Mason, 1978; Chalupa and Abramson, 1988, 1989; Hutchins and Updyke, 1989) or by studying the connections between the cortical maps and the thalamic maps with neuroanatomical tracers (Updyke, 1977, 1981). A single small, localized injection of an anterograde tracer like tritiated proline into area 17, or in one of the other visual areas, produces a number of localized spots of terminal label in the thalamus, one in each nucleus or nuclear subdivision. By varying the position of the area 17 injection it can be shown that each spot forms a part of a more or less complete map of the visual field in the thalamus.

The way in which these several thalamic and cortical representations of the visual field are interconnected is likely to be of vital importance to the functioning of the thalamocortical visual system, and to our understanding of it, but currently we know relatively little about the precise pattern of the interconnections. Although multiple cortical areas have been defined in rats, mice, cats, and monkeys (Olavarria and Montero, 1984, 1989; Tusa *et al.*, 1979; Tusa and Palmer, 1980; Zeki, 1969; Van Essen *et al.*, 1990, 1992), the extent to which cortical areas in different species can be treated as functionally equivalent or developmentally homologous is currently largely unknown, and we have yet to find out whether there is a generalization that can be made across species about the way in which the two way connections between these several cortical areas and the thalamus are organized.

2. MIRROR REVERSALS OF MAPS AND PATHWAYS

One important and common feature of thalamocortical pathways that include more than one cortical or thalamic map is that in many of the accounts of multiple maps for a single modality like vision or touch one commonly sees that adjacent maps are mirror reversals of each other. The developmental mechanisms that underlie these many mirror reversals are not understood, nor is it known whether the mirror reversals have any functional significance in the adult. However, from the point of view of this discussion they demonstrate the capacity of the thalamocortical and corticothalamic pathways to form the complex crossings that are needed within each hemisphere in order to produce orderly connections between a thalamic relay and two or more of such mirror-reversed cortical maps (see Adams *et al.*, 1997). It is important to stress that no amount of twisting of axonal pathways can produce the connections that are required. An actual crossing of axons in one dimension of the map is required. In the pathways that link thalamus and cortex, a complex system of crossing axons can be seen in two regions quite early in development. One is immediately beneath the cortex in a region that corresponds to the *subplate* of early development (Allendoerfer and Shatz, 1994; Ghosh and Shatz, 1994), and the other is in the thalamic reticular nucleus and in a region that lies close to it laterally, the *perireticular* nucleus, which lies in the region of the smaller dashed arrow in Figure 3 (Mitrofanis, 1994;

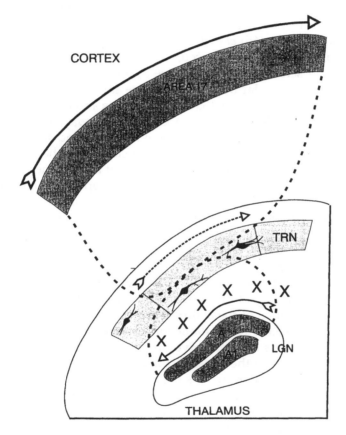

Figure 3
Schematic representation of the pathways that connect the visual cortex and lateral geniculate nucleus. Only two geniculate layers (labeled A and A1) are shown. For the two maps to be topographically interconnected a pathway crossing is necessary as shown here for the corticothalamic axons, which are shown crossing in the region of the thalamic reticular nucleus. Three cells of the reticular nucleus are shown with their dendritic arbors stretched out in the plane of the reticular nucleus. The two continuous arrows represent the horizontal meridian of the visual hemifield in the cortex and in the lateral geniculate nucleus. The dashed arrow represents the visual field representation that might originally have been expected within the visual sector of the thalamic reticular nucleus; further details are in the text. TRN, thalamic reticular nucleus; LGN, lateral geniculate nucleus; XXX, position of the cells of the perigeniculate nucleus.

Earle and Mitrofanis, 1996). Both of these cell groups are present as the axons are growing to link thalamus and cortex, but are largely lost because of heavy cell death in later development.

The expected pattern of axons crossing each other is present in the pathways from the lateral geniculate nucleus to the first visual cortical area (area 17) and also in that going from the ventral posterior nucleus to the first somatosensory cortical area (see Adams *et al.*, 1997). For the visual pathways the mirror reversal represents a reversal of the horizontal meridian about an axis formed by the vertical meridian. The theoretical need for a pathway crossing in the geniculocortical pathway was recognized by Connolly and Van Essen (1984) and demonstrated experimentally by Nelson and LeVay (1985). They showed that the crossing of the thalamocortical fibers occurs in the white matter underlying the visual cortex. In contrast, Lozsádi *et al.* (1996) showed that the crossing for the corticogeniculate axons in a rat occurs within the thalamic reticular nucleus (as shown schematically in Figure 3) and also just lateral to this nucleus in the perireticular nucleus. Evidence about where the crossings occur for most of the pathways that link thalamus and cortex is currently not yet available. Possibly the thalamocortical pathways all cross in the subcortical regions that develop from the subplate and the corticothalamic axons all cross in the perireticular nucleus and the thalamic reticular nucleus (see below); that would be a neat arrangement in accord with the fact that the two pathways take quite independent courses. The perireticular nucleus and the subplate are two very similar cell groups. They are present transiently next to cortex and next to the reticular nucleus, they share many immunohistochemical staining properties, and are most evident at the developmental stages when the thalamocortical and corticothalamic axons are growing through these regions (Mitrofanis and Guillery, 1993). They are likely to play a significant role in establishing the complex crossings. In the adult, the region of the thalamic reticular nucleus, where much of the complex pattern of axon crossing occurs, serves as a crucial nexus for thalamocortical and corticothalamic pathways because both sets of axons give off collateral branches to the thalamic reticular nucleus as they pass through.

In the thalamic reticular nucleus there is not only the pattern of axon crossing necessitated by the mirror reversals, but there is also a significant amount of convergence and divergence in the pathways that link thalamic nuclei and cortical areas. Although in primates almost all current evidence (Fries, 1981; Yukie and Iwai, 1981) shows that only a few of the geniculocortical axons go to extrastriate visual cortex, in other species there are projections from the lateral geniculate nucleus to several visual cortical areas. Each of these areas in turn sends axons to the lateral geniculate nucleus and to the pulvinar and lateralis posterior nuclei. So far as we know at present many of these several converging and diverging thalamocortical and corticothalamic pathways send branches to the reticular nucleus and connect to the same portion or sector of the reticular nucleus, described more fully below.

The multiplicity of cortical areas and thalamic nuclei for any one modality produces a combination of crossing pathways, divergent and convergent pathways that link thalamus and cortex (see Figure 6 in Chapter I). Each pathway can give off collateral branches to the same sector of the reticular nucleus, and as several systems pass through the reticular nucleus, they can relate several different thalamocortical and corticothalamic pathways concerned with a single modality to the cells of the reticular nucleus. In the remainder of this chapter we first consider abnormal crossings of the thalamocortical pathways that pass through the reticular nucleus, and then we consider the organization of connections in the normal reticular nucleus.

E. Abnormal Maps in the Visual Pathways

Abnormal visual pathways can be produced by naturally occurring mutations or by surgical interference with the developing system. The mutations that we will consider here act like a delicate piece of experimental surgery, causing a particular group of axons, some of the normally uncrossed axons from the temporal retina, to take an abnormal, crossed pathway. The experimental manipulations involve an early postnatal enucleation in hamsters or ferrets. These are both species who have young born at very immature stages, and the enucleation can be done at an early developmental stage before the thalamocortical pathways are fully formed. The results are comparable and show that the abnormal retinal input influences the topography of the thalamocortical pathways. Both produce rather similar abnormal thalamocortical pathways, but because the details have been worked out more fully for the mutants, we will give an account of these results and then consider the experimental results more briefly in the light of what is known for the mutants.

1. ABNORMAL PATHWAYS IN ALBINOS

Much of the work that has attempted to define the developmental basis of the gene action in the mutants or to analyze the functional capacities of the adult visual pathways is beyond the scope of this book. Stent (1978) has written a stimulating analysis, and more recent summaries can be found in Guillery *et al.* (1995) and Guillery (1996). Here we are concerned to use the mutant systems to look at what happens when the abnormalities produce disrupted maps like Cajal's broken arrow, or produce nonmatching maps within adjacent geniculate layers, and to compare thalamus and cortex because there appears to be an important difference in the way that the two react to these abnormal connections.

Albino animals and many other mutants with an abnormal distribution of melanin in the retina early in development have an abnormal crossing of some of the retinofugal axons. Figure 4 shows the normal and abnormal pathways in a cat. The abnormal pathways have been worked out in most detail for Siamese cats, a variant of cat that is homozygous for an allele of the albino series and that only makes melanin in the colder

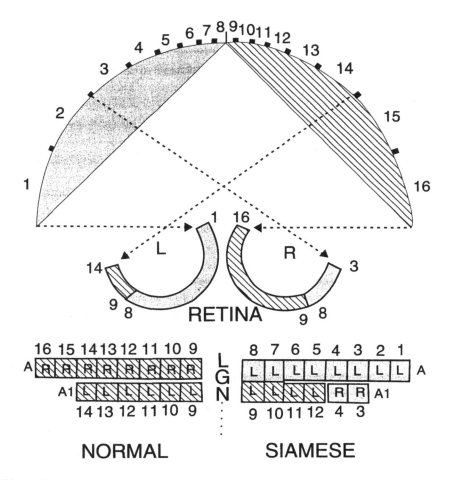

Figure 4
Schema to show the layout of the retinogeniculate pathways in normal and Siamese cats. The top half applies to both types of pathway. It is only the connection from the retina to the lateral geniculate that differs in the two. The normal pattern is shown for the lateral geniculate nucleus on the left side, and the abnormal Siamese pattern is shown on the right. R, L, right and left; LGN, lateral geniculate nucleus.

parts of its body late in life. In normal cats, whose visual pathways are represented in the left part of the figure, all the layers of the lateral geniculate nucleus are receiving inputs from the contralateral visual hemifield. The abnormal pathways of Siamese cats are shown in the right part of the figure, and here there is a segment of the lateral geniculate nucleus that is receiving from the sectors numbered 9 to 12 of the visual hemifield on the same side. This is due to an abnormal crossed pathway from the retinal sectors numbered 9 to 12 in the temporal retina of the left eye. These axons go to the appropriate locus in the lateral geniculate nucleus, but on the wrong side, the right side instead of the left side in the figure. The result is that geniculate segments that would normally receive, in

E. Abnormal Maps in the Visual Pathways

mediolateral sequence, from visual field segments 8, 7, 6, and 5 now receive from 9, 10, 11, and 12. Not only are the retinal axons from a part of the temporal retina terminating on the wrong side of the brain, but they are also terminating in a sequence that is a mirror reversal of the normal representation of the visual field. Further, within the layer labeled A1 there is now a disruption not unlike Cajal's broken arrow, which can be appreciated by following the numerical sequence within layer A1 on the right side of the figure: 9, 10, 11, 12, // 4, 3. Finally, whereas normally the numbers match across layers, in the abnormal pathways the numbers only match for the most lateral part of the nucleus (sectors 3 and 4). We are concerned with looking at what effect the "broken arrow" has on the system and with demonstrating the extent to which the mapping of the thalamocortical pathways reflects the map formed by the afferents that innervate the thalamic relay cells.

Electrophysiological recordings in Siamese cats (Hubel and Wiesel, 1971; Guillery and Kaas, 1971; Kaas and Guillery, 1973) show that the normally and abnormally innervated cells in the major geniculate layers that receive from the temporal retina (the layers that are labeled A1 in the figure) respond well to visual stimuli, and the same has been recorded for the superior colliculus (Berman and Cynader, 1972; Lane et al., 1974). That is, contrary to Cajal's suggestion (see Figure 1), these cell groups can accept a disrupted representation of the visual field.

In the visual cortex the situation is different. Cajal's suggestion seems correct. A broken arrow does not make for a functional connection. Figure 5A shows the normal geniculocortical pathway, with the reversal that is also shown in Figure 3. It has been experimentally demonstrated (Kaas and Guillery, 1973) that in normal and Siamese cats both sets of geniculate layers, those receiving crossed axons and those receiving uncrossed axons, send axons to visual cortex since both undergo the classical retrograde degeneration after a lesion of the visual cortex. This connection is indicated by the two sets of numbers shown in area 17 and in an adjacent, mirror-reversed part of area 18 in Figures 5A–C. For the abnormal cats, two distinctly different patterns of axonal connections have been defined. They were first described on the basis of the visual receptive fields mapped in visual cortex, and subsequently this was confirmed anatomically on the basis of the retrograde degeneration. They were labeled Boston and Midwestern patterns, which indicated where each was first described. In some abnormal cats (Midwestern, Kaas and Guillery, 1973; see Figure 5B) the whole abnormal visual field representation arising from the temporal retina, although present in the lateral geniculate nucleus, is simply lost within the cortex and correspondingly in the figure all of the numbers relating to the abnormally innervated layer have been crossed out in the cortex. In spite of the anatomical evidence that the relevant geniculocortical fibers reach the cortex, records of neural activity in the cortex show very few cortical cells that respond to any of the inputs from the temporal retina. Almost all of the cortical cells are driven by inputs from the nasal retina alone. It is important to note that all of the inputs from the temporal retina are lost in the cortex, not just those from the

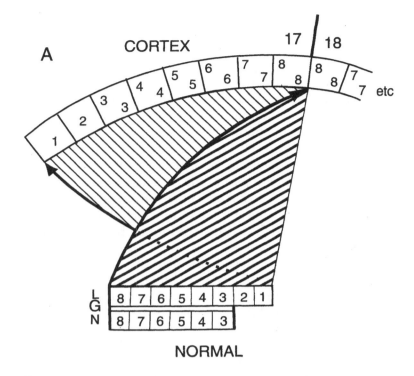

Figure 5
(A) Normal. Schema for a normal cat to show the reversed connections to visual cortex (area 17 or V1) from two geniculate layers in which the visual field representations are in register as indicated by the numbers; 17/18 shows the boundary between areas 17 and 18; LGN, lateral geniculate nucleus. (B) Midwestern. Schema using the same conventions as in A to show that in the Midwestern geniculocortical pathways of Siamese cats none of the inputs from the abnormally innervated layer of the lateral geniculate nucleus have effective cortical connections. That is, crossed inputs shown in Figure 5 coming from retinal sectors numbered 9, 10, 11, and 12, and uncrossed inputs numbered 4 and 3 are not received in cortex. Essentially, the temporal retina fails to reach cortex. In contrast, the crossed inputs coming from the nasal retina (sectors 1–8) make normal cortical connections. (C) Boston. Schema to show the correction that occurs in the Boston geniculocortical pathways of Siamese cats. Inputs from nasal and temporal retina provide effective input to cortex; the continuous sequence of visual field representation is recreated by the reversal of a sector of the geniculocortical pathway (9, 10, 11 and 12) to its own cortical area adjacent to the 17/18 boundary. Further details are in the text.

abnormally connected sector of the temporal retina. That is, the cortex (in contrast to the subcortical centers) appears to be unable to deal with a disrupted representation like Cajal's broken arrow. Further confirmation that cortical cells are not responding to inputs from the temporal retina comes from experiments showing that these cats are unable to respond to stimuli falling on any part of the temporal retina (Elekessy *et al.*, 1973; Guillery and Casagrande, 1977). That is, the parts of the visual fields that

E. Abnormal Maps in the Visual Pathways 217

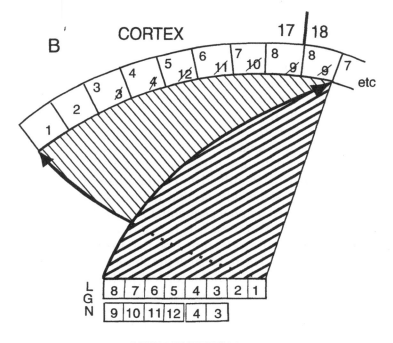

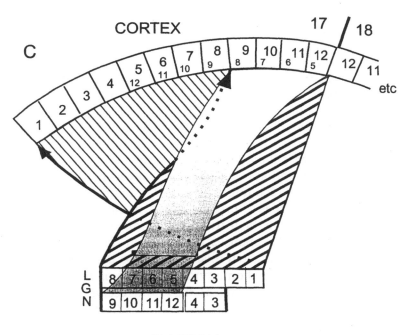

Figure 5 (continued)

are not able to drive cortical cells also fail to evoke visual orientation when tested behaviorally.

In other abnormal cats (Boston, described by Hubel and Wiesel, 1971; see Figure 5C), it has been found that there is a partial reversal of the geniculocortical projection for the abnormally innervated geniculate segment. The abnormal segment of the geniculate (9, 10, 11, 12) is now connected to cortex without the normal mirror reversal and is inserted next to the 17/18 border, which in a normal animal represents the vertical meridian but in the Boston pathways represents a position about 20 degrees into the visual field on the same side. This reversal corrects the disruption and produces an essentially orderly, continuous representation of the normal, contralateral visual hemifield and of the adjacent, additional part of the hemifield on the same side, so that the numbering in the cortex shown by the large figures is now sequential from 1 through 12. Cajal's arrow has been repaired! And these cats respond to visual stimuli in the temporal retina as well as the nasal retina, so that the repaired arrow can be regarded as functional.[4] The reversal of the abnormal sector in the Boston pathways is of interest for thalamocortical pathways in general because it demonstrates that the map in the thalamocortical pathway depends on the map that arrives at the thalamic relay from the driver afferents. Further, it shows that a visual field reversal very like the normally occurring visual field reversals discussed in the previous sections, has been induced in the abnormal brains in what appears to be a response to the abnormal inputs. The experimental results described in the next section confirm that a change in the retinal input pattern can lead to a change in the mapped projection of the thalamocortical pathway. Finally, it is important to note that the map reversal must occur somewhere on the route of the thalamocortical fibers, between the thalamus and the cortex; it will affect the mapped connections of the modulatory connections coming from cortex and, to the extent that it is likely to involve the connections of the thalamic reticular nucleus, it will also affect those coming from the reticular nucleus.

Figure 5B shows that there are two strikingly abnormal features of the visual field representation in the geniculate layers of the Midwestern pathways, either one of which might be expected to produce a visual abnormality. One is that there is a "broken arrow" in the abnormally innervated layer, and this may be the feature that produces the abnormalities in cortical and behavioral responses to stimuli in the temporal retina. However, there is also an unusual mismatch across the layers. Whereas these normally receive from homonymous points of the two retinas (Walls' toothpick in the club sandwich), they now receive from disparate parts of the visual field. Whereas the broken arrow effect can be expected to affect the whole of the relevant geniculate layer (layer A1 in Figure 5), the mismatch affects only the abnormally innervated segment of layer A1.

---

[4] Hubel and Wiesel (1971) also described a small number of cortical cells having two receptive fields in mirror position to each other about the vertical meridian. These have not been shown in the figure.

E. Abnormal Maps in the Visual Pathways

From this it would appear that the abnormal cortical and behavioral responses are due to the first and not the second effect.

This interpretation, that the mismatch across geniculate layers does not produce an abnormal visual response, is confirmed by two lines of evidence. One is that in some cats, in which the normally connected sector of the temporal retina is extremely small (Leventhal and Creel, 1985; Ault et al., 1995) and in albino monkeys, which also have a very small uncrossed component (Guillery et al., 1984), there appears to be no significant loss of responsiveness in the cortex. That is, there is no significant broken arrow effect within the individual geniculate layer that normally receives the uncrossed afferents (A1). The second is that there is no loss of responsiveness for the normal eye in Siamese cats in which one eye has been sutured from before eye opening (Guillery and Casagrande, 1977). Figure 4 shows that in a Siamese cat with such an early monocular suture, layer A1, when tested in the adult with both eyes open, should be able to receive inputs from either the right eye (labeled R in the right part of the figure) or the left eye (L) only. In these cats no single geniculate layer receives a disrupted, broken arrow representation of the temporal retina: one side of the brain receives the normal inputs from the temporal retina and the other side receives the abnormal inputs. The two appear not to be in conflict, and the cats react normally to visual stimuli falling on the temporal retina. This behavioral response to visual stimuli on one side can be abolished by a cortical lesion on the other side, showing that one is dealing with behavioral responses that depend on the cortical mechanisms.

The right side of Figure 4 shows that after a unilateral lid-suture of the right eye, there is still a mismatch of visual field representations across layers A and A1. Inputs coming from the left eye, and marked "L" in layers A and A1 in the figure, do not match across the layers. Although there is no longer a broken arrow in any one lamina, the parts of the visual field viewed by adjacent parts of two geniculate layers still do not match. Again, this mismatch appears not to disrupt the behavioral responses that have been tested. This may seem odd in view of what we said earlier about the match of visual field maps across layers. If this match is important for the modulatory pathways, then are there subtle abnormalities in such monocularly lid-sutured Siamese cats that remain to be defined? Or has the mismatch been dealt with in some other way? The answer is not known because we do not have appropriate behavioral tests of the function of the modulatory pathways. A partial answer may be in two sets of experiments, one relevant to the Boston, the other to the Midwestern pattern, showing that each has a distinct modification of the corticothalamic pathways, each apparently matching the pattern of the thalamocortical pathway (Montero and Guillery, 1978; Shatz and LeVay, 1979). It seems probable that there would be corresponding abnormalities of the reticulothalamic and thalamoreticular pathways, but these have not been studied. The results that have been obtained from these genetically abnormal animals provide striking evidence that the continuity of Cajal's arrow is important for the development of functional representations of sensory surfaces in the cortex. The Boston animals have also shown that there is a capacity for the developing thalamocortical pathways to produce reversals that serve to recreate a continuous arrow from a discontinuous one. It would seem probable that this capacity reflects a part of the normal developmental repertoire of the thalamocortical system and is not something that the mutant cats have developed *de novo* in response to the retinofugal abnormalities.

2. Experimental Modifications of the Thalamocortical Pathway

Evidence that the map formed by the thalamocortical pathways depends on the afferents reaching the thalamus comes from recent experiments by Trevelyan and Thompson (1992) and Krug *et al.* (1998). They studied the effects of a very early postnatal monocular enucleation in hamsters or ferrets. This leaves one lateral geniculate nucleus, the one innervated by the eye on the same side, with an abnormal, partially reversed projection in the retinogeniculate pathway (see also Schall *et al.*, 1988). By making small injections of different retrograde markers in visual cortex at various stages of development and in the adult, they were able to show that there is a mirror reversal of the geniculocortical projection on the side of the enucleation, reminiscent of the Boston abnormality, and that this develops early in postnatal life. These experiments demonstrate clearly that the arrangement of sensory maps that feed into the thalamus plays a significant role in the development and in the functioning of the mapped thalamocortical pathways. That is, the thalamocortical pathways have the capacity to produce modifications and mirror reversals in the early postnatal brain (Krug *et al.*, 1998, define the time course), and it is probable from the limited evidence we have from Siamese cats that the modulatory pathways that innervate the thalamic relay cells from cortex, probably also from the reticular nucleus, are correspondingly modified. These are issues that remain open for further study. The important point for the next section is that one should start to think of the maps that have been demonstrated in the thalamic reticular nucleus not only as part of a modulatory system that can act locally, but also as one whose local action is probably dependent on the pattern of the mapped inputs that are arriving from the afferent drivers. This is a view that bears serious exploration no matter whether these drivers represent a sensory surface or a higher cortical area and no matter whether we know or are entirely ignorant about the functional variable that is mapped in the topographically organized pathways that link thalamus and cortex.

F. Maps in the Thalamic Reticular Nucleus

The thalamic reticular nucleus is a relatively narrow sheet of cells wrapped around the rostral, lateral, and dorsal aspects of the thalamus (Figure 2 in Chapter I). Both the thalamocortical and the corticothalamic axons have to pass through this nucleus, and as they do, they give off branches that

F. Maps in the Thalamic Reticular Nucleus

provide mostly excitatory innervation to the reticular cells (Jones, 1985; Murphy and Sillito, 1996; Cox and Sherman, 1999). The reticular cells, in turn, provide inhibitory innervation for the thalamic nuclei from which they receive afferents (Jones, 1985; Ahlsén et al., 1985). When the connections of the reticular nucleus were first defined (see Jones, 1985), several sectors of the nucleus, each corresponding to a major thalamic nucleus or group of nuclei, were recognized. However, it was thought that each of these sectors received a diffuse, nonmapped input from the thalamus and from the corresponding area of cortex. The idea that there were no detailed maps within the pathway for any one modality such as the visual, the auditory, or the somatosensory system was reinforced by the observations made some years earlier (Cajal, 1911; Scheibel and Scheibel, 1966) that the individual cells of the reticular nucleus stretched their dendrites in the plane of the reticular sheet, extending as a discoid over large parts of any one sector or even between sectors (Figure 3). One expected, wrongly as it turned out, that the reticular sheet would correspond to the cortical sheet and that, if there were maps, these would be laid out in the plane of the reticular sheet, as they are in the cortex (see the dashed arrow in Figure 3) so that the parts of the map would be "smeared out" over the long axis formed by the dendrites of each reticular cell.

During the past decade it has become clear that there are relatively accurate maps in the thalamic reticular nucleus, but that they do not follow the line of the dashed arrow in Figure 3 (Crabtree and Killackey, 1989; Crabtree, 1992a,b; Cicirata et al., 1990; Cornwall et al., 1990; Conley et al., 1991). They lie at right angles to this line, formed as the corticogeniculate axons cross each other on their way through the nucleus. These maps can be defined by plotting the terminals of corticoreticular axons, by plotting the terminals of thalamoreticular axons, or by looking at the retrograde labeling of reticular cells after local injections of tracer into the dorsal thalamus. For any one reticular sector, related cortical areas along one axis of the cortex are mapped perpendicular to the expected plane. Such representations of small cortical areas are shown as half-cheeses (labeled A and B in the reticular nucleus) in Figure 6. These representations are stacked on top of each other for the horizontal meridian (A, B) and next to each other for the vertical meridian. The latter would be represented by (B) in the section of the reticular nucleus illustrated in Figure 6 and by a region representing (C), which would lie in a more rostral section that is not shown in the figure. That is, a small cortical injection produces a narrow slab (a whole cheese) of label within a small fraction of the thickness of the reticular sheet, and these slabs extend along the dendrites of the reticular cells (compare Figures 3 and 6), so leading to a reinterpretation of what the spread of reticular dendrites can mean for the capacity of the nucleus to carry reasonably accurate maps. Further, a reticular sector concerned with one modality receives inputs from the several cortical areas and thalamic nuclei concerned with that modality.

When evidence for a visual field mapping in the thalamic reticular nucleus first came to light (Montero et al., 1977), there was a good reason for expecting this pattern of representation in the reticular nucleus to

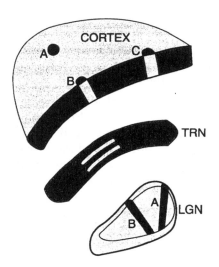

Figure 6
The organization of cortical, geniculate, and reticular representations of the visual field in the left hemisphere. The cortex is shown as flattened, with cortical area C rostral to area B. The thalamic reticular nucleus (TRN) and the lateral geniculate nucleus (LGN) are shown as they would appear in a coronal section, with medial to the right. At the top, A, B, and C are three small cortical areas; A and B represent small parts of the visual field along the horizontal meridian; B and C represent small parts along the vertical meridian. At the bottom of the figure the lateral geniculate nucleus is shown with the layers not indicated, so that the figure could represent a rat, rabbit, or bush baby. Cortical areas A and B are connected by thalamocortical and corticothalamic axons to columns of cells that run through all of the layers of the lateral geniculate nucleus as shown for A and B. Cortical area C would be connected to parts of the reticular and geniculate nuclei at more rostral levels, and its representations are not shown. Based on data described in the text.

correspond to the dashed arrow in Figure 3. This is the pattern of representation seen in the perigeniculate nucleus, which is indicated by the Xs in Figure 3. So far we have treated the perigeniculate nucleus as a part of the thalamic reticular nucleus, characteristically found only in cats and other carnivores. It lies between the lateral geniculate nucleus and the reticular nucleus. From almost everything we know about the nucleus, including its apparent shared developmental origin with the reticular nucleus (Mitrofanis, 1994), the shape and orientation of its dendritic arbors, and its transmitters, receptors, thalamic, cortical, and brain stem connections (Sherman and Guillery, 1996), there appear to be no differences, and the perigeniculate nucleus has long been regarded as simply a part of the reticular nucleus of carnivores that is slightly displaced toward the lateral geniculate nucleus. However, there is one important known difference between, on the one hand, the cat's perigeniculate nucleus and, on the other hand, the sectors of the reticular nucleus that deal with vision in other species or with somatosensory or auditory pathways in any species including the cat. This is a difference in the mapping. The cells of the perigeniculate nucleus have receptive fields that are in register with those of the lateral geniculate nucleus (see the lower continuous arrow in Figure 3; Sanderson, 1971), whereas the other maps are oriented perpendicular to this as shown in Figure 6.

The difference is not trivial. It is possible to entertain the notion that the perigeniculate nucleus may be developmentally related to the lateral geniculate nucleus and distinct from the reticular nucleus. This can be based on a developmental stage when the perigeniculate nucleus appears to be included with the lateral geniculate nucleus within the terminal field of retinal afferents, which later retreat from the region of the perigeniculate nucleus (Cucchiaro and Guillery, 1984). However, in the rest of this book we treat the perigeniculate nucleus as a part of the reticular nucleus in accord with the developmental account of Mitrofanis (1994) and interpret the layout of the map as indicative of the fact that where the corticogeniculate fibers reach the perigeniculate nucleus they have completed their crossings and are aligned in accord with their geniculate termination.

Although the reticular nucleus is relatively thin, there is now evidence for several species that primary visual, somatosensory, motor, and auditory cortex are each mapped in a distinctive sector of the reticular nucleus and that within each of these sectors small cortical areas are represented by slabs that are stacked next to each, as are A and B in Figure 6, from the inner toward the outer surface of the nucleus (see especially Crabtree and Killackey, 1989; Crabtree, 1992a, 1998; Conley et al., 1991). The afferents from the relevant thalamic nuclei also appear to show the same pattern of mapping. The mapped relationships that are established within the reticular nucleus will need to be understood before it is possible to determine how the several cortical and thalamic afferents that feed into any one reticular sector relate to each other and interact to produce the modulatory, inhibitory inputs of the reticulothalamic pathway.

Not all afferents to the reticular nucleus from cortical areas or thalamic nuclei show this relatively simple mapping. The anterior thalamic nuclei of the rat and cingulate cortex are mapped in a distinctive pattern (Lozsádi, 1995), and the mediodorsal thalamic nucleus and frontal cortex have reticular connections that are topographically mapped but that do not follow the slablike arrangement illustrated for the major sensory modalities in Figure 3 (Cornwall et al., 1990). The visual pulvinar and parts of the medial geniculate nucleus of the bush baby relating to cortical areas other than the first auditory area (Conley and Diamond, 1990; Conley et al., 1991) show no evidence of an orderly map within their sectors of the reticular nucleus. For some cortical areas a single injection of tracer produces not one but two or three slabs running parallel to the plane of the reticular nucleus within the same sector (Symonds and Kaas, 1978; Conley and Diamond, 1990; Cicirata et al., 1990). The significance of these multiple mappings is not clear and merits further study. The degree to which the several cortical representations within any one reticular sector are mapped appears to vary (see Guillery et al., 1998), and many of the details of corticoreticular and thalamoreticular pathways remain undefined at present. This is an area where details of connectivity are seriously in need of attention, although it seems likely that a detailed account may have to wait until descriptive research is once more recognized as worthy of financial support.

Three points must be stressed about the maps in the reticular nucleus. The first concerns the accuracy of the maps where they have been defined. Even though the slabs extend roughly parallel to the dendritic arbors of the reticular cells, these arbors occupy a fair proportion of the thickness of the reticular nucleus, and the reticular nucleus itself is thin relative to the degree of localization that would be needed if the reticular nucleus were to be able to pass well-localized information back to the thalamus. Although there clearly are maps in the reticular nucleus, the receptive fields of the reticular cells are larger than those of the thalamic or cortical cells (So and Shapley, 1981; Murphy et al., 1994). It is probable that reticular activity can focus on particular parts of the thalamic relay, but it is likely that dendrites of single reticular cells do extend across borders of experimentally defined slabs. That is, in accord with its relatively large receptive field a single reticular cell will be subject to afferents from a relatively large sensory area while nonetheless having clear focus within the reticular map. It may be important to recognize a distinction between a thalamic cell, which signals the presence or absence of an incoming driving stimulus, and a reticular cell, which signals a region within which modulatory activity is focused.

The second point concerns the surprising orientation of the reticular maps and the relationship of these to the complex crossing in the reticular nucleus. The pattern of crossing fibers in the region of the reticular nucleus was already recognized by Kölliker (1896). He called it the "Gitterkern" because he saw its appearance in an axon preparation as a lattice (Gitter) of complexly intercrossing axons. The comparison to a net in the current name has the same implication, but the multiple crossings produced by passing axons have often not been appreciated, and the idea that the axons in the thalamic radiation actually pass from thalamus to cortex in a direct, radiating pattern has been seriously entertained in spite of the obvious crossings in the reticular nucleus (Frost and Caviness 1980). If one looks at a preparation in which the axons are well stained, one can recognize that the axons approach the thalamus from the cortex by running roughly parallel to each other, and that then, some distance external to the outer or lateral border of the reticular nucleus, the corticothalamic and thalamocortical axons begin the formation of a complex latticework of intertwining axons. This is the site of the embryonic perireticular nucleus, only a few of whose cells survive in the adult. The latticework continues right through to the inner border of the reticular nucleus and then stops abruptly as the axons continue their course into the substance of the thalamus. As soon as the axons enter the thalamus, they run in strikingly straight, parallel bundles to their final thalamic destination or origin. In a good preparation these straight, parallel bundles running into the thalamus from the complex intertwining plexus of the perireticular and reticular nuclei look like a rainstorm descending from a cloud (see Mitrofanis and Guillery, 1993, their Figure 3). It appears as though the corticothalamic axons, having made the complex traverse of the reticular nucleus, are now directly on the right path and can proceed without further deviation. It would appear that the mapped connections to the reticular nucleus are formed as the several corticothalamic pathways that go through any one sector become lined up (one can think of them as stacked up in register) on their way through the nucleus.

The third point is that within any one sector of the reticular nucleus, and the sectors that we understand best are the ones concerned with the major sensory modalities, one sees a representation of several thalamic nuclei or nuclear subdivisions and also of several cortical areas concerned with the modality to which that sector is primarily dedicated. Where there are a multiplicity of cortical areas concerned with one modality, as for the visual pathways of primates, we do not yet know the details of how each cortical area relates to the reticular nucleus, but generally one finds that each corticothalamic pathway, whatever its origin, has a terminal zone in the thalamic reticular nucleus. That is, each major reticular sector receives afferents from thalamocortical and corticothalamic axons concerned with both the first and higher order circuits relating to the modality processed in that sector.

This arrangement of axons can now be related to the convergence, divergence, and reversal of maps between several cortical areas and thalamic nuclei, all concerned with one modality, that we noted earlier. It begins to look as though the representations of cortical areas and thalamic nuclei for any one modality are brought into relation to each other in one reticular sector as the relevant thalamocortical and corticothalamic axons relate to each other within the latticework of that sector of the reticular nucleus. Most of the details of how first and higher order circuits interrelate within any one sector of the nucleus are still to be defined, but the important point is that there is an opportunity for several first and higher order circuits to innervate cells in the same parts of the reticular nucleus and for these cells then to provide inhibitory inputs to thalamic relay cells in the related first and higher order circuits, some by axons that branch to innervate more than one of the related nuclei, some by axons that do not branch (see Chapter III). In the rabbit, the reticular representation of cortical area V1 lies in the outer two thirds of the visual sector of the reticular nucleus, and the representation of V2 lies in the inner two thirds. In terms of the visual field maps each is probably a mirror reversal of the other, and where the two meet, the junction itself represents the midline vertical meridian (Crabtree and Killackey, 1989). The somatosensory representations in the cat have shown a different pattern, with S1 and S2 representations overlapping entirely within the reticular nucleus (Crabtree, 1992a,b). To produce these patterns of alignment of cortical maps within the reticular nucleus there have to be significant rearrangements of neighborhood relationships among the corticofugal and thalamofugal pathways, and it is reasonable to conclude that the latticework represents the means by which these rearrangements are produced and that the axonal rearrangements within the reticular nucleus are the means whereby cortical maps are converted to thalamic maps and vice versa.

One key to understanding what may be happening in the reticular nucleus is to recognize not merely that many of the connections of the reticular nucleus have local sign (that is, they are mapped) but also that they provide a site where several different maps related to a single modality can be brought into relation with each other. That is, in any one sector of the reticular nucleus, several related higher order circuits are brought into close relationship to each other and to the related first order circuit; all innervate the reticular cells within that sector and the reticular cells serve as a "final common pathway" sending inhibitory afferents to the thalamus so that these inhibitory pathways can represent all of the thalamoreticular and corticoreticular circuits concerned with the relevant modality. There is a further point about the reticular nucleus that has not been addressed so far. Not only does it receive the thalamocortical and corticothalamic inputs discussed above, but it also receives ascending afferents from the brain stem, hypothalamus, and basal forebrain, which were described in Chapter III. So far as we know, most of these afferents lack local sign and probably act globally within any one reticular sector, and even, for some of the afferents, across all of the sectors. That is, the reticular nucleus can act with local sign, can relate several distinct maps to each other, or it can act globally to modify transmission through the thalamus as a whole.

G. *Summary*

The importance of mapped projections in the pathways to and from the thalamus has been explored. There is evidence that most, possibly all, driver afferents to the thalamus are mapped, although our knowledge of what is mapped is limited largely to relays concerned with sensory surfaces. Large parts of the thalamocortical pathways are known to have a topographically ordered projection but we lack information about the functional variable that is mapped. Some of the modulatory pathways are also mapped, whereas others show no topographic organization, and this suggests that some modulation is likely to be global, either for a particular modality or else for the whole thalamus, but that modulation limited to a small part of a projection can also occur.

Each major modality is represented by several maps. These show varying degrees of topographic order and are often mirror reversals of each other. The connections required between thalamus and cortex to link such mirror-reversed maps involve a great deal of fiber crossing, much of which occurs in the thalamic reticular nucleus, forming a complex latticework there.

The extent to which the brain can function with maps that are disrupted or abnormal is explored by consideration of mutants in which such disrupted maps are produced by a misrouting of pathways. These mutants and some simple experimental procedures have also served to illustrate the capacity of the thalamocortical projections to produce mirror reversals of the normal pattern of topographically ordered projections.

As the corticothalamic and thalamocortical axons pass through the complex axonal latticework that characterizes the thalamic reticular nucleus, they give off branches to innervate the cells of the reticular nucleus, influencing the modulatory effects of the reticular cells on the thalamocortical relay. For any one modality the pathways that link several cortical areas to the thalamus pass through this lattice in the same sector of the thalamic reticular nucleus, so that the reticular nucleus can serve as a nexus for interactions between the many thalamocortical and corticothalamic pathways concerned with that modality.

H. Some Unresolved Questions

1. Are all driver pathways to thalamus mapped?
2. Are all driver outputs from thalamus mapped?
3. For those pathways that are not concerned with visual, auditory, somatosensory, or motor pathways (e.g., pathways to frontal or cingulate cortex), what is the variable, if any, that is mapped? And where the topographic accuracy of sensory or motor maps in higher cortical areas is relatively crude or absent, is there another variable that is mapped?
4. Which modulatory pathways are mapped?
5. Is the alignment in the lateral geniculate nucleus of visual field representations that come through the left and the right eye related to the local, topographically organized action of modulators, or is there some other way of understanding the functional significance of Walls' toothpick?
6. Are the mirror reversals of thalamocortical maps and the consequent crossings of pathways that occur in the thalamic reticular nucleus an essential part of the functional organization of thalamocortical circuitry?
7. How do first order and higher order circuits relate to each other in the thalamic reticular nucleus (a) in terms of the maps that can be defined? (b) in terms of their synaptic connections to reticular cells?
8. (Closely related to question 7) Since first and higher order pathways for any one modality share a reticular sector, should one look for evidence that in the modulation of thalamocortical relays they generally act together, acting in unison, complementing (or perhaps opposing) each other?

CHAPTER VIII

Two Types of Thalamic Relay

A. The Basic Categorization of Relays

Earlier chapters introduced the distinction between first and higher order thalamic relays. One type receives its driving afferents from ascending pathways and transmits messages, which the cortex has not seen before, to cortex. This was called a first order relay (FO; hatched in Figure 1, which is a repeat of Figure 2 of Chapter I) because the thalamus here provides the first approach to cortical processing. The second type of relay, bringing driver messages to the thalamus from cortex for transmission from one cortical area to another was called a higher order relay (HO) because the thalamus is here processing messages that have already reached cortex and been processed in at least one cortical area (Guillery, 1995). Such a higher order relay, even though it involves thalamic cells, which are generally seen as representing a lower, subcortical stage of forebrain processing, must be regarded as part of higher cortical circuitry because it serves to reflect a part of the output of one cortical area onto another. That is, from the point of view of thalamocortical organization, this is a second (or third or more) run through the thalamocortical circuitry. These relays were not called second order relays because it is reasonable to expect that there will prove to be third and higher order loops going through thalamus and transmitting information from one higher cortical area to others. In this chapter, we will summarize the evidence that is currently available about these two functionally distinct types of relay, going over some of the material that has appeared briefly in earlier chapters and looking at how the recognition of these two types of thalamic relay relates to our view of the organization of the thalamic relay in general.

Nuclei that appear to be largely or entirely first order relays are shown in diagonal hatching in Figure 1. They include the ventral posterior

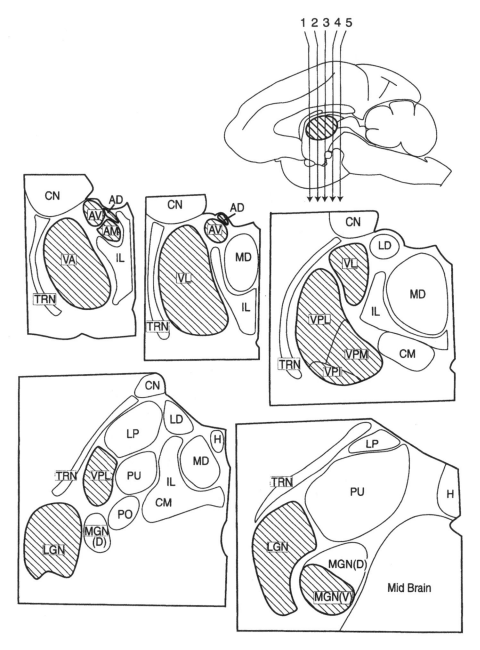

Figure 1
Schematic view of five sections through the thalamus of a monkey. The sections are numbered 1 through 5 and were cut in the coronal planes indicated by the arrows in the upper right midsagittal view of the monkey brain. The major thalamic nuclei in one hemisphere for a generalized primate are shown. The nuclei that are outlined by a heavier line and filled by diagonal hatching are described as first order nuclei (see text) and the major functional connections of these, in terms of their afferent (input) and efferent (output) pathways to cortex are indicated in Figure 3 in Chapter I. Abbreviations: AD, anterodorsal nucleus; AM,

A. The Basic Categorization of Relays

nucleus, the ventral part of the medial geniculate nucleus, the lateral geniculate nucleus, the anterior thalamic nuclei, and the ventral lateral and ventral anterior nuclei, receiving somatosensory, auditory, visual, mamillary, cerebellar, and pallidal afferents, respectively.

It is possible to argue that the last three afferent pathways are all subject to cortical influences that act on the mamillary bodies from the hippocampus, on the cerebellum by way of the pons, and on the pallidum by way of the striatum, and that even the first three can also be influenced by cortical pathways that descend to the gracile and cuneate nuclei, to the auditory relays, and to the superior colliculus (which in turn projects to some of the laminae of the lateral geniculate nucleus). We have stressed that we regard the thalamus as serving to relay information from a set of driver afferents to cortex. From this point of view there is an important difference between driving afferents that may have been influenced by prior corticofugal modifications and driving afferents that come from cortex itself, and it is this distinction that is important for identifying whether a particular relay is first or higher order; for a higher order relay the driver afferents must come from cortex itself. The GABAergic pathways from the globus pallidus and the substantia nigra to the thalamus are perhaps the most problematic of the relays considered above from the point of view of the proposed classification. There are two related issues. One is whether these afferents are drivers or modulators. This issue will be considered in Chapter IX, where it is argued that an inhibitory pathway is unlikely to be an efficient driver of a cell that communicates by generating action potentials. The second issue would only arise if the pallidal and nigral afferents were drivers. Then one would have to regard the pallidal relay to the thalamus as a part of a somewhat questionable first order relay because it can be regarded as bringing cortical outputs, that have come through the striatum, to the thalamus. Insofar as this striatal circuitry can be expected to introduce an entirely new, noncortical activity to the pathway, we would, however, hesitate to regard it as a typical higher order relay. Once we understand the nature of the messages that they bring to the thalamus, it may prove easier to categorize the pallidal and nigral afferents. It may prove that they represent an entirely distinct category of thalamic afferent, not fitting easily into any current category.

A second point that needs to be clarified concerns the identification, by hatching, in Figure 1, of certain *nuclei* as first order. For each of the nuclei so identified in the figure all of the driving afferents, or, in some of

---

anteromedial nucleus, AV, anteroventral nucleus; CM, center median nucleus; CN, caudate nucleus; H, habenular nucleus; IL, intralaminar (and midline) nuclei; LD, lateral dorsal nucleus; LGN, lateral geniculate nucleus; LP, lateral posterior nucleus; MGN, medial geniculate nucleus; PO, posterior nucleus; PU, pulvinar; TRN, thalamic reticular nucleus; VA, ventral anterior nucleus; VL, ventral lateral nucleus; VPI, VPL, VPM are the inferior, the lateral, and the medial parts of the ventral posterior nucleus or nuclear group.

these nuclei all but a very small proportion[1] of the driving afferents, are ascending afferents, so that it is appropriate to think of these as first order relay *nuclei*. However, the identification of higher order nuclei is more problematic because for most we cannot exclude the possibility that some first order afferents may also have a relay in them. There may well be several of such "mixed" nuclei; possibly, when all the evidence is in, it will turn out that there are no pure higher order nuclei at all, that all contain a mixture of first and higher order relays, and for this reason it is clearer to speak of first and higher order *relays* or *circuits* rather than to identify *nuclei* as being necessarily entirely one or the other. For example, the evidence, outlined below, that the pulvinar and lateralis posterior nuclei receive driving afferents from visual cortex indicates that there are higher order relays in these nuclei. However, there is also a tectal input to the pulvinar and the lateralis posterior nucleus and the interpretation of these axons as driver or modulator afferents is uncertain. Light and electron microscopic accounts differ as to their appearance (Ling *et al.*, 1997; Robson and Hall, 1977; Mathers, 1971), depending on species and probably also on the particular subdivision of the pulvinar or lateralis posterior nucleus that is under study; there are several distinct functionally organized zones in the pulvinar and lateralis posterior nuclei (e.g., Abramson and Chalupa, 1988; Luppino *et al.*, 1988). Thus, receptive field properties of some cells in the rabbit's lateralis posterior nucleus appear to be dependent on a collicular input (Casanova and Molotchnikoff, 1990), whereas in the pulvinar of the rabbit the effect of tectal inactivation produces an augmentation or a diminution of responses, not their abolition (Molotchinoff *et al.*, 1988). Bender (1988) reports that in the primate brain: "the colliculus contributes rather little to the neuronal response properties in the pulvinar, in contrast to what one would expect if the pulvinar served a major role as a 'relay nucleus'." That is, some tectal afferents appear to establish first order relays through the pulvinar or lateralis posterior nucleus, so probably making some parts of these nuclei mixed first and higher order relays, whereas other tectal afferents may serve a modulatory role comparable to the cortical afferents that come from cortical layer 6, and we would be inclined to regard the regions innervated by these as a pure higher order relays until some other, noncortical, candidate for a driving input came to light.

Comparably, in the mediodorsal nucleus, which receives RL terminals from prefrontal cortex (Schwartz *et al.*, 1991) and thus should be regarded as having higher order relays, there are some first order relays as well. Afferents come from the amygdala (Aggleton and Mishkin, 1984; Price, 1986; Groenewegen *et al.*, 1990), the piriform cortex (Kuroda *et al.*, 1992), and from several brain stem structures (Russchen *et al.*, 1987; Kuroda and Price, 1991). Although many of these afferents have been described by methods that do not allow a distinction between drivers and

---

[1] Evidence that a few driver afferents may be coming from the cortex to the ventral anterior and to the ventral posterior nuclei was considered in Chapter III (Section C.3).

A. The Basic Categorization of Relays

modulators, some are known to form RL or type 2 terminals in the mediodorsal nucleus (Kuroda and Price, 1991; Kuroda *et al.*, 1992) and these are likely to be drivers. It remains to be determined precisely how the higher order relay coming from the frontal cortex relates to the first order relays receiving from these ascending afferents. They may involve different subdivisions of the mediodorsal nucleus or overlap in some parts of the nucleus.

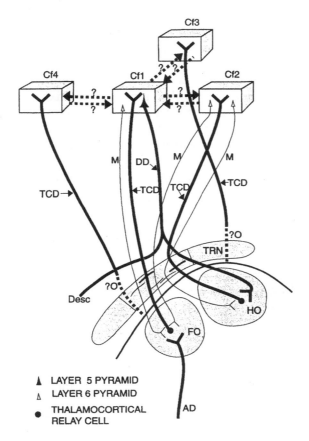

Figure 2
Schema to show the thalamocortical, corticothalamic, and corticocortical interconnections related to first order (FO) and higher order (HO) thalamocortical circuits. The inhibitory connections that go from the thalamic reticular nucleus (TRN) and from interneurons to the thalamocortical relay cells are not shown. Cf1, Cf2, Cf3, and Cf4 show interconnected cortical areas. TCD indicates thalamocortical drivers, M shows corticothalamic modulators, DD shows a descending driver; the ?s relating to the interrupted lines that link cortical areas indicate that we have no evidence regarding their driver or modulator functions. ?O shows thalamocortical axons innervating cortical areas Cf3 and Cf4, the precise thalamic origin of which is undefined for this schema, and often unknown for the real brain. AD, Ascending driver. More details are in the text.

We have mentioned that the magnocellular part of the medial geniculate nucleus is a strong candidate for a mixed first and higher order relay, with type 2 axons that have large RL terminals coming from cortex, but also with a direct afferent component coming in the brachium of the inferior colliculus (Bartlett et al., 2000). The intralaminar nuclei are also very likely to represent a mixture of first and higher order relays. We saw in Chapter III (Section E.3) that injections of horseradish peroxidase into the centre median nucleus show that it receives afferents from cortical layer 5, suggesting that there might be some corticothalamic terminals in this nucleus having the character of RL terminals. The possibility that there are higher order circuits in the intralaminar nuclei, which also receive ascending afferents (Jones, 1985) that may well prove to be driving afferents, suggests that there may be a mixture of first and higher order circuits in these nuclei as well.

On the basis of current evidence it thus seems highly probable that there will be several cell groups that represent a mixture of first and higher order circuits. Where such mixed inputs are identified, an important next question will be whether first and higher order relays ever share a thalamic relay cell or whether within one nucleus they are kept as separate parallel pathways going through the nucleus independently like the X and Y pathways in the A-laminae of the cat (see Chapter III).

B. *The Evidence That There Are Two Distinct Types of Afferent That Go from Neocortex to Some Thalamic Nuclei*

One of the major approaches we have used in this book is to treat the thalamus, including all of its nuclei, as a structure that has a common developmental history and that, therefore, morphological features and functional relationships that can be seen in one nucleus will reappear in others. The argument that there are cell groups in the thalamus that receive their major driver afferents from the cerebral cortex has three parts. The evidence for each part has been considered in Chapter III, but it will be summarized here to focus specifically on the distinction between first and higher order circuits.

The first part of the argument is that well-established driver afferents have been identified in the major first order relay nuclei for visual, auditory, and somatosensory afferents, and that these driver afferents have a characteristic light and electron microscopic appearance (type II axons with RL terminals, see Chapter III) as well as a common basic pattern of synaptic organization, and generally or perhaps invariably, no branches to the thalamic reticular nucleus. We regard axons having such a characteristic light and electron microscopic appearance as candidate driver afferents even if they come from pathways whose driver function is not yet established experimentally. It is important to stress that this is a hypothesis, but a testable one about the function of afferent pathways to the thalamus, and that currently there is no clear counter evidence to show that any type II axons with RL terminals have only a modulatory and no

B. The Evidence That There Are Two Distinct Types of Afferent

driver function in any thalamic nucleus. The type II afferents having RL terminals that come from the cerebral cortex are, therefore, reasonably regarded as drivers, determining the nature of the message that is to be passed to another cortical area from the thalamic nucleus in which they terminate.

The second part of the argument is that there are two distinct types of corticothalamic afferent identifiable on the basis of their structure. One, coming from cortical layer 6 appears to provide afferents to all thalamic nuclei, including those that have well-defined ascending, sensory driver afferents. On the basis of their terminal structure these are type I axons with RS terminals and are quite unlike any known driver afferents. The second type of cortical afferent comes from output pyramidal cells in cortical layer 5, has type II axons with RL terminals, and goes primarily to nuclei that have few, poorly defined, or no ascending driver afferents recognizable as type II axons with RL terminals.

The third part of the argument is that in the few situations where it has been possible to test the effects of silencing the candidate corticothalamic driver afferents it has been shown that the basic receptive field properties of the relevant nucleus depend on the cortical input that comes from cortical layer 5, but does not depend on the layer 6 input. We look at the structural evidence supporting the first two parts of the argument first, and then consider the functional evidence.

1. THE STRUCTURE AND LAMINAR ORIGIN OF THE CORTICOTHALAMIC AXONS

Evidence that corticothalamic afferents can have the same fine structural appearance as afferents coming from the retina or the medial lemniscus has been available for a long time. Mathers (1972) made lesions in the visual cortex of squirrel monkeys and showed that two distinct types of axon terminal in the pulvinar degenerate after such a lesion. One resembled the corticothalamic terminals that had previously been described in the main visual auditory and somatosensory nuclei (RS terminals; see Chapter III), and the other resembled the ascending driver afferents to these nuclei (the RL terminals). This observation was confirmed for the pulvinar of squirrels and primates by Robson and Hall (1977), Ogren and Hendrickson (1979), and Feig and Harting (1998); for the rat lateral posterior nucleus by Vidnyánszky et al. (1996a); and for the mediodorsal nucleus in the macaque monkey by Schwartz et al. (1991).

The light microscopic evidence came much more recently than the fine structural evidence. This may seem odd but has a simple explanation, since early methods of tracing axons such as axonal degeneration or axonal labeling with radioactive materials or horseradish peroxidase did not allow identification of the morphological characteristics of individual terminals, whereas electron microscopy did. Observations made in several species (rat, mouse, cat, monkey) on small groups of axons or single axons that had been filled with an anterograde marker such as biocytin or phaseolus lectin (see Bolam, 1992, for these and related methods) showed the structure of the axons going from somatosensory cortex to the posterior nucleus (PO) (Hoogland et al., 1991; Bourassa et al., 1995), from the auditory cortex to the magnocellular part of the medial geniculate nucleus (Ojima, 1994; Bartlett et al., 2000), and from the visual cortex to the pulvinar or lateralis posterior nucleus (Bourassa and Deschênes, 1995; Rockland, 1996, 1998). It was found, again, that there are two characteristic types of axon coming from the cortex, one resembling the type I axons and the other resembling the type II axons described in Chapter III for the major sensory relay nuclei, that is, the corticothalamic and the ascending afferents, respectively. It was further shown on the basis of single cell injections that the former come from layer 6 cells in the cortex, whereas the latter come from layer 5 cells. In addition, in some instances these laminar origins were in accord with earlier experiments that had shown the distribution of retrogradely labeled cortical cells in layers 5 or 6 after injections of horseradish peroxidase into some of the relevant thalamic nuclei (Gilbert and Kelly, 1975; Abramson and Chalupa, 1985).

These experiments, involving injections of one or a few cells, also showed two other potentially important features of the axons that were coming from the cortex. One is that layer 6 cells commonly send branches to the thalamic reticular nucleus, whereas layer 5 cells do so rarely or not at all (see Chapter III), and the other is that the axons from layer 5 cells, but not from the layer 6 cells, can sometimes be seen to be branches of long descending axons that go to more caudal parts of the brain stem (Deschênes et al., 1994). The pattern of some corticothalamic axons as branches of corticotectal axons can also be demonstrated by recording experiments (Casanova, 1993); it may be that there are some layer 5 corticothalamic axons that have no descending branches and that other descending axons have no thalamic branches, but resolving these two issues depends on the interpretation of negative results from difficult experiments, so that both questions must be seen as open. However, we can conclude that, at least some, possibly all, of the type II corticothalamic axons represent branches of cortical output axons. Possibly, there is a more general statement that can be made about driver afferents to thalamus: no matter whether they come from cortex or from lower centers, they are likely to send branches to brain stem, as opposed to the modulator pathways, which, as we have seen, send branches to the thalamic reticular nucleus, but not to centers below the thalamus.

One line of evidence that is currently available only for the ventral posterior nucleus (Hoogland et al., 1991) and the medial geniculate nucleus (Bartlett et al., 2000) is the demonstration that in the higher order nuclei the type I and the type II axons identified light microscopically correspond in fine structural terms to the RS and the RL terminals, respectively. We can feel reasonably secure about the correspondence in first order nuclei like the lateral geniculate nucleus on the basis of the evidence presented in Chapter III, but for the higher order circuits the evidence is generally less secure and indirect. On the basis of the terminal sizes this is a reasonable expectation for most parts of the thalamus because the small drumstick-like side branches of the type I terminals could hardly form characteristic RL terminals, anymore than the large, complex type II terminals coming from layer 5 that have been illustrated and described in the studies cited earlier, could appear as RS terminals in electron microscopic sections. However, the sizes of the terminals of type I and type II cannot be regarded as diagnostic. Although they show no overlap in the A-laminae of the cat's lateral geniculate nucleus (Van Horn et al., 2000) they have not been shown to form distinct populations throughout the thalamus, and it would be reassuring to have a great deal more electron microscopic evidence available about corticothalamic axons that have the light microscopic appearance of type II or type I axons and that can be shown to arise from either cortical layer 5 or 6, respectively.

2. The Functional Evidence for Two Distinct Types of Corticothalamic Afferent

We have proposed, on the basis of evidence from the retino-geniculo-cortical pathway, that the thalamic relay serves as a modulatory gate, not an integrator. That is, for any one relay cell the driver afferents are few and generally represent a single function; the function of the relay cell is to pass the driver message to cortex. The message may be passed to cortex in tonic or in burst mode, or it may not be passed to cortex at all in certain sleep states. There may be slight modifications of the message in the relay, but essentially our proposal is that the thalamus will not produce a significant change in the message.

The issue of identifying the drivers in any thalamic nucleus and distinguishing them from modulators is considered in the next chapter, where the problem of making this distinction is also raised in general for any part of the nervous system. We have argued that, in the thalamus, drivers can be identified by certain morphological and functional features (Sherman and Guillery, 1998). Where information about receptive field properties is available, one can use the transmission of receptive field properties for identifying drivers. However, before we look at this functional argument and explore its use for distinguishing corticothalamic drivers from modulators, it is necessary to recognize two important provisos. One is that for many pathways other than the retinogeniculate pathway, manipulation of cortex can act either directly on the thalamic relay or it may act on the driver afferents that innervate this relay. For example, there are pathways that go from somatosensory cortex to the gracile and cuneate nuclei and from auditory cortex to the inferior colliculus, but there is no cortical innervation of the retina. Whereas modifications of activity in visual cortex can have no direct action on the retinogeniculate pathway, the same is not true for the auditory and somatosensory pathways. The second proviso arises when we extend our argument to nuclei that are not obviously on the route to one of the primary sensory cortical areas. Here, defining the critical properties that characterize the nature of the driver input becomes important but, as yet, largely unexplored. We lack information about the nature of the message that is delivered by the putative driver either because many of the pathways that pass through the thalamus are not concerned with sensory messages, and the concept of a "receptive field" is not applicable or because, for most higher order circuits, the nature of the receptive fields has not been defined, even though the circuit is concerned with a particular sensory modality. The distinction between the "nature" of the driver input defined above, which refers to the message that it carries, and the action that the driver has on its postsynaptic cell, which is the contribution that it makes to the discharge of the postsynaptic cell, and which is discussed in the next chapter on drivers and modulators, is of crucial importance. Here we have to recognize that currently for all but a few of the thalamic relays our primary clue for identifying an afferent as a driver is knowledge about the message that it carries and a demonstration that this message is passed on through the relay.

a. Distinguishing Corticothalamic Drivers from Modulators in Sensory Pathways

Studies of receptive field properties have provided some useful evidence about the action of corticothalamic axons for first and higher order relays of visual and somatosensory pathways. In the visual pathways, it has been shown that relay cells in the lateral geniculate nucleus, which receive corticothalamic type I (RS) axons but no corticothalamic type II (RL) axons, have receptive fields that survive cortical lesions or cooling in cat or monkey (see Kalil and Chase, 1970; Schmielau and Singer, 1977; Baker and Malpeli, 1977; Geisert *et al.*, 1981), whereas receptive field properties in the primate higher order pulvinar, which receives corticothalamic type II axons, are lost after lesions of visual cortex (area 17; Bender, 1983). Comparably, for the somatosensory pathways of rats, the experiments of Diamond *et al.* (1992) have shown that the receptive field properties of cells in the ventral posterior nucleus (first order) survive inactivation of the somatosensory cortex, whereas the receptive field properties of cells in the posterior nucleus (POm, higher order) are lost after such cortical inactivation. It should be noted that here we are not discussing subtle changes in receptive field sizes or in the temporal pattern of the thalamic responses (see, for example, Krupa *et al.*, 1999), but a much more dramatic loss of the receptive field. That is, these experiments show that the type I axons having RS endings and arising from layer 6, which are the only or the major cortical afferents tested in the first order nuclei, can serve to modify the receptive field properties, but that they differ fundamentally from the primary afferents to first order nuclei and the type II axons from layer 5 to higher order nuclei, which on the basis of these experiments should be regarded as the drivers, bringing the impulse traffic that defines the receptive field of the relay.

There may be an important difference in the effects, for different first order relays, of cortical lesions, stimulation, or silencing. In all nuclei, the elimination of the modulatory input from layer 6 would have rather similar but subtle effects. In a relay like the lateral geniculate nucleus this is the only effect, but in certain relays, such as the medial geniculate and ventral posterior nuclei, the ascending driver inputs themselves may be affected by elimination of corticofugal inputs to lower level relays. For example, because the inferior colliculus and the dorsal column nuclei both receive inputs from cortex, cortex can be expected to act over these pathways on the ascending driver pathways, so that silencing cortex could produce a change in the thalamus that is not due to the interruption of any direct corticothalamic input.

b. Defining the Functional Nature of Driver Afferents in First Order Nuclei

Identifying the sorts of stimuli that are likely to drive thalamic relay cells receiving first order visual, auditory, or somatosensory afferents may seem relatively straightforward, but even that often proves to be a challenge of finding the right stimulus variable for any one particular cell type. For first order nuclei that are not in receipt of a major sensory pathway, clearly the problem is more difficult, and it becomes increasingly difficult as we move to higher order nuclei. Thus, although there have been careful studies of the discharge properties of nerve cells in the deep cerebellar nuclei and in the globus pallidus (Thach et al., 1992; Middleton and Strick, 1997; Mushiake and Strick, 1995), it is not yet understood precisely how the activity in a particular cerebellar or pallidal output neuron relates to its thalamic relay. Mushiake and Strick (1995) writing about the pallido-thalamo-cortical circuit have said: "What each stage in this circuit contributes . . . to motor behavior remains to be determined," and this question can be repeated for the contribution made by most thalamic cells, other than those on a sensory pathway, to the input of the cortical areas in which their axons terminate. For example, recent observations of the anterior thalamic relay have shown that head orientation in space, "head direction," is a signal that is passed through the lateral mamillary nucleus and the anterodorsal thalamic nucleus (Taube, 1995; Stackman and Taube, 1998). Since the lateral mamillary nucleus receives afferents from the fornix and from the midbrain and sends bilateral efferents to the anterodorsal nucleus, this can be seen as a thalamic relay that transmits a definable message relating to the animal's spatial orientation on to the limbic cortex. However, exactly how this message is transmitted through the thalamus, or what the other two, larger, anterior thalamic nuclei may be doing, remains unknown. To say for any one thalamic nucleus that the afferents come from the mamillary nucleus, the cerebellum, the pallidum, tectum or a particular area of cortex is like identifying a ship by its port of origin, not by its cargo, which is often more important for those waiting to welcome, unload, or consume the cargo. We need to know what aspect of the relevant prethalamic functional organization the messages represent so that we can understand what it is that is being passed on to the cortex. This is a serious and difficult problem for some of the first order nuclei in the thalamus and is largely unresolved for higher order circuits.

3. WHAT IS THE FUNCTIONAL SIGNIFICANCE OF A HIGHER ORDER THALAMIC INPUT?

Defining the functional role of a higher order thalamic relay will be difficult. We suggest that all higher order relays serve to pass messages from one area of cortex to another through the constraints imposed by the thalamic "gate" and that these constraints are basically similar for all thalamic nuclei. That is, just as sensory information passes through first order nuclei *en route* to cortex, and there is no direct pathway to cortex, so perhaps most, possibly all, corticocortical messages must pass through a higher order thalamic relay, with the direct corticocortical pathways performing some other, perhaps modulatory, function. The advantage of information reaching a cortical area by first passing through thalamus is not entirely clear, but lies at the heart of understanding the thalamus. It seems probable that the thalamic gating functions relating to burst and tonic firing properties of the thalamic relay neurons, discussed in Chapter VI, may be equally important for information transmitted by first order and higher order relays. We do not expect a thalamic relay, whether first or higher order, to modify the content of the message or to act as an integrator of two or more messages, but this is an expectation that, although it is in accord with what we know about some first order relays, has not been tested for any higher order relay.

We can consider the lateralis posterior and pulvinar nuclei as examples of higher order relays for which we know a fair amount about the connections of the relay cells and their functional properties and ask what sorts of information may be needed for us to understand the functional role played by these higher order relays. These nuclei send thalamocortical efferents to several different cortical visual areas (Niimi *et al.*, 1985; Lysakowski *et al.*, 1988; Dick *et al.*, 1991; Rockland *et al.*, 1999) and also receive cortical afferents from several cortical visual areas (Updyke, 1977; Wall *et al.*, 1982; Yeterian and Pandya, 1997). In order to understand how these relays function in sending messages to cortex it will be necessary first to define which of the several groups of corticothalamic afferents represent type II axons with RL terminals, and whether all of these take origin in cortical layer 5; then to define the functional characteristics of the cortical layer 5 cells that send axons to the pulvinar. Our expectation is that all corticothalamic type II axons projecting to these higher order nuclei will have RL terminals and originate in layer 5. Further, we expect that there may well be some cortical areas that send only type I axons from layer 6 to some subdivisions, comparable to the relationship between area 17 and the lateral geniculate nucleus, and that there will be other cortical areas that send only type II axons from layer 5, comparable to the relationship of area 17 to the lateral posterior nucleus. However, most of the details of such connections for most of the higher order nuclei are not yet established, and defining them represents an important challenge.

A further, and perhaps more important, point is that we expect to find that the messages that the relay cells in the pulvinar and lateralis

B. The Evidence That There Are Two Distinct Types of Afferent

posterior nucleus send on to peristriate cortical areas will depend *entirely or to a significant extent* upon the nature of the activity received from the layer 5 type II cortical afferents that reach the nucleus. Not many studies define the message passed from cortex to the thalamus by these cells. Casanova (1993) recorded from nerve cells in area 17 that project to the lateralis posterior nucleus in the cat. These, on the basis of Abramson and Chalupa's (1985) earlier study of cat corticothalamic pathways, are probably layer 5 cells and were shown by Feig and Harting (1998) to form RL axon terminals in the pulvinar. Casanova tested these cells with drifting sine wave gratings and showed them to be like cortical complex cells: orientation and directionally selective, with a tendency to favor horizontal or vertical gratings, mostly binocularly driven, with some monocular cells innervated from the contralateral eye. The response characteristics were similar to those of corticotectal cells, and 6 of 40 cells that were studied were demonstrated to have tectal branches. These observations provide an insight into the type of information that is needed to understand the messages that higher order nuclei receive from their layer 5 innervation. If these are driver inputs to parts of pulvinar, we predict that some pulvinar cells should have receptive fields similar to these complex cells, just as geniculate relay cells have receptive fields much like their retinal drivers.

The next step will be to define the processing that occurs in the cortical areas receiving from these higher order relay cells and to determine the extent to which the features that have been defined for the cortical afferents to these thalamic relay cells are the ones that are relevant to this processing. As pathways are followed further from the first order nuclei, we can expect that defining the crucial properties (receptive field or other) will become increasingly more difficult and will depend more on an intuition about what the higher cortical area may be doing than on a clear view of how the first order receiving cortex processes its thalamic afferent messages. The choice of response properties that are to be tested must depend on choices that the experimenter makes as to what are the relevant stimuli to be examined. The assumption that these are the particular properties that fit the specific functional role of the relay is one that is not always easy to justify in a sensory system like the visual or somatosensory pathways; it is even more difficult to establish for higher order relays like the mediodorsal or anterior thalamic nuclei where we have very few clues as to the particular properties that are relevant to the functional role of the relay. Another potential problem is that many receptive field studies have been performed in anesthetized animals, and anesthetics seem to have relatively little effect on first order relays and their drivers, such as the lateral geniculate nucleus and retina. Higher order relays and their cortical drivers may be significantly suppressed by anesthetics, so that such studies may have to be carried out in unanesthetized, behaving animals.

If we now compare first and higher relays, we have to ask whether our proposal that the two are functionally similar is, indeed, a viable proposal. Are the thalamic relay cells in each type of nucleus simply passing on messages whose characteristics are already established in the ascending afferents or in the layer 5 cells of one of the connected cortical areas? Or are the higher order nuclei generating new messages either on the basis of their intrinsic, intrathalamic connections or by integrating two or more afferent (driver) pathways? After that, the really interesting, and currently unanswered, question is about the action of these thalamocortical axons on the cortical areas that they innervate. Are they the effective drivers for the cortical cells that they innervate? Is the laminar distribution of their cortical terminals relevant to determining whether their cortical action is that of driver or modulator? Do they, like their counterparts from the first order nuclei, dominate the nature of the message that the relevant cortical area processes (see Figure 2)? Or is there a fundamental difference between cortical cells that receive from first order relays and those that receive from higher order relays? Are the former heavily dominated by their thalamic afferents as would appear from currently available evidence (e.g., for striate cortex, Hubel and Wiesel, 1977; Reid and Alonso, 1995; Ferster *et al.*, 1996), and the latter dominated by corticocortical pathways, as would appear from most current speculations about corticocortical connections (e.g., Zeki and Shipp, 1988; Van Essen *et al.*, 1992; Salin and Bullier, 1995)? It is possible, although it seems unlikely to us, that higher cortical areas, which all receive thalamic afferents, depend less or not at all upon these thalamic afferents and instead are driven by corticocortical connections. However, it would be more attractive and more in accord with a view of neocortex as having a basic structural pattern for all cortical areas, if it could be shown that all areas of cortex resemble the primary visual, auditory, and somatosensory cortical areas in receiving their primary driver input from the thalamus. That is, it would seem reasonable to consider that all of neocortex receives primary driving afferents from the thalamus, and that the corticocortical connections serve modulatory but not driving functions. If this could be demonstrated, then the search for the origin of cortical receptive field properties would be dramatically redirected. At present the question is an open one, waiting for critical experimental evidence to define which axonal pathways are drivers and which are modulators.

This possibility runs directly counter to the prevailing assumption that corticocortical pathways convey the main message for corticocortical organization. The idea here is that information initially arrives at cortex after being relayed through first order relays, and then once in cortex, it remains strictly at the cortical level, being analyzed and communicated among cortical areas only by way of corticocortical pathways, until some instruction is ready to be sent to lower centers. For vision, for example, information first gets to cortex by way of the geniculostriate pathway and is then analyzed strictly within cortex among the various areas and sent out to centers concerned with producing the necessary responses. The same argument has been made for auditory and somatosensory processing. This view provides no real function for what we call the higher order relays that represent such a large volume of the thalamus in primates. In contrast, our very different hypothesis, that corticocortical information transfer is relayed through these higher order thalamic nuclei invests these relays with a critical function. Not only does it place the thalamus in a key role in corticocortical communication, but it also links that key role to the outputs that the relevant cortical areas are sending to the brain stem and spinal cord. At present we lack the critical data to choose between these competing views because we know of no experiments that define either the thalamic or the cortical afferents as the drivers for any higher cortical area; however, we do know that the drivers come from thalamus for primary cortical receiving areas and have no good reason for thinking that cortical areas differ so dramatically from each other that the component acting as a driver in one cortical area is not likely to do the same for another.

The important question raised at the beginning of this section, as to what the functional significance of a higher order relay may be, has not been answered in any of the points raised so far. Perhaps the closest we can yet come to a useful answer is that having a thalamic relay in corticocortical pathways imposes the complex "gating" functions discussed in earlier chapters upon these corticocortical pathways. Of the several possible functions for a thalamic gate, the one we understand best is the one that involves the switch from the burst to the tonic mode. We have argued that either a particularly salient afferent stimulus, or a generalized wakeup call coming from the brain stem, can serve to switch a relay from burst to tonic mode. It should now be clear that this may happen for messages passing through the thalamus from one cortical area to another, just as it can for afferent messages on their way to the cortex. We have to think of the brain stem sending its general wakeup calls not just to the afferent pathways so that the perceptual mechanisms will be transmitting an accurate, linear copy of the afferent activity to cortical receiving areas, but also involving afferents to all cortical areas so that these will all be transmitting efficiently. That is, when it becomes important for a higher cortical area to receive an accurate, linear representation of what other cortical areas are sending out along the axons of their layer 5 pyramidal cells, then one must expect that the thalamic relays to these higher cortical areas will also be in tonic mode. That is, when we are in a fully alert mode we expect all of our cerebral processes to be performing well, although we may focus on one or another aspect as circumstances demand. The possibility that one or another corticothalamic or corticoreticular pathway can serve to generate a *local* wakeup call, will apply to higher order circuits just as it does to first order circuits. These may be circuits concerned primarily with one modality, but they need not be.

When we are trying to find the best words with which to express difficult thoughts, our sensory inputs coming through first order relays are barely getting through to cortical levels, but there are active cerebral processes going on, and it would be a reasonable bet to say that to a significant extent these must involve higher order circuits going through the thalamus and linking cortical areas to each other. Then, when the writing is done and we can relax with some music or conversation, the tonic mode in the first order circuits becomes more important for what we are doing and the corticothalamic and reticulothalamic afferents will serve to ensure that a new set of thalamocortical pathways is functioning in the tonic mode. Also, of course, there is the important possibility of interactions between the first and higher order circuits, which can occur, perhaps through type I axons going from higher order cortical areas to first order thalamic relays or through connections in the thalamic reticular nucleus because it is in this nucleus that one finds the closest relationships between first and higher order circuits for any one modality. Finally, if we next go to sleep, then, if messages from one cortical area to another make an obligatory pass through thalamus, the rhythmic bursting seen throughout thalamus during slow-wave sleep would largely disrupt corticocortical communication, and this would be an important aspect of such sleep. It is interesting that dreaming may require corticocortical integration, and dreaming is not seen in slow-wave sleep but is seen in REM sleep during which rhythmic bursting is prevented in thalamus by strong activation of brain stem afferents (Steriade and McCarley, 1990).

The preceding account could be taken to imply that higher order circuits and the links between first and higher order circuits are significant for abstract thought but that a simple relay through thalamus to cortex is all that is required for simpler processes. There are good reasons for thinking that thalamic pathways to neocortex in all mammals, even in those with relatively simple brains and rather limited amounts of neocortex, involve first and higher order circuits. This is because all mammals that have been studied from this point of view show a reduplication of cortical areas (Kaas 1995; Krubitzer, 1998). They have two or more visual, auditory, or somatosensory areas, not just one. It has been shown that even a simple brain like that of the hedgehog has two somatosensory and two visual areas. It looks as though one of the significant advantages that the evolution of thalamocortical pathways offered our ancestors was the opportunity for having a higher order thalamocortical circuit that functioned on the basis of receiving driving afferents that themselves were already the output stage of cortical processing. That is, right from the start, the thalamus could serve as the organ of reflection of cortical activity back upon itself. The reiteration of thalamocortical circuitry may well prove to be one of the special advantages that the evolution of neocortex provided for mammals. This says nothing about the function of multiple cortical areas but addresses the importance of the way in which they are interconnected, allowing one area to depend for a significant part of its input on the output of another area. One tends to think of the neocortex sitting on top of the brain and doing all of the really difficult things that need to be done. However, it is worth bearing in mind that thalamus and neocortex have evolved in parallel, and that neither would amount to much without the other. If thalamic circuitry had developed as the deepest cortical layer, we would call it layer 7 and treat it as an integral part of cortex, and it could possibly have developed to do everything that the thalamus and reticular nucleus now do. We don't know why the diencephalic and telencephalic parts of the circuitry evolved as separate entities, but whatever the reasons, it is likely to be useful to treat the links between the two, particularly the reiterative links formed by higher order circuits, as an essential part of the neocortical mechanisms.

C. The Relationship of First and Higher Order Relays to the Thalamic Reticular Nucleus

There are several sectors in the thalamic reticular nucleus and each of these sectors relates to one modality or to one functionally related group of thalamic nuclei (see Chapter VII). Corticothalamic and thalamocortical axons send branches to the sector of the reticular nucleus related to their functional grouping, and this applies both to first and also to higher order relays (Crabtree and Killackey, 1989; Conley and Diamond 1990; Cicarata et al., 1990; Cornwall et al., 1990; Conley et al., 1991; Crabtree, 1992a,b; Lozsadi, 1995; Coleman and Mitrofanis, 1996). Further, the cells of the thalamic reticular nucleus send their axons back to the thalamus, relaying on first and higher order relay cells, and in some instances actually sending two branches of a single axon back to the first and the related higher order nucleus (Crabtree, 1992a; Pinault et al., 1995b). Exactly how first and higher order circuits relate to each other within a single sector of the thalamic reticular nucleus is still poorly defined. The few studies that are available show different relationships for different sectors and show some differences between species (summarized in Guillery et al., 1998). For example, the studies cited above show that in the visual pathways of rat, rabbit, and bush baby the first order pathways connect to the outer third of the reticular sector, whereas the higher order pathways connect to the inner third. The same pattern has been described for the somatosensory pathway of the rat, but in the cat the first and higher order pathways of the somatosensory pathways that have been studied both connect to the full thickness of the relevant reticular sector. In the auditory sector of the bush baby the relationships are more complex (Conley et al., 1991).

Whatever the precise relationships between first and higher order circuits within a sector of the reticular nucleus, there are likely to be significant interactions. Where first and higher order circuits are confined to the outer and inner parts of the nucleus, respectively, it is likely that the dendrites of reticular cells do not stay exactly within the boundaries defined by these connections. Further, there is evidence for local connections within the reticular nucleus by dendrodendritic junctions and local axonal branches that were considered in Chapter III, and we have seen that some of the reticulothalamic axons are distributed to first and higher order thalamic nuclei by single axons that branch. The reticular nucleus can be seen as an important site for several possibilities for interactions between first and higher order circuits, so that a locally salient aspect of ascending afferent or of descending cortical activity can influence all of the circuits related to the same modality or thalamic grouping. That is, the thalamic reticular nucleus can act as a nexus for the interactions of several thalamocorticothalamic circuits, and the functional significance of such a nexus may be most important for the interactions among first and higher order circuits that belong to the same modality grouping. Certainly it is here that one sees the connectivity patterns most suggestive of functional interactions.

If we imagine an afferent stimulus (say visual) that is passed to cortex through a bursting geniculate relay and that produces no change to the tonic mode, perhaps because the stimulus is not salient, or threatening, or else because the burst mode is too well entrenched, then it may be that it is only after reaching one or another higher cortical area that the stimulus acquires the salience that can produce the change from burst to tonic in the relevant thalamic relay, or in the relevant part of the relay. The extent to which each of the several visual cortical areas has type 1 axons with RS terminals that go back to the several relevant thalamic first and higher order relays has not been defined in detail for any species, but we know that there is considerable overlap of these corticothalamic pathways, that many of them have a topographic map, and that each has a synapse in the reticular nucleus, which also shows a rough mapping. The possibilities that such a system creates for any one pathway to activate the thalamic gate for all related pathways is intriguing, and the role of the shared reticular relay is likely to provide an important key to understanding how the thalamic gate functions for first and higher order relays.

We have pointed out that having duplicate or multiple cortical areas for any one modality is a very general mammalian feature, and that it is the duplication of cortical areas, especially when, as is usually the case, two cortical areas carry mirror-reversed maps, that produces the complex latticework characteristic of the thalamic reticular nucleus. This latticework is seen not only in mammalian brains but also in the chick and in the turtle (Adams *et al.*, 1997). We know very little about the functional organization of the nonmammalian thalamic reticular nucleus, but the characteristic "reticular" structure suggests that wherever this part of the diencephalon is recognizable there will be an interchange between different thalamocortical and (probably) corticothalamic pathways, suggesting that the thalamus and cortex depend on complex interactions with each other and that the notion of the thalamus simply providing a supply of afferent messages to cortex through a neatly ordered thalamic "radiation" is dead.

D. *Summary*

In this chapter we have focused on the evidence that allows a distinction between first and higher order relays, the former transmitting messages to cortex from subcortical structures and the latter transmitting messages to cortex from cortex. A crucial point about the higher order relays is that these thalamic relays are innervated by axons that serve as cortical output pathways to lower centers, so that a higher order relay tells one cortical area about the instructions that another cortical area is sending to lower centers. It has been noted that there may be many parts of the thalamus that are not purely first order or higher order, but that there may be an intermingling of first and higher order circuits in some parts of the thalamus. The possibility that two such pathways might interact in the thalamus remains completely unexplored, although we suggest that the two pathways may well be almost entirely independent of each other as they pass through the thalamus, each, however, subjected to the functions of the same thalamic gating mechanisms. One of the important gating mechanisms involves the thalamic reticular nucleus, and we stress the importance of the close relationship that first and higher order circuits concerned with the same modality (or with closely related motor mechanisms) may establish within the reticular nucleus.

E. Some Unresolved Questions

1. Are there thalamic nuclei that have mixed, first order, and higher order relays?
2. If there are, do the two types of relay pass through the nucleus as parallel, independent pathways, comparable to the X and Y pathways of the cat visual system, or is there an interaction within the thalamic relay?
3. Do the axons that arise in cortex and that form RL terminals in the thalamus always arise from layer 5 cells, and are they always (or only sometimes) branches of long descending axons?
4. Can it be demonstrated that the axons considered in question 3 are always the drivers of the relay cells in the nucleus that they innervate? Where this cannot be achieved by the study of receptive fields, can it be done with the methods proposed in the next chapter?
5. Do higher order relay cells resemble the cells of the lateral geniculate nucleus in passing messages on to cortex without significantly altering the nature of the message, and do they have the same "gating" functions as first order relay cells?
6. Do higher order cortical areas receive their major driver afferents from higher order thalamic relays or from other cortical areas?
7. What is the nature of the action that direct corticocortical pathways transmit from one cortical area to another?
8. Is there a good example in any species of a first order relay that does not have a functionally corresponding or related higher order relay?

CHAPTER IX

Drivers and Modulators

We have in earlier chapters introduced the distinction between "drivers" and "modulators" and have discussed this concept previously (Sherman and Guillery, 1998). Examples of the former are retinal inputs to the lateral geniculate nucleus and medial lemniscal inputs to the ventral posterior nucleus; examples of the latter are various brain stem inputs and the feedback pathway from cortical layer 6. Perhaps the most important aspect of this distinction is the now well-documented (see Chapter VI) but often overlooked point that the many synaptic inputs to a neuron have distinct postsynaptic actions and that one cannot hope to understand the functional organization of a circuit just by mapping all of the inputs and outputs of that circuit. The nature of synaptic inputs, and in particular whether they are drivers or modulators or perhaps even either one or the other under differing conditions, is key to this understanding. For example, if we could not identify drivers versus modulators among inputs to relay cells of the lateral geniculate nucleus, we might easily be misled by consideration of the relative numerical strengths of inputs. Because, as we have shown in Chapter III, the brain stem afferents provide more synaptic contacts on relay cells than do the retinal afferents, it would be easy to conclude, in the absence of other information, that the lateral geniculate nucleus relays brain stem information to cortex and not retinal information. Here we shall attempt to explain more formally what we mean by drivers and modulators and consider the extent to which this distinction applies throughout thalamus and perhaps even to other parts of the brain. We start by looking specifically at relationships in the lateral geniculate nucleus because drivers and modulators are most clearly understood there.

A. Drivers and Modulators in the Lateral Geniculate Nucleus

It may seem unnecessary to define the driver input for the lateral geniculate nucleus because it seems a tautology to accept retinal afferents for this role, but if we stop to consider why we so readily believe this for the lateral geniculate nucleus, it will be useful in thinking about driver inputs to other, less well-defined thalamic relays. There are two major reasons to be confident that retinal input is the driver for the lateral geniculate nucleus, and both relate to the concept that the driving input conveys to the thalamic nucleus the main form of information to be relayed to cortex.

First, it is clear from clinical and experimental evidence that striate cortex is primarily, or exclusively, involved with vision and that its main thalamic innervation arrives from the lateral geniculate nucleus; also, of the potential sources of subcortical afferent systems innervating all relay cells of the lateral geniculate nucleus (which includes various brain stem sources described in Chapter III), only the retinal afferents are frankly visual. The tectal inputs, which innervate some geniculate laminae, fail the next criterion for identifying a driver, which is based on receptive field properties, because the directionally selective receptive fields of tectal cells cannot readily account for the center/surround properties of the geniculate cells that they innervate. Second, one way to determine the nature of information processed by a sensory neuron is by studying its receptive field properties, and where the receptive field properties can be mostly accounted for by any subset of afferents, these afferents must be considered the drivers. For geniculate relay cells, retinal input is clearly the driver, because the retinal afferents pass their receptive field properties to geniculate relay cells with only minor changes (reviewed in Cleland *et al.*, 1971; Sherman and Spear, 1982; Cleland and Lee, 1985; Sherman, 1985; Shapley and Lennie, 1985; Usrey *et al.*, 1998). It should be noted here that it is not necessary that the postsynaptic cell have virtually the same receptive field properties as their driver inputs, only that the drivers can account for these properties. It happens that geniculate relay cells have virtually the same center/surround receptive fields as do their retinal afferents (see Chapter II, Section A.3). However, as we describe later in this chapter, cells in visual cortex have receptive fields that are quite different from their geniculate afferents, but these afferents can nonetheless account for most of the cortical receptive field properties, which is one reason we argue that the geniculate afferents are the drivers for their postsynaptic target cells in cortex.

Although retinal inputs are the only plausible subcortical source of driver input to the lateral geniculate nucleus, a major input to this structure derives from the visual cortex itself. Could corticogeniculate axons provide significant driver input? The best evidence that they do not comes from studies of ablating or otherwise silencing the visual cortex and thus this input. Such studies, and there have been many, have found rather subtle effects of removing this pathway on receptive fields of geniculate

A. Drivers and Modulators in the Lateral Geniculate Nucleus

TABLE 1 DRIVERS VERSUS MODULATORS IN THE LATERAL GENICULATE NUCLEUS

	Driver (retina)	Modulator
Terminal size	Large	Small
TRN innervation	No	Yes
Metabotropic receptors	No	Yes
Afferent convergence	Small	???

neurons (Kalil and Chase, 1970; Richard *et al.*, 1975; Schmielau and Singer, 1977; Baker and Malpeli, 1977; Geisert *et al.*, 1981; McClurkin and Marrocco, 1984; McClurkin *et al.*, 1994). More recent studies have suggested that the pathway affects temporal properties of relay cell discharges (McClurkin *et al.*, 1994; Godwin *et al.*, 1996b) or establishes correlated firing among nearby relay cells with similar receptive field properties (Sillito *et al.*, 1994). Overall, the available evidence indicates that the corticogeniculate pathway is modulatory and not a driver input.

If we accept retinal afferents as the drivers, we can begin to explore what other features distinguish the drivers from the modulators. Table 1 summarizes some of the differences between retinal and nonretinal afferents for the lateral geniculate nucleus of the cat. We shall go on to suggest that these differences help to distinguish drivers from modulators throughout thalamus.

1. GENERAL DIFFERENCES BETWEEN DRIVERS AND MODULATORS

The drivers in the lateral geniculate nucleus are the retinal afferents and these form the characteristically large axon terminals known as type 2 axons on the basis of their light microscopic appearance and as RL terminals on the basis of their electron microscopic appearance. They make synaptic contacts on dendrites close to the cell body of relay cells and are described in Chapter III. On many relay cells, they form triadic synaptic complexes typically in glomeruli. From the point of view of the functional distinctions we shall draw, it is of special interest that retinal afferents use glutamate as a neurotransmitter and activate only ionotropic receptors (see Chapter VI). There is rather little convergence in retinogeniculate contacts because relay cells typically receive most of their input from one or a very small number of retinal axons (Cleland *et al.*, 1971; Sherman and Spear, 1982; Cleland and Lee, 1985; Sherman, 1985; Shapley and Lennie, 1985; Usrey *et al.*, 1998). Where they have been documented, these features characterize presumed driver input to other sensory thalamic relays like the auditory and the somatosensory afferents to the medial geniculate nucleus and ventral posterior nucleus, respectively (see Chapter III). All of these inputs contact proximal dendrites of relay cells with RL terminals, often in triadic arrangements within glomeruli, and they use glutamate as a neurotransmitter (reviewed in Sherman and Guillery, 1996), although the ionotropic or metabotropic nature of their associated postsynaptic receptors has not yet been identified. Also, because of the similarity in receptive field structure between these afferents and target relay cells, there seems to be little convergence within these inputs. These features are not generally known for other thalamic nuclei (see Chapter III), but it seems a reasonable step to suggest that some or all of these features characterize drivers throughout thalamus, and that some will also apply in other parts of the brain.

The modulators in the lateral geniculate nucleus include several distinct components, each with rather distinct properties. However, they do share certain features (see Table 1). They include glutamatergic afferents from layer 6 of cortex, cholinergic and noradrenergic afferents from brain stem, and histaminergic afferents from the hypothalamus. Other modulatory afferents to relay cells come from local GABAergic cells, including interneurons and cells in the adjacent thalamic reticular nucleus. Serotonergic afferents from the dorsal raphé nucleus also innervate the lateral geniculate nucleus, but available evidence summarized in Chapter VI suggests that these innervate interneurons and not relay cells. For reasons given below, we are inclined to regard them together with other inhibitory afferents (see also below) as modulators of the relay cells. Some geniculate regions, containing W and K cells in the lateral geniculate nuclei of the cat and monkey, respectively, also receive input from the superior colliculus, but as we have argued earlier, these lack a critical feature that would lead us to regard them as drivers.

Synaptic terminals from modulator inputs are relatively small and most make only a single synaptic contact. Modulator inputs from layer 6 of cortex form synapses onto distal dendrites, those from brain stem are primarily found proximally, adjacent to retinal (driver) inputs, and those from GABAergic sources make contacts throughout the dendritic arbor (Wilson et al., 1984; Erişir et al., 1997). Modulators as a total population outnumber drivers significantly because only about 5 to 10% of synapses on geniculate relay cells derive from retina (Van Horn et al., 2000). Related to this is the observation that at least some modulators exhibit a high degree of convergence in contacting relay cells. This is clearest for the corticogeniculate axons from layer 6. These axons outnumber geniculate relay cells by a factor of 10 to 100 (Sherman and Koch, 1986). Because corticogeniculate axons branch richly and probably contact many geniculate cells, the amount of convergence is likely to be much greater than this.

Perhaps most importantly, all of the modulator input pathways to geniculate relay cells activate metabotropic receptors, and finally, whereas all of the modulator pathways innervating the lateral geniculate nucleus also innervate the thalamic reticular nucleus, many of them by means of branching axons (see Chapter III), the retinal (driver) afferents do not send any branches to innervate the thalamic reticular nucleus.

Thus in the lateral geniculate nucleus, the retinal drivers may be distinguished from modulators by a number of criteria, of which the main ones are shown in Table 1: the retinal driver inputs have relatively few, large axon terminals making contacts close to the cell body, whereas modulators have many more, smaller terminals dominating on peripheral dendrites; afferent convergence onto relay cells from retinal drivers is small, whereas for at least some modulators (i.e., corticogeniculate afferents), such convergence is high; retinal drivers activate only ionotropic receptors, whereas modulators activate metabotropic receptors often in addition to ionotropic ones; retinal drivers do not branch to innervate the thalamic reticular nucleus, whereas modulators do. However, these distinctions are not generally diagnostic for drivers versus modulators, and if the classification is to be useful in other brain regions, further criteria are needed. It will be very important to define how many of these features are identifiable for other thalamic nuclei and other parts of the brain. For thalamus, some features, such as terminal morphology, are more readily identified than others, such as relative convergence. Also, although drivers in thalamus may turn out to be relatively homogeneous across various nuclei in terms of these features, modulators may be quite variable. For instance, it appears that histaminergic inputs to the lateral geniculate nucleus, unlike other modulatory inputs, do not operate through conventional synapses but instead release transmitter into the neuropil, so questions of afferent convergence are moot.

2. Receptive Fields and Cross-correlograms

As noted previously, retinal inputs dominate geniculate relay cell responses in terms of their receptive field properties. That is, the center/surround receptive field of a geniculate relay cell is more like that of its retinal afferent(s) than of afferents from visual cortex or thalamic reticular nucleus. It is very likely quite different from that of any brain stem or hypothalamic afferent, all of which almost certainly lack classical visual center/surround receptive fields. Thus, in terms of receptive field properties, retinal afferents dominate the output of a geniculate relay cell when it is transmitting visual information.[1] In contrast, all the other afferents have relatively little obvious effect on basic, qualitative receptive field properties of the relay cell but instead modulate the input/output relationships to control quantitative features of the relay.

Thus for relay cells of the lateral geniculate nucleus, we can define drivers as primary transmitters of receptive field properties and modulators as inputs that do not provide the basic receptive field properties to the relay cell. This may also serve for some of the other first order sensory thalamic relays (see Chapter VIII), such as the ventral posterior nucleus and the medial geniculate nucleus, where receptive field properties of the relay cells are well understood. However, this criterion will not serve to

---

[1] This refers to receptive fields seen in the lightly anesthetized or awake preparations because geniculate relay cells respond quite differently during certain phases of sleep (see Section E of this chapter and also Ramcharan et al., 2000).

distinguish drivers from modulators in other thalamic nuclei where receptive fields have not been defined. Examples include the medial dorsal nucleus, midline and intralaminar nuclei, and much of the pulvinar. We must consider other features that distinguish drivers from modulators to develop more general criteria and can again start by considering the lateral geniculate nucleus.

The functional linkage between retinal afferents and their target geniculate relay cells involves more than the structure of their receptive fields. During simultaneous recording between a geniculate relay cell and one of its retinal afferents, a close temporal correspondence is seen in the action potentials in these connected cells (Cleland and Lee, 1985; Usrey et al., 1998): most retinal action potentials are followed with a fixed latency by one in the relay cell. This is graphically evident when a cross-correlogram is constructed showing the temporal relationship between the action potentials in the two cells.[2] Such a cross-correlogram has a relatively narrow peak, with a latency of several milliseconds, and a relatively low, flat baseline (Figures 1A and 1C).

There are several criteria that the EPSPs from an afferent must meet to produce a sharp, narrow peak in the cross-correlogram. One criterion is that there be relatively little latency variation in the EPSP because this would tend to smear the peak. Another concerns the relationship between the EPSP duration and frequency. As long as the EPSP duration is briefer than the interspike interval of the afferent, there will be no temporal summation in the evoked response. With no temporal summation, the peak of the cross-correlogram will be closely related to the duration of individual EPSPs, but if temporal summation occurs, the peak will again be smeared. Therefore, the briefer EPSP related to activation of ionotropic receptors (see Chapter VI) allows the production of a narrow peak with moderate rates of afferent firing, whereas the longer EPSPs related to activation of metabotropic receptors would smear already lengthy EPSPs to produce an even broader peak.

From the point of view of signal transmission, there is an important distinction between a narrow peak and less temporal summation and a broader peak with more temporal summation. With more temporal summation, individual input spikes are no longer clearly resolved in the transmitted signal because summation creates a single, broader EPSP. In fact, only if individual action potentials in the retinal afferent occur with

---

[2] Such a cross-correlogram is constructed by computing for each retinal action potential the probability that an action potential is seen in the relay cell as a function of time, typically looking for several hundred milliseconds before and after the retinal action potential, which is set at time zero. This is repeated for every retinal action potential, and an average histogram is then constructed. This histogram is the cross-correlogram. Consider what would be expected if most retinal spikes evoked one spike at a fairly constant latency in the postsynaptic relay cell, and very few other spikes were seen postsynaptically. This cross-correlogram would have a large peak at a time after zero that reflects transmission time of the retinal action potential from the recording site to the terminals in the lateral geniculate nucleus plus synaptic delay (e.g., several milliseconds, depending on the recording site), and the width of the peak would reflect synaptic "jitter." There would be very few events in the cross-correlogram outside of this peak. See Fetz et al. (1991) and Nelson et al. (1992) for a further discussion of cross-correlograms.

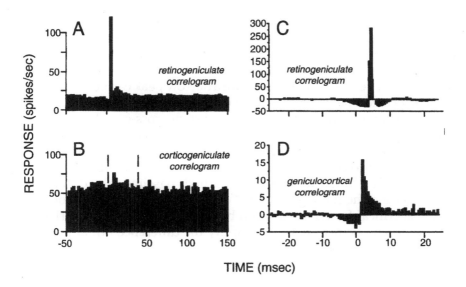

Figure 1
Cross-correlograms displaying the difference between drivers and modulators. Each is based on simultaneous recording in cats from two neurons, one presynaptic to the other. The cross-correlograms represent the firing of the postsynaptic cells relative to a spike at time zero for the presynaptic cell. (A) Retinogeniculate cross-correlogram based on spontaneous activity in both the retinal and geniculate neurons. Note the narrow peak rising out of a flat, low baseline that marks this as a driver connection. (B) Corticogeniculate cross-correlogram based on spontaneous activity in both the layer 6 cell in area 17 and geniculate neuron. Glutamate was applied to cortex to enhance the spontaneous firing of the afferent cell. Between the vertical, dashed lines it is possible to discern a very gradual, prolonged, and small peak arising from a noisy, high baseline that marks this as a modulator connection. (C and D) Cross-correlograms taken from the same laboratory using identical techniques for easier comparison. Both are based on visually driven activity and involve a "shuffle correction" (Perkel et al., 1967), and they are normalized against the firing level of the afferent, which is why some bins fall below zero. Both represent driver inputs and include another retinogeniculate pair (C) plus a geniculocortical pair (D). Note the difference in vertical scale, indicating that the retinal input accounts for more postsynaptic spikes in the geniculate cell (C) than does the geniculate input to the layer 4 cell of striate cortex (D). Note also that the time represented by these cross-correlograms is much briefer than that for A and B. Nonetheless both cross-correlograms have narrow peaks rising from a flat, low baseline, marking them as driver inputs.

long enough interspike intervals (i.e., during low-frequency firing) will each be correlated with a specific postsynaptic action potential and thus be individually recognized. What this all boils down to is the longer the EPSP, as with metabotropic responses, the more high-frequency information is lost. Conversely, the fact that retinal input activates only ionotropic receptors preserves high-frequency information in the relay to cortex. This is akin to saying that, compared to the fast postsynaptic responses to the driver input that are produced by ionotropic receptors, the slower metabotropic responses act like a low-pass temporal filter that fails to pass higher frequencies.

This "driver" cross-correlogram with a narrow peak can be contrasted with the expected one between a relay cell and its modulator input. Unfortunately, examples from the literature of such cross-correlograms are exceedingly rare, and indeed only one is known to us. This is shown in Figure 1B, and it is a cross-correlogram obtained from a concurrent recording of a layer 6 cortical cell and a target geniculate cell. This shows a small peak with a broad foundation on a high baseline. The difference between this and the retinogeniculate "driver" cross-correlogram of Figure 1A,C is critical for the distinction between modulators and drivers. It may eventually be necessary to quantify the difference but at present there is too little relevant evidence for this. We would also predict that other modulatory inputs to geniculate relay cells, such as from brain stem or local GABAergic cells, will exhibit cross-correlograms with relay cells that resemble that in Figure 1B much more than those in Figures 1A and 1C.

The difference between driver and modulator cross-correlograms shown in Figures 1A–1C probably relates to at least two factors. First, as noted earlier, the sharp peak of Figures 1A and 1C depends in part on there being no (slow) metabotropic receptors activated by retinal afferents on relay cells, only ionotropic receptors, whereas in Figure 1B the corticogeniculate afferents from layer 6 activate metabotropic receptors as well as the ionotropic ones (see Chapter VI). This will result in a prolonged EPSP with a long and relatively variable latency, and one would expect such an EPSP to promote prolonged firing of the relay cell with individual action potentials in the postsynaptic cell not clearly related to specific action potentials in the cortical afferent. Second, there are likely to be many modulators, but few drivers for any one relay cell. With little convergence of the driver retinogeniculate afferents, the postsynaptic responses of the relay cell to any one driver will be relatively strong and well synchronized to it. In contrast, the modulators, such as layer 6 corticogeniculate afferents, show significant convergence (see earlier), and the effect of any one modulator, possibly one of hundreds, may be minuscule. Furthermore, unless the modulator inputs from cortex fire in a highly synchronized manner, which seems unlikely in the normal, behaving animal, the postsynaptic action potentials will not be synchronized to those of any one corticogeniculate afferent. Either or both of these two factors can contribute to the differences seen between the driver cross-correlogram of Figures 1A and 1C and the modulator example of Figure 1B.

A driver, if it is to drive, must produce a distinct, measurable effect. The quantitative relationships of putative drivers to their postsynaptic neurons will almost certainly prove important, and the extent to which any one driver in any other thalamic relay can actually produce a cross-correlogram as sharp as that in Figures 1A and 1C is untested and is likely to be a useful feature to explore. Because modulatory inputs may often far outnumber driver inputs in thalamus, a numerically strong afferent pathway, which is sometimes interpreted as a pathway that must be important in information transfer, may instead often indicate modulatory influences. That is, one cannot simply assume a dominant (i.e., driver) input on the basis of large numbers, and understanding the functioning of thalamic circuits requires identifying drivers and modulators by criteria other than the numerical strength of the inputs.

This use of the cross-correlogram as a criterion for the driver/modulator distinction provides a relationship based on individual action potentials that can be applied to thalamic (and other) relays where receptive fields cannot be defined. Where, as in the transmission of many receptive fields, critical temporal relationships are a key function of the driver, any transmission not producing a sharp cross-correlogram (i.e., a narrow peak arising from a flat baseline) would lose all temporal information not contained in the lowest frequencies. This raises an important qualification for this criterion. Clearly, higher frequencies are important to vision, touch and kinesthesis, and hearing, so this criterion makes sense when applied to the thalamic relays for these sensory pathways. High frequencies may also be important for the transmission of most other types of information by way of thalamic relays, but this needs to be established. Smell, for instance, or taste, are examples of sensory systems that may involve only slow processes and thus very low temporal frequencies, and perhaps the cross-correlogram reflecting the driver input relayed through the thalamic nuclei associated with smell or taste [i.e., respectively a part of the mediodorsal nucleus and a small-celled group related to the ventral posterior nucleus called VMb by Jones (1985)] may not resemble those of Figures 1A and 1C. This is an issue that remains to be experimentally defined, and that may well differ for thalamic cells concerned with smell compared with those that deal with taste. That is, the thalamic relay for taste is reasonably treated as a first order relay on the pathway to the cortical taste area (see Benjamin and Burton, 1968), where the ascending afferents must be the functional drivers. In contrast to this, the functional significance of the olfactory afferents to the medial dorsal nucleus is less clear because these come from pyriform cortex (Kuroda *et al.*, 1992), and we know of no morphological or functional evidence that would establish this pathway as a driver or a modulator of medial dorsal relay cells.

3. WHY ARE DRIVER INPUTS SO SMALL IN NUMBER?

There is a logical explanation for the large number of modulators and the smaller number of driver afferents impinging on any one relay cell. Neuromodulation is likely to be produced by afferents that come from many different sources and that act together to produce a finely graded effect linked to many aspects of the behavioral state such as alertness and attention; this implies a large number of inputs, each contributing a relatively small effect. Drivers, on the other hand, need a relatively small number of units to carry the basic message to the target level. This is analogous to the situation in many brain structures where the main output message of complex neural computations is carried by very few cells. For instance, in cortex, the number of cells reflecting the output of a column (e.g., the subset of layer 5 cells carrying the results of the columnar computation to subcortical targets) represents a minority of cells in the column.

The conclusion that retinal input dominates geniculate relay cell response properties may at first seem at odds with the observation that retinal synapses are such a small percentage of inputs to these geniculate cells. It is likely that the strength or efficacy of retinal synapses is much stronger, on average, than that of other synapses on the relay cells. There are at least two ready explanations for this. First, sizes of the retinal contact zones on relay cells tend to be much larger than those of other inputs, and larger contacts might lead to more transmitter release and larger EPSPs. Second, as noted in Chapter VII, synapses can vary greatly with respect to the probability that an action potential invading the presynaptic terminal leads to transmitter release (reviewed in Lisman, 1997). If retinal synapses, on average, have a much higher probability of release than other synapses, then the effective EPSP in a relay cell from an action potential in a retinal axon would consequently be relatively larger than expected from a source providing only 5 to 10% of the inputs to that relay cell. For example, if a retinal axon produced, say, 600 synapses on a relay cell, but each had a probability of release of 0.9, then an action potential along that axon would generate release from 540 synapses. If several cortical axons together produced 600 synapses onto the same cell and fired together, but the probability of release of their synapses was only 0.1 (and as reviewed by Lisman, 1997, release probabilities of 0.1 or lower are common in hippocampus and neocortex), then only 60 synapses would be activated. Thus before we can begin to relate relative synaptic numbers to functional properties, we must know more about synaptic physiology of these inputs than we do at present.

It is also interesting that some other putative driver inputs to neurons outside of thalamus seem to be a small minority as well. For instance, only 5 to 10% of the synapses to layer 4 cells in striate cortex come from the lateral geniculate nucleus, and the geniculate afferents can be regarded as the major (or only) source of driving afferents (Ahmed et al., 1994; Latawiec et al., 2000). The remarkable similarity in the relative synaptic numbers that retinogeniculate and geniculocortical drivers contribute to their postsynaptic cells may be a coincidence, but perhaps not, and only a more widespread survey of drivers and the relative number of synapses they form on their postsynaptic targets will address the generality of these numbers. Interestingly, morphological studies of spinal motoneurons suggest that Ia afferents, which constitute a major "driver" input, provide <5% of the synaptic terminals to these cells (reviewed on p. 462 of Henneman and Mendell, 1981).

B. The Geniculate Input to Cortex as a Driver

Relay cells of the lateral geniculate nucleus and other main sensory thalamic relays have an unusual property that may make them poor exemplars for a general definition of drivers versus modulators. We pointed out earlier that the lateral geniculate relay appears to be the only relay, from retinal receptor to higher cortical visual areas, that produces no significant spatial change in receptive field properties. Geniculate receptive fields are essentially like those of retinal ganglion cells, which makes it particularly easy to identify retinal inputs as the drivers. This is not typical for other relays in the visual pathways, which are presumably also innervated by drivers and modulators. Where new receptive field structures are synthesized, as happens in retina and cortex, we must expect a more complex grouping of convergent driving afferents.

For example, several geniculate axons converge to innervate a single cell in layer 4 of striate cortex. Even here, however, the convergence is relatively small compared, say, with the corticogeniculate pathway. If the synaptic influences sum linearly, the postsynaptic receptive field will reflect all of the receptive fields of the inputs. This seems to be the case for the geniculocortical pathway (Reid and Alonso, 1995; Ferster *et al.*, 1996), and the observation that the receptive field of the postsynaptic cell is strongly determined by the input from geniculate axons marks these inputs as drivers. However, action potentials can occur in the postsynaptic cell in relation to action potentials from *any* of its inputs. If the geniculate afferents fired independently of each other, the cross-correlogram based on one of these afferents could still be fairly sharp, but the peak would be smaller and the baseline would be higher and noisier because of firing of the other afferents. A cross-correlogram can identify the driver as long as the number of convergent, independently firing afferents does not prevent the baseline from obscuring the peak that each alone would produce. Indeed, the relatively sharp cross-correlogram seen for geniculocortical synapses (Figure 1D) identifies this as a possible driver identifiable at a synapse in the cortex that takes our argument beyond the confines of the thalamus.

Clearly convergence of many independently firing drivers must be limited if a diagnostic, sharp cross-correlogram is to be produced. Large numbers of convergent inputs could produce a sharp cross-correlogram *only* if their firing were highly correlated. This is well shown in geniculocortical connections: cross-correlograms for geniculate and cortical cells indicate a driver input (Figure 1D), but they are sharper when the relevant geniculate cells fire in synchrony (see Alonso *et al.*, 1996).

The cross-correlogram for the geniculocortical input (Figure 1D) further suggests that the synapses must activate mainly ionotropic receptors because if metabotropic receptors were activated strongly, the result would be a broad peak in the cross-correlogram similar to that seen in Figure 1B. We can predict, then, that, like the retinogeniculate synapse (a driver), and unlike the corticogeniculate synapse (a modulator), the geniculocortical synapse activates predominantly ionotropic receptors. This hypothesis could be readily tested, but direct evidence is not yet available.

C. Tonic and Burst Modes in Thalamic Relay Cells

Because geniculate relay cells generally show both tonic and burst firing (see Chapters V and VII) in varying degrees under conditions like those for which the cross-correlograms of Figures 1A and 1C were obtained, it is likely that some of the responses forming the peaks in these cross-correlograms were due to bursts. A question that has not been experimentally asked is: What is the difference in the form of the retinogeniculate cross-correlogram for the burst or tonic response modes of the relay cell? Action potentials in the tonic firing mode result directly from EPSPs, but during bursting, they result from the Ca^{2+} spike and are thus indirectly linked to the EPSP. This would affect the cross-correlograms. During tonic firing, an action potential in the retinal afferent may evoke an action potential in the relay cell with a tight one-to-one coupling, resulting in a cross-correlogram with an extremely narrow peak only a millisecond or so across. During burst firing, an action potential in the retinal afferent may activate a Ca^{2+} spike, and the resultant burst of several action potentials lasts for 20 msec or so. Thus the coupling between input and output action potentials is no longer one-to-one and would produce a broader peak than expected during tonic firing. Nonetheless, one would expect that the peak during burst firing would still be quite sharp compared with that produced by modulators.

D. Key Differences between Drivers and Modulators

We can now summarize our definition for drivers versus modulators innervating relay cells of the lateral geniculate nucleus. The key functional feature is the nature of cross-correlograms (see Figure 1). For driver inputs, they have a sharp peak rising out of a low, flat baseline. For modulators, they would have a broad, gradual peak, if any, set against a noisy baseline. The sharp cross-correlogram conforms to the need of the relay cell to convey the signal of its driver input to cortex, and do so faithfully for all firing frequencies in the driver input; a lack of such cross-correlograms for the modulator inputs means that the higher frequency signals they carry are not relayed to cortex. These critical features of cross-correlograms are merely products of a variety of circuit and cellular properties that characterize the contacts made by these inputs. Thus a geniculate relay cell receives afferents from relatively few retinal afferents that terminate near the cell body, show little convergence, and activate only ionotropic receptors. In contrast, the modulatory inputs come from many more different sources, they are generally further from the cell body and can show considerable convergence; and they all activate metabotropic in addition to ionotropic receptors.

We have shown that there are several characteristic features of drivers in the lateral geniculate nucleus that distinguish them from modulators. These include their fine structural appearance, their synaptic relationships, their degree of convergence on relay cells, their relatively small number relative to the modulators, their absence of connections to the thalamic reticular nucleus, their transmitter and receptor characteristics, and the nature of the cross-correlogram they produce when stimulated. Many of these properties are seen in all first order thalamic nuclei where the drivers can be identified. Other properties remain to be studied in these other nuclei, and some, such as the cross-correlograms, have not been empirically defined for modulators in any other thalamic nucleus. This summary for the lateral geniculate nucleus leads to a number of clear questions about other thalamic nuclei, where current knowledge of their functional organization does not make clear which is the driver, although in some the morphological evidence has provided a useful clue. In extending the definition of drivers and modulators to other brain centers, such as the cortex and possibly even to the spinal cord, the characteristics we have identified may prove to be relevant.

E. The Sleeping Thalamus

The functioning of the thalamus during wakefulness seems very different from its functioning during sleep, a topic that will be briefly considered here in the context of drivers and modulators. The reader is referred to other sources for a fuller account of sleep (Favale *et al.*, 1964; Dagnino *et al.*, 1965, 1966, 1971; Ghelarducci *et al.*, 1970; Marks *et al.*, 1981; Llinás and Pare, 1991). Sleep actually has two very different major components, each of which can be further subdivided. One is slow wave sleep (or synchronized sleep), which is characterized in electroencephalographic (EEG) recording by low-frequency, high-amplitude oscillations. The other is desynchronized sleep, which is characterized in EEG recording by high-frequency, low-amplitude oscillations. The latter is also sometimes called REM (for rapid eye movement) sleep, because bouts of rapid eye movements occur during this phase of sleep. As one passes from wakefulness to drowsiness to sleep, slow wave sleep is entered first, and from there, REM sleep can be initiated.

1. Slow Wave Sleep

During slow wave sleep, the cholinergic parabrachial inputs to the thalamus become much less active, and this undoubtedly enhances the propensity of relay cells to fire in burst mode. Indeed, bursting based on low Ca^{2+} threshold spikes is most frequently seen during slow wave sleep, and when it is seen, it tends to be rhythmic and synchronized among thalamic neurons. As noted in Chapter V, Section A, this synchronized, rhythmic bursting depends on interactions among reticular cells and between reticular and relay cells. It is important to note that not all thalamic cells show rhythmic bursting during slow wave sleep and that such bursting is typically interspersed with periods of tonic firing (McCarley et al., 1983; Ramcharan et al., 2000). The current dogma (Livingstone and Hubel, 1981; McCarley et al., 1983; Steriade and McCarley, 1990; Steriade et al., 1990, 1993; Steriade and Contreras, 1995) suggests that rhythmic bursting represents a period during which the thalamic relay of normal driver information is blocked, because the rhythmic bursting does not correspond to the firing pattern of the drivers.

However, it is not clear why rhythmic bursting should interrupt thalamic relays (see also Chapter VI). Perhaps the synchronized rhythmic volleys established in the thalamic reticular nucleus come to so dominate thalamic relay cell-firing patterns that their normal driver inputs become ineffective (e.g., Sherman and Guillery, 1998). However, recent evidence suggests that rhythmic bursting may not always dominate responses of all thalamic cells during sleep (Ramcharan et al., 2000), and other evidence suggests that transmission of driver inputs may only be slightly depressed during sleep (Ghelarducci et al., 1970; Dagnino et al., 1971; Meeren et al., 1998). This issue needs further study. We need much better evidence for the status of thalamic relays during the various phases of sleep and specifically during slow wave sleep.

If the dogma does prove to be mostly correct, that synchronized, rhythmic bursting indeed reflects a breakdown of relay of normal driver inputs, and we emphasize the "If," then we can suggest the following. During this synchronized bursting, input from the thalamic reticular nucleus dominates relay cells, and EPSPs generated by driver inputs may be insufficient to break the stranglehold of reticular inputs on thalamic relay cell responses. The relay is interrupted not by silencing relay cells, but rather by forcing them to burst rhythmically and independently of normal driver input. Thus, instead of no signal, cortex receives a clear, positive signal that the relay is disrupted. Silence alone would be ambiguous: the absence of a driver-carried message would be indistinguishable from a disruption of an effective relay of such a message. The rhythmic bursting, by signaling the "no-relay" alternative, avoids this ambiguity.

In any case, because reticular input seems to dominate relay cell responses during synchronous, rhythmic bursting, it is interesting to imagine what the cross-correlograms between a reticular cell and its postsynaptic relay cell would look like. We would expect there to be a fairly sharp peak with little baseline in the cross-correlogram during the synchronized, rhythmic bursting. The peak would have a fairly long latency because the direct synaptic effect of the reticular activity (a burst) would be powerful inhibition, silencing the relay cells for hundreds of milliseconds, and the relay cell would then fire a burst only after it depolarized, partly caused by I_h and partly caused by cessation of reticular cell firing. Nonetheless, the interval between the reticular firing and the relay cell firing, while hundreds of milliseconds, would be fairly regular, resulting in a peak. This begins to look like one of our major criteria for a driver input. One could argue that the thalamic relay cells are responding to a "message" sent by the reticular cells. The more general issue of whether an inhibitory input like the reticular input is likely to be a driver is considered in the following. However, this form of "driving" by reticular input during slow wave sleep occurs because of the special relationships that produce the highly correlated firing of the synchronized reticular cells. Without such correlation, there would be no driving. It seems that the action of these afferents has to be distinguished from drivers considered earlier, but that it cannot be regarded as modulation. We suggest that this action be treated as a "disrupting" action that is distinct from driving and modulation and that may be special to the thalamic reticular nucleus and its thalamic connections.

2. REM Sleep

During REM sleep, cholinergic, parabrachial inputs to thalamus become highly active, which makes thalamic cells fire mostly or perhaps even exclusively in tonic mode. Some very limited evidence (Dagnino et al., 1965, 1966; Ghelarducci et al., 1970; Marks et al., 1981) suggests that relay functions are much greater during REM sleep than during slow wave sleep and may actually be equivalent to the relay seen during the awake state. In fact, there are as yet no clear criteria that distinguish thalamic relay functions during REM sleep and the alert, awake state. A question often asked and not yet answered (e.g., Llinás and Pare, 1991) is: Why, if sensory information is effectively relayed through thalamus to cortex during REM sleep, are we generally unaware of it?

F. Can Extradiencephalic GABAergic Inputs to Thalamus Be Drivers?

In the preceding section, we considered the role of the input from the thalamic reticular nucleus, which seems clearly to be a modulator under most conditions. However, during slow wave sleep, these reticular inputs seem to play a different role, something other than driver or modulator, because of the very special condition that they then fire synchronously. We have suggested that this role might be considered as a disrupter. But what of other GABAergic inputs to thalamic relay cells, particularly those of extradiencephalic origin described for some thalamic nuclei (see Chapter III)?

Possibly the most interesting question regarding these extradiencephalic GABAergic afferents is whether they can act functionally as drivers, or whether they all operate as modulators. In Chapter III, we showed that the morphology and connections are not decisive on this issue for all of these afferents. Here we add the point that the functional evidence would suggest that inhibitory afferents are most unlikely to be acting as drivers. It is important to recognize that inhibitory and excitatory afferents to spiking cells are not mirror images of each other functionally. That is, a synapse creating an EPSP or IPSP might convey the same information to a nonspiking postsynaptic cell, as commonly happens in retina, but not to a cell that passes on information based on action potentials activated from a voltage threshold depolarized with respect to rest. In the extreme case where the depolarized cell has zero baseline or spontaneous activity, an IPSP will not readily affect the spiking of a cell with no spontaneous activity, and so the message conveyed by an inhibitory afferent is lost,[3] whereas an EPSP can activate an action potential, and so the presynaptic message can be transmitted further. If, however, the postsynaptic cell has enough spontaneous activity, both an EPSP, by elevating the firing rate, and an IPSP, by lowering it, can transmit information through the postsynaptic cell. Nonetheless, for moderate levels of spontaneous activity (i.e., less than roughly 20–50 Hz), the action of EPSPs in creating extra spikes is more likely to be detected in the postsynaptic spike train than is the action of an IPSP in removing occasional spikes. Possibly at very high baseline firing levels the action of IPSPs can influence the firing of the postsynaptic cell strongly and with a high degree of temporal resolution (i.e., the flip side of an EPSP). However, for the IPSP to be effective as a driver, the baseline rate must be high (well over 50 Hz, and the higher the better) because the brief IPSP must have a high likelihood of canceling an action potential and not fall between them, as would often happen with lower firing rates; but such baseline rates are rarely, if ever, observed in thalamic relay cells under normal conditions. For these reasons, we regard inhibitory, GABAergic inputs to thalamus as unlikely candidates to be drivers and suggest they are all modulators. It is interesting in this regard that baseline firing rates of neurons in the deep cerebellar nuclei, which are innervated from the cerebellar cortex by GABAergic Purkinje cells, tend to be fairly high (Gardner and Fuchs, 1975; Armstrong and Edgley, 1988).

However, it should be clearly understood that this conclusion is at best provisional and based on limited and indirect evidence. We need functional studies of the extradiencephalic inhibitory inputs to thalamus for a more direct test of the hypothesis that they are all modulators. We also need a better understanding of the different roles played by EPSPs and IPSPs in information processing.

G. Summary

The distinction between drivers and modulators (and possibly disrupters) is important for understanding thalamic relays. Perhaps the most

---

[3] There is an interesting exception to this argument. If the inhibitory input sustains strong and long enough inhibition in a previously depolarized thalamic relay cell, either by temporal summation of $GABA_A$-mediated IPSPs or by $GABA_B$-mediated IPSPs, this would de-inactivate I_T. When the IPSP(s) ceased, the relay cell would passively depolarize to its previous value, activating I_T and thus a burst of action potentials. Inhibitory input could thereby lead to burst firing. However, burst firing alone would limit the nature of information relayed to cortex (see Chapter VI), there would be a long ($\leq$50–100 msec) and variable delay before a burst could be activated beause of the time needed to de-inactivate I_T, and much temporal information in the input would be lost by a relay limited to burst mode. This, then, would severely limit the information relayed by an inhibitory driver input if it depended on activating burst firing in the relay cell.

G. Summary

important question to be answered for any thalamic nucleus is how its inputs are divided among drivers and modulators. Where receptive fields can be defined, as in the main, first order sensory relays, identifying the drivers versus modulators is fairly straightforward, but there is much to learn about drivers and modulators for most of the rest of thalamus. We have outlined in this chapter some features other than receptive field properties that can be used to help identify drivers versus modulators. Possibly this distinction can be applied much more broadly to cerebral cortex as suggested by Crick and Koch (1998), and possibly to other cerebral centers as well, and perhaps generally to broad areas of the CNS, such as spinal cord. Within cortex, we think it likely that thalamocortical axons, especially those going to cortical layer 4, and possibly all of them, will prove to be drivers, whether from a sensory relay nucleus like the lateral geniculate nucleus or a higher order nucleus like the pulvinar.

It is important in this context to appreciate that the classification of corticocortical pathways is largely untested in terms of the criteria proposed here. The current dogma is that information flow within cortex is carried exclusively, or nearly so, by direct corticocortical connections (see Felleman and Van Essen, 1991; Bullier et al., 1996), but this seems largely to be an implicit assumption based on the sheer anatomical number of these connections. Indirect corticocortical routes, such as higher order cortico-thalamo-cortical ones, have been largely ignored in terms of information flow, partly because of their relatively small size anatomically, and partly because they have not been recognized. A re-investigation of direct and indirect corticocortical pathways could significantly affect the current dogma that all information here travels in the direct connections. By analogy with the anatomy of driver and modulator inputs to thalamic nuclei, such as retinogeniculate input, drivers may generally represent a very small input anatomically, whereas the modulators are the anatomically dominant input. One distinct and intriguing possibility is that the major source of a functional drive for corticocortical communication actually goes through the thalamus and derives from cells in layer 5 of one cortical area, which then provide a driver input to relay cells in a higher order thalamic nucleus such as the pulvinar (as suggested in Chapter VIII; see Figure 1 there). These thalamic cells then, in turn, send their axons as drivers to layer 4 of another cortical area. An extreme corollary might be that most, and perhaps even all, direct corticocortical pathways serve as modulators. This would mean that information flowing from one cortical area to another, by passing through a thalamic relay, would be subject to the same control of information flow as exists for information coming into cortex from subcortical sources. The thalamus would thus serve as a "gate" not only in the control of information to particular cortical areas about sensory events, but also of information passed to other cortical areas about the descending outputs emanating from layer 5.

If the categorization of inputs as driver or modulator (or possibly as disrupter) is to have an agreed general significance, or if one is to determine whether disrupters are unique to thalamus or can also be identified in other parts of the brain, it becomes important that experimental criteria for identifying the class of an input be clearly understood. We have tried to provide an introduction to the problems that need to be addressed if a classification that has wide applicability is to be used. Possibly it will prove that there are too many problem areas and intermediate positions for the distinction to be of any service outside the thalamus. The observations remain to be made.

H. Some Unresolved Questions

1. Can any afferent to the thalamus act as a driver under some conditions and as a modulator under other conditions?

2. How clear are distinctions between drivers and modulators in other parts of the brain?

3. Are there types of afferent that need to be considered other than "drivers" and "modulators"?

4. Do all driver inputs in the brain use glutamate as a transmitter and activate only ionotropic glutamate receptors? For instance, might some be cholinergic, activating nicotinic receptors?

5. Can cross-correlograms generally serve to distinguish drivers from modulators for the thalamus? Can they also do so for other parts of the brain?

6. Does a metabotropic receptor with its concomitant long timecourse of transmission represent an identifier of a modulator in the thalamus? Does this apply to other parts of the brain? Does this also apply where timing is less critical, as in olfactory or gustatory pathways?

7. Do modulators outnumber drivers in all (most) parts of the brain. Is the discrepancy generally of the same scale as it appears to be in the thalamus?

CHAPTER X

Overview

In the first part of the book we have presented the thalamus as it was seen historically: as a group of several different nuclei, each carrying a functionally distinct set of afferents to the cerebral cortex. We have stressed that although this may be telling us something about the messages that pass to the cortex through some parts of the thalamus, it tells us very little about what these parts of the thalamus actually do and tells us nothing about the function of the rest of the thalamus, represented by the nuclei that are entirely or largely higher order relays. To explore what it is that the thalamus does, we have used the visual pathways, primarily of cat and primate, to examine some of the functional properties of the thalamic relay in general. This has been a dual approach. On the one hand it raises a number of questions about what exactly the thalamus is doing for the visual relay itself, and on the other hand it asks whether some of the functional properties that can be defined or hypothesized for the visual relay may also apply to other thalamic relays, possibly to all thalamic nuclei. Insofar as our exploration has an expected outcome, it should serve to guide an investigation of the particular thalamic features that can be treated as general properties found in all parts of the thalamus and of those that are special to one nucleus or to one group of nuclei. It can be argued that it is unlikely, for example, that auditory, cerebellar, visual, and mamillothalamic relays will all show the same thalamic organization in all details. Yet it is clear that in terms of the morphological features seen light and electron microscopically or in terms of membrane properties and transmitter functions, there are many shared features across most, possibly all, thalamic nuclei. That is, there is a basic common pattern to thalamic relays; the challenge for future studies is to define exactly what these are and to show where the functional organization of any one thalamic nucleus differs from that of another. That is, although one can reasonably expect to find differences between thalamic nuclei, a systematic search for differences has not been undertaken, and those differences that have been documented represent a small and as yet poorly defined part of our knowledge of the thalamus.

The differences between thalamic nuclei can be treated as either intrinsic or extrinsic. That is, they may depend on external connections or on the internal organization of the particular relay. The most important of the extrinsic differences are represented by the origins and the functional properties of the driver afferents to each nucleus, and these are the differences explored by the classical approach, except that this approach did not recognize driver afferents coming from the cortex. Adding the drivers that come from cortex to our analysis of thalamus opens many new doors to the way in which we can look at the functional organization of many previously puzzling thalamic nuclei. However, a key step in such an analysis, defining the nature of the message that is sent by the corticothalamic drivers, must play a crucial role in such an analysis in the future, as must the determination of whether all, or only some, of these axons are branches of long descending axons going to brain stem or spinal cord. Further, we have argued that a critical unknown currently is the extent to which any one cortical area depends for its major driver afferents on either the thalamus or on other cortical areas, and we have suggested that the functional importance of thalamic afferents to higher order cortical areas may in the past have been seriously underestimated. Other extrinsic differences include the particular combination of modulator afferents that a nucleus is receiving from, for example, the brain stem cholinergic pathways or from noradrenergic, serotonergic, histaminergic, or other systems. In addition to these several distinct modulatory pathways, the extradiencephalic inhibitory afferents, which we have argued are likely to be modulators, represent another important known extrinsic difference between thalamic nuclei; only a few thalamic nuclei appear to receive such afferents and the function of these afferents very likely differs from one nucleus to another, but at present we have no clear picture of the role that these afferents play. As the functional role of the several modulatory pathways becomes clear, it will be important to understand how each relates to any one thalamic nucleus so that the patterns of modulation that can be imposed on that nucleus can be appreciated.

Although probably every thalamic nucleus receives modulatory (type 1) afferents from cortex, there may be important differences that relate to whether these come from just one cortical area, or from several different cortical areas; further the extent to which these corticothalamic axons arise from cortical areas that are receiving afferents from the same nucleus or from cortical areas that do not receive afferents from the same nucleus is often not clearly defined for any one thalamic relay, and the details will almost certainly prove relevant to understanding the conditions that produce modulation in the thalamic relay. There are rare examples of modulatory corticothalamic axons having a bilateral distribution in the thalamus, and, further, we have presented evidence that the layer 6 cells that give rise to the modulatory corticothalamic afferents are not a uniform population so that understanding the particular pattern of modulation received by any one thalamic nucleus may well prove important.

A further difference in the extrinsic connections of thalamic nuclei relates to the distribution of their thalamocortical axons. We have seen that there are differences in the cortical layers within which the axons terminate, although there is rather little evidence as to exactly how this relates to the actions they have within the receiving cortical area, and there are also differences in the extent to which any one nucleus or relay cell sends its axons to just one cortical area or to more than one cortical area. Possibly, from the point of view of understanding any one cortical area, all one needs to know is where the thalamocortical afferents take origin, but clearly in terms of understanding overall functional relationships between cortical areas, it will be important to understand where the thalamic afferents to two cortical areas are or are not shared. Finally, in terms of the extrinsic connections that are relevant to understanding differences among thalamic nuclei, there is the difference between the thalamic cells that do and those that do not send axonal terminals to the striatum. Does cortical processing of messages that come from afferents having branches to the striatum differ from that of messages that go to cortex alone? Clearly, whether the overall action does or does not include the striatum will have major functional significance for the brain as a whole, but exactly what relationship, if any, this has to cortical processing itself remains to be defined.

We have discussed the importance of mapped projections to and from thalamic nuclei and have suggested that there is likely to be an important difference between modulators like those coming from cortex that are topographically mapped, and thus can be expected to be able to act quite locally, and many other modulators that are not mapped and so may prove to have a more global action. Where maps can be defined for any one of the extrinsic connections it may be obvious what those maps represent, as it is for example in the early stages of visual or somatosensory processing, but in many instances we see a pathway that is topographically organized but have no idea about the feature that is mapped. There may prove to be a critical distinction between one thalamic nucleus and another in terms of how maps are organized, but it is probable that once the nature of the several different maps is understood, this may reveal a basic common pattern in thalamic organization.

We have summarized some of the current knowledge about pathways that link the thalamic reticular nucleus and thalamic relay nuclei. At the level of individual thalamic, cortical, and reticular cells details of connectivity patterns are likely to prove critical to our understanding of what these pathways are doing. At the level of the major reticular segments and their related thalamic nuclei and cortical areas, we need to learn much more about the connections that are established within the reticular nucleus among first and higher order circuits, as well as the mapped (or unmapped) connections that they form. On the basis of the close connections that it establishes with the thalamic relays, the reticular nucleus is neither entirely intrinsic nor really extrinsic to the thalamic relay. It is an integral part of the links that connect thalamus and cortex. It accommodates many of the complex fiber interchanges that are an essential part of thalamocorticothalamic communication, and although we can anticipate that every thalamic relay has a reticular connection, it is reasonable to expect that the details will not be the same for each relay and may differ, in particular when first order relays are compared with higher order relays.

In terms of the intrinsic organization of thalamic relays there are many significant common features. It would appear probable that all thalamic nuclei receive glutamatergic driver afferents as well as local GABAergic inhibitory afferents coming from interneurons, from cells of the thalamic reticular nucleus, or from both. Perhaps more importantly it would appear that the capacity to function in either tonic or burst mode is characteristic of all thalamic relay cells. We have treated this as a key to understanding at least some of the aspects of thalamic gating functions and are inclined to treat it as a general property of thalamus until a group of thalamic relays that fail to show these two modes are demonstrated. We have summarized the several distinct conductances and receptor mechanisms that have been demonstrated for the lateral geniculate nucleus. Only a few have been studied systematically in more than one thalamic nucleus. It remains to be defined exactly how each relates to the relay functions of the thalamus and the extent to which their distribution differs from one thalamic nucleus to another.

The distribution of metabotropic receptors is likely to relate to the distribution of modulator afferents in many, possibly all, thalamic relays; it seems unlikely that a driver that is transmitting a message that requires accurate timing could function with metabotropic receptors, but the possibility that there are driver afferents that function on a slower time base remains to be explored for systems such as taste or olfaction and, by implication, for thalamic relays whose functions are currently completely unknown.

Interneurons represent an important component of some, but not all, thalamic nuclei, and their distribution in different nuclei and in different species represent perhaps the strongest argument against treating all thalamic relays as equal. We know that there are differences between nuclei that have almost no interneurons and nuclei that have 20% or more thalamic cells serving as interneurons. It is not yet clear exactly what functional difference the presence or absence of interneurons implies for the function of a particular relay. The fact that thalamic somatosensory relays in cat and rat differ markedly in this respect, whereas the visual pathways in the two species seem to resemble each other, needs to be understood in terms of how each of these relays can affect the messages that are passed on to cortex. However, the difference within the visual pathways of cats, where the X cells appear to be involved in a relay heavily linked to interneuronal connections whereas the Y cells are not, suggests that it may be more appropriate to compare relay cells that are or are not connected to interneurons instead of simply comparing thalamic nuclei. Possibly there are two types of relay characteristic of the thalamus in general, and perhaps they relate to the two types of nerve cell originally described by Kölliker, although on present evidence that is far from an established conclusion. We have seen that the issue of classifying neurons and so distinguishing functionally distinct classes is a difficult one for relay cells and also for interneurons. It is clear that for both of these types of thalamic neuron there are likely to be functionally distinct classes. However, at present we have rather poor criteria for establishing differences and almost no insights as to the functional significance of the differences that can be identified.

We are still far from understanding exactly what it is that the thalamus does. Perhaps the most promising leads will come from recognizing the difference, in terms of the information transmitted, between the tonic mode on the one hand and the burst mode, either rhythmic or not, on the other. This can provide a clue as to what cortex may be receiving from the thalamus at any one time, provided that we know, in turn, what it is that the relevant part of the thalamus is receiving from its drivers. Often, as for some first order and most higher order relays we have all too little information about the nature of the message that the thalamic relay cells are receiving. Defining the functional characteristics of the afferents may prove extremely difficult and may depend on intuitions about the significant variables and on a certain amount of luck. However, knowing what message is carried by the drivers to any one nucleus will prove to be an essential part of understanding that nucleus and the messages that it is sending to cortex. Where a thalamic nucleus receives more than one driver afferent there is a crucial question about how these pathways relate. Is there some integrative interaction in the nucleus or do the two sets of messages pass through with relatively little or no interaction? If it is a higher order relay, then an important next step will be to define how the messages that are being sent to the cortical receiving area relate to the messages sent to the same area through corticocortical pathways. Which are drivers? Which are modulators? If there is more than one driver reaching cortex, what is the nature of the integration, if any, that occurs in the cortex?

We have presented many questions about the thalamus in the course of this book and in this brief overview. The reader will almost certainly be able to add others and may know of some answers that we have missed in our review of the relevant literature. We are inclined to think of the thalamus as central to all cortical functions and to believe that a better understanding of the thalamus will lead to a fuller appreciation of cortical function. It has become fashionable in journals that deal with neuroscience for authors to introduce their subject, whatever it may be, as "the central problem of neuroscience"; or occasionally, for slightly more modest authors, as "one of the central problems of neuroscience." We are not tempted to make such a claim about the thalamus, although it does occupy a strikingly central position in the brain. However, we suggest that cerebral cortex without thalamus is rather like a great church organ without an organist: fascinating, but useless.

Bibliography

Abramson, B. P., and Chalupa, L. M. (1985). The laminar distribution of cortical connections with the tecto- and cortico-recipient zones in the cat's lateral posterior nucleus. *Neuroscience* **15**, 81–95.

Abramson, B. P., and Chalupa, L. M. (1988). Multiple pathways from the superior colliculus to the extrageniculate visual thalamus of the cat. *J. Comp. Neurol.* **271**, 397–418.

Adams, N. C., Lozsádi, D. A., and Guillery, R. W. (1997). Complexities in the thalamocortical and corticothalamic pathways. *Eur. J. Neurosci.* **9**, 204–209.

Adams, P. R. (1982). Voltage-dependent conductances of vertebrate neurones. *Trends Neurosci.* **5**, 116–119.

Aggleton, J. P., and Mishkin, M. (1984). Projections of the amygdala to the thalamus in the cynomolgus monkey. *J. Comp. Neurol.* **222**, 56–68.

Ahlsén, G., Lindström, S., and Lo, F.-S. (1984). Inhibition from the brain stem of inhibitory interneurones of the cat's dorsal lateral geniculate nucleus. *J. Physiol. (Lond.)* **347**, 593–609.

Ahlsén, G., Lindström, S., and Lo, F.-S. (1985). Interaction between inhibitory pathways to principal cells in the lateral geniculate nucleus of the cat. *Exp. Brain Res.* **58**, 134–143.

Ahmed, B., Anderson, J. C., Douglas, R. J., Martin, K. A. C., and Nelson, J. C. (1994). Polyneuronal innervation of spiny stellate neurons in cat visual cortex. *J. Comp. Neurol.* **341**, 39–49.

Allendoerfer, K. L., and Shatz, C. J. (1994). The subplate, a transient neocortical structure: its role in the development of connections between thalamus and cortex. *Annu. Rev. Neurosci.* **17**, 185–218.

Allman, J. M., and Kaas, J. H. (1974). A crescent-shaped cortical visual area surrounding the middle temporal area (MT) in the owl monkey (*Aotus trivirgatus*). *Brain Res.* **81**, 199–213.

Allman, J. M., and Kaas, J. H. (1975). The dorsomedial cortical visual area: a third tier area in the occipital lobe of the owl monkey (*Aotus trivirgatus*). *Brain Res.* **100**, 473–487.

Alonso, J. M., Usrey, W. M., and Reid, R. C. (1996). Precisely correlated firing in cells of the lateral geniculate nucleus. *Nature* **383**, 815–819.

Angel, A., Magni, F., and Strata, P. (1967). The excitability of optic nerve terminals in the lateral geniculate nucleus after stimulation of visual cortex. *Arch. Ital. Biol.* **105**, 104–117.

Angelucci, A., Clascá, F., Bricolo, E., Cramer, K. S., and Sur, M. (1997).

Experimentally induced retinal projections to the ferret auditory thalamus: development of clustered eye-specific patterns in a novel target. *J. Neurosci.* **17**, 2040–2055.

Arcelli, P., Frassoni, C., Regondi, M. C., De Biasi, S., and Spreafico, R. (1997). GABAergic neurons in mammalian thalamus: a marker of thalamic complexity? *Brain Res. Bull.* **42**, 27–37.

Armstrong, D. M., and Edgley, S. A. (1988). Discharges of interpositus and Purkinje cells of the cat cerebellum during locomotion under different conditions. *J. Physiol. (Lond.)* **400**, 425–445.

Asanuma, C. (1989). Axonal arborizations of a magnocellular basal nucleus input and their relation to the neurons in the thalamic reticular nucleus of rats. *Proc. Natl. Acad. Sci. USA* **86**, 4746–4750.

Asanuma, C. (1992). Noradrenergic innervation of the thalamic reticular nucleus: a light and electron microscopic immunohistochemical study in rats. *J. Comp. Neurol.* **319**, 299–311.

Asanuma, C. (1994). GABAergic and pallidal terminals in the thalamic reticular nucleus of squirrel monkeys. *Exp. Brain Res.* **101**, 439–51.

Asanuma, C., and Porter, L. L. (1990). Light and electron microscopic evidence for a GABAergic projection from the caudal basal forebrain to the thalamic reticular nucleus in rats. *J. Comp. Neurol.* **302**, 159–72.

Ault, S. J., Leventhal, A. G., Vitek, D. J., and Creel, D. J. (1995). Abnormal ipsilateral visual field representation in areas 17 and 18 of hypopigmented cats. *J. Comp. Neurol.* **354**, 181–92.

Aumann, T. D., Ivanusic, J., and Horne, M. K. (1998). Arborisation and termination of single motor thalamocortical axons in the rat. *J. Comp. Neurol.* **396**, 121–130.

Baas, P. W. (1999). Microtubules and neuronal polarity: lessons from mitosis. *Neuron* **22**, 23–31.

Bacci, A., Verderio, C., Pravettoni, E., and Matteoli, M. (1999). The role of glial cells in synaptic function. *Phil. Trans. Roy. Soc. Lond. B.* **354**, 403–409.

Baker, F. H., and Malpeli, J. G. (1977). Effects of cryogenic blockade of visual cortex on the responses of lateral geniculate neurons in the monkey. *Exp. Brain Res.* **29**, 433–444.

Bal, T., and McCormick, D. A. (1993). Mechanisms of oscillatory activity in guinea-pig nucleus reticularis thalami in vitro: a mammalian pacemaker. *J. Physiol. (Lond.)* **468**, 669–691.

Bal, T., Von Krosigk, M., and McCormick, D.A. (1995). Synaptic and membrane mechanisms underlying synchronized oscillations in the ferret lateral geniculate nucleus *in vitro*. *J. Physiol. (Lond.)* **483**, 641–663.

Balercia, G., Kultas-Ilinsky, K., Bentivolglio, M., and Ilinsky, I. A. (1996). Neuronal and synaptic organization of the centromedian nucleus of the monkey thalamus: a quantitative ultrastructural study, with tract tracing and immunohistochemical observations. *J. Neurocytol.* **25**, 267–288.

Bartlett, E. L., and Smith, P. H. (1999). Anatomic, intrinsic, and synaptic properties of dorsal and ventral division neurons in rat medial geniculate body. *J. Neurophysiol.* **81**, 1999–2016.

Bartlett, E. L., Stark, J. M., Guillery, R. W., and Smith, P. H. (2000). A comparison of the fine structure of axon terminals in the rat medial geniculate body. *Neuroscience*, in press.

Beck, P. D., Pospichal, M. W., and Kaas, J. H. (1996). Topography, architecture, and connections of somatosensory cortex in opossums: evidence for five somatosensory areas. *J. Comp. Neurol.* **366**, 109–133.

Belluscio, L., Koentges, G., Axel, R., and Dulac, C. (1999). A map of pheromone receptor activation in the mammalian brain. *Cell* **97**, 209–220.

Bender, D. B. (1983). Visual activation of neurons in the primate pulvinar depends on cortex but not colliculus. *Brain Res.* **279**, 258–261.

Bender, D. B. (1988). Electrophysiological and behavioral experiments on the primate pulvinar. *Progr. Brain Res.* **75**, 55–65.

Benjamin, R. M., and Burton, H. (1968). Projection of taste nerve afferents to anterior opercular-insular cortex in squirrel monkey *(Saimiri sciureus)*. *Brain Res.* **7,** 221–231.

Bentivoglio, M., Macchi, G., and Albanese, A. (1981). The cortical projection of the thalamic intralaminar nuclei, as studied in cat and rat with multiple fluorescent retrograde tracing technique. *Neurosci. Lett.* **26,** 5–10.

Berendse, H. W., and Groenewegen, H. J. (1991). Restricted cortical termination fields of the midline and intralaminar thalamic nuclei in the rat. *Neuroscience* **42,** 73–102.

Berman, A. L., and Jones, E. G. (1982). "The Thalamus and Basal Telencephalon of the Cat: A Cytoarchitectonic Atlas with Stereotaxic Coordinates." University of Wisconsin Press, Madison, WI.

Berman, N., and Cynader, M. (1972). Comparison of receptive-field organization of the superior colliculus in Siamese and normal cats. *J. Physiol. (Lond.)* **224,** 363–389.

Bernander, O., Douglas, R. J., Martin, K. A. C., and Koch, C. (1991). Synaptic background activity influences spatiotemporal integration in single pyramidal cells. *Proc. Natl. Acad. Sci. USA* **88,** 11569–11573.

Bickford, M. E., Günlük, A. E., van Horn, S. C., and Sherman, S. M. (1994). GABAergic projection from the basal forebrain to the visual sector of the thalamic reticular nucleus in the cat. *J. Comp. Neurol.* **348,** 481–510.

Bickford, M. E., Günlük, A. E., Guido, W., and Sherman, S. M. (1993). Evidence that cholinergic axons from the parabrachial region of the brainstem are the exclusive source of nitric oxide in the lateral geniculate nucleus of the cat. *J. Comp. Neurol.* **334,** 410–430.

Bickford, M. E., Breckinridge, C. W., and Patel, N. C. (1999). Two types of interneurons in the cat visual thalamus are distinguished by morphology, synaptic connections, and nitric oxide synthase content. *J. Comp. Neurol.* **413,** 83–100.

Birnbacher, D., and Albus, K. (1987). Divergence of single axons in afferent projections to the cat's visual cortical areas 17, 18 and 19: a parametric study. *J. Comp. Neurol.* **261,** 543–561.

Bishop, G. (1959). The relation between nerve fiber size and sensory modality: phylogenetic implications of the afferent innervation of cortex. *J. Nerv. Ment. Dis.* **128,** 89–114.

Bishop, P. O., Jeremy, D., and McLeod, J. G. (1953). Phenomenon of repetitive firing in lateral geniculate nucleus of cat. *J. Neurophysiol.* **16,** 437–447.

Black, M. M., and Baas, P. W. (1989). The basis of polarity in neurons. *Trends Neurosci.* **12,** 211–214.

Blasdel, G. G., and Lund, J. S. (1983). Termination of afferent axons in macaque striate cortex. *J. Neurosci.* **3,** 1389–1413.

Bloomfield, S. A., Hamos, J. E., and Sherman, S. M. (1987). Passive cable properties and morphological correlates of neurones in the lateral geniculate nucleus of the cat. *J. Physiol. (Lond.)* **383,** 653–692.

Bloomfield, S. A., and Sherman, S. M. (1989). Dendritic current flow in relay cells and interneurons of the cat's lateral geniculate nucleus. *Proc. Natl. Acad. Sci. USA* **86,** 3911–3914.

Bodian, D. (1962). The generalized vertebrate neuron. *Science* **137,** 323–326.

Bolam, J. P. (1992). "Experimental Neuroanatomy: A Practical Approach," pp. 1–273. IRL Press, Oxford, UK.

Bourassa, J., and Deschênes, M. (1995). Corticothalamic projections from the primary visual cortex in rats: a single fiber study using biocytin as an anterograde tracer. *Neuroscience* **66,** 253–263.

Bourassa, J., Pinault, D., and Deschênes, M. (1995). Corticothalamic projections from the cortical barrel field to the somatosensory thalamus in rats: a single-fibre study using biocytin as an anterograde tracer. *Eur. J. Neurosci.* **7,** 19–30.

Bowling, D. B., and Michael, C. R. (1984). Terminal patterns of single, physiologically characterized optic tract fibers in the cat's lateral geniculate nucleus. *J. Neurosci.* **4,** 198–216.

Boycott, B. B., and Wässle, H. (1974). The morphological types of ganglion cells of the domestic cat's retina. *J. Physiol. (Lond.)* **240,** 397–419.

Boyd, J. D., and Matsubara, J. A. (1996). Laminar and columnar patters of geniculocortical projections in the cat: relationship to cytochrome oxydase. *J. Comp. Neurol.* **365**, 659–682.

Britten, K. H., Shadlen, M. N., Newsome, W. T., and Movshon, J. A. (1993). Responses of neurons in macaque MT to stochastic motion signals. *Visual. Neurosci.* **10**, 1157–1169.

Brown, D. A., Abogadie, F. C., Allen, T. G., Buckley, N. J., Caulfield, M. P., Delmas, P., Haley, J. F., Lamas, J. A., and Selyanko, A. A. (1997). Muscarinic mechanisms in nerve cells. *Life Sci.* **60**, 1137–1144.

Bullier, J., Kennedy, H., and Salinger, W. L. (1984). Bifurcation of subcortical afferents to visual areas 17, 18, and 19 in the cat cortex. *J. Comp. Neurol.* **228**, 309–328.

Bullier, J., and Norton, T. T. (1979). X and Y relay cells in cat lateral geniculate nucleus: quantitative analysis of receptive-field properties and classification. *J. Neurophysiol.* **42**, 243–273.

Bullier, J., Schall, J. D., and Morel, A. (1996). Functional streams in occipito-frontal connections in the monkey. *Behav. Brain Res.* **76**, 89–97.

Bullock, T. H. (1959). Neuron doctrine and electrophysiology. *Science* **129**, 997–1002.

Cajal, S., Ramón y. (1911). "Histologie du Système Nerveux de l'Homme et des Vertébrés." (Translated by L. Azoulay.) Maloine, Paris, France.

Cajal, S., Ramón y. (1995). "Histology of the Nervous System of Man and Vertebrates." (Translated from the French by N. and L. W. Swanson.) Oxford University Press, Oxford, UK.

Carey, R. G., Fitzpatrick, D., and Diamond, I. T. (1979). Layer I of striate cortex of *Tupaia glis* and *Galago senegalensis*: projections from thalamus and claustrum revealed by retrograde transport of horseradish peroxidase. *J. Comp. Neurol.* **186**, 393–437.

Casagrande, V. A. (1994). A third parallel visual pathway to primate area V1. *Trends Neurosci.* **17**, 305–309.

Casagrande, V. A., and Kaas, J. H. (1994). The afferent, intrinsic and efferent connections of primary visual cortex in primates. *In* "Cerebral Cortex" (A. Peters and K. S. Rockland, eds.), Vol. 10, pp. 201–259. Plenum Press, New York, NY.

Casagrande, V. A., and Norton, T. T. (1991). Lateral geniculate nucleus: A review of its physiology and function. *In* "Vision and Visual Dysfunction" (A. G. Leventhal, ed.), pp. 41–84. MacMillan Press, London, UK.

Casanova, C. (1993). Response properties of neurons in area 17 projecting to the striate-recipient zone of the cat's lateralis posterior-pulvinar complex: comparison with cortico-tectal cells. *Exp. Brain Res.* **96**, 247–59.

Casanova, C., and Molotchnikoff, S. (1990). Influence of the superior colliculus in visual responses of cells in the rabbit's lateral posterior nucleus. *Exp. Brain Res.* **80**, 387–396.

Castro-Alamancos, M. A., and Connors, B. W. (1997). Thalamocortical synapses. *Progr. Neurobiol.* **51**, 581–606.

Catsman-Berrevoets, C. E., and Kuypers, H. G. J. M. (1978). Differential laminar distribution of corticothalamic neurons projecting to the VL and the center median. An HRP study in the cynomolgus monkey. *Brain Res.* **154**, 359–365.

Caviness, V. S. Jr., and Frost, D. O. (1983). Thalamocortical projections in the reeler mutant mouse. *J. Comp. Neurol.* **219**, 182–202.

Chalupa, L. M., and Abramson, B. P. (1988). Receptive-field properties in the tecto- and striate-recipient zones of the cat's lateral posterior nucleus. *Progr. Brain Res.* **75**, 85–94.

Chalupa, L. M., and Abramson, B. P. (1989). Visual receptive fields in the striate-recipient zone of the lateral posterior-pulvinar complex. *J. Neurosci.* **9**, 347–357.

Churchill, L., Zahm, D. S., and Kalivas, P. W. (1996). The mediodorsal nucleus of the thalamus in rats—I. forebrain GABAergic innervation. *Neuroscience* **70**, 93–102.

Cicirata, F., Angaut, P., Serapide, M. F., and Panto, M. R. (1990). Functional organization of the direct and indirect projection via the reticularis thalami nuclear complex from the motor cortex to the thalamic nucleus ventralis lateralis. *Exp. Brain Res.* **79,** 325–337.

Cleland, B. G., Dubin, M. W., and Levick, W. R. (1971). Sustained and transient neurones in the cat's retina and lateral geniculate nucleus. *J. Physiol. (Lond.)* **217,** 473–496.

Cleland, B. G., and Lee, B. B. (1985). A comparison of visual responses of cat lateral geniculate nucleus neurones with those of ganglion cells afferent to them. *J. Physiol. (Lond.)* **369,** 249–268.

Coleman, K. A., and Mitrofanis, J. (1996). Organization of the visual reticular thalamic nucleus of the rat. *Eur. J. Neurosci.* **8,** 388–404.

Collingridge, G. L., and Bliss, T. V. P. (1987). NMDA receptors: Their role in long-term potentiation. *Trends Neurosci.* **10,** 288–293.

Colonnier, M. (1968). Synaptic patterns on different cell types in the different laminae of the visual cortex. An electron microscope study. *Brain Res.* **9,** 268–287.

Colonnier, M., and Guillery, R. W. (1964). Synaptic organization in the lateral geniculate nucleus of the monkey. *Z. Zellforsch.* **62,** 333–355.

Conley, M., and Diamond, I. T. (1990). Organization of the visual sector of the thalamic reticular nucleus in *Galago*. Evidence that the dorsal lateral geniculate and pulvinar nuclei occupy separate parallel tiers. *Eur. J. Neurosci.* **2,** 211–226.

Conley, M., Fitzpatrick, D., and Diamond, I. T. (1984). The laminar organization of the lateral geniculate body and the striate cortex in the tree shrew (*Tupaia glis*). *J. Neurosci.* **4,** 171–197.

Conley, M., Kupersmith, A. C., and Diamond, I. T. (1991). The organization of projections from subdivisions of the auditory cortex and thalamus to the auditory sector of the thalamic reticular nucleus in *Galago*. *Eur. J. Neurosci.* **3,** 1089–1103.

Conn, P. J., and Pin, J. P. (1997). Pharmacology and functions of metabotropic glutamate receptors. *Ann. Rev. Pharmacol. Toxicol.* **37,** 205–237.

Connolly, M., and Van Essen, D. (1984). The representation of the visual field in parvicellular and magnocellular layers of the lateral geniculate nucleus in the macaque monkey. *J. Comp. Neurol.* **226,** 544–564.

Connor, J. A., and Stevens, C. F. (1971). Prediction of repetitive firing behaviour from voltage clamp data on an isolated neurone soma. *J. Physiol. (Lond.)* **213,** 31–53.

Cornwall, J., Cooper, J. D., and Phillipson, O. T. (1990). Projections to the rostral reticular thalamic nucleus in the rat. *Exp. Brain Res.* **80,** 157–171.

Cotman, C. W., Monaghan, D. T., and Gamong, A. H. (1988). Excitatory amino acid neurotransmission: NMDA receptors and Hebb-type synaptic plasticity. *Annu. Rev. Neurosci.* **11,** 61–80.

Coulter, D. A., Huguenard, J. R., and Prince, D. A. (1989). Calcium currents in rat thalamocortical relay neurones: kinetic properties of the transient, low-threshold current. *J. Physiol. (Lond.)* **414,** 587–604.

Cowan, W. M., and Powell, T. P. S. (1954). An experimental study of the relation between the medial mamillary nucleus and the cingulate cortex. *Proc. Roy. Soc. Lond.* B **143,** 114–125.

Cox, C. L., Huguenard, J. R., and Prince, D. A. (1996). Heterogeneous axonal arborizations of rat thalamic reticular neurons in the ventrobasal nucleus. *J. Comp. Neurol.* **366,** 416–430.

Cox, C. L., Huguenard, J. R., and Prince, D. A. (1997). Nucleus reticularis neurons mediate diverse inhibitory effects in thalamus. *Proc. Natl. Acad. Sci. USA* **94,** 8854–8859.

Cox, C. L., and Sherman, S. M. (1999). Glutamate inhibits thalamic reticular neurons. *J. Neurosci.* **19,** 6694–6699.

Cox, C. L., and Sherman, S. M. (2000). Control of dendritic outputs of inhibitory interneurons in the lateral geniculate nucleus. *Neuron,* in press.

Cox, C. L., Zhou, Q., and Sherman, S. M. (1998). Glutamate locally activates dendritic outputs of thalamic interneurons. *Nature* **394,** 478–482.

Crabtree, J. W. (1992a). The somatotopic organization within the cat's thalamic reticular nucleus. *Eur. J. Neurosci.* **4,** 1352–1361.

Crabtree, J. W. (1992b). The somatopic organization within the rabbit's thalamic reticular nucleus. *Eur. J. Neurosci.* **4**, 1343–1351.

Crabtree, J. W. (1996). Organization in the somatosensory sector of the cat's thalamic reticular nucleus. *J. Comp. Neurol.* **366**, 207–22.

Crabtree, J. W. (1998). Organization in the auditory sector of the cat's thalamic reticular nucleus. *J. Comp. Neurol.* **390**, 167–182.

Crabtree, J. W., and Killackey, H. P. (1989). The topographic organization and axis of projection within the visual sector of the rabbit's thalamic reticular nucleus. *Eur. J. Neurosci.* **1**, 94–109.

Cragg, B. G. (1969). The topography of the afferent projections in the circumstriate visual cortex of the monkey studied by the Nauta method. *Vis. Res.* **9**, 733–747.

Crair, M. C., Gillespie, D. C., and Stryker, M. P. (1998). The role of visual experience in the development of columns in cat visual cortex. *Science* **279**, 566–570.

Crick, F. (1984). Function of the thalamic reticular complex: The searchlight hypothesis. *Proc. Natl. Acad. Sci. USA* **81**, 4586–4590.

Crick, F., and Koch, C. (1998). Constraints on cortical and thalamic projections: the no-strong-loops hypothesis. *Nature* **391**, 245–250.

Cucchiaro, J. B., Bickford, M. E., and Sherman, S. M. (1991). A GABAergic projection from the pretectum to the dorsal lateral geniculate nucleus in the cat. *Neuroscience* **41**, 213–226.

Cucchiaro, J., and Guillery, R. W. (1984). The development of the retinogeniculate pathways in normal and albino ferrets. *Proc. Roy. Soc. Lond. B* **223**, 141–164.

Cucchiaro, J. B., Uhlrich, D. J., and Sherman, S. M. (1988). Parabrachial innervation of the cat's dorsal lateral geniculate nucleus: an electron microscopic study using the tracer *Phaseolus vulgaris leucoagglutinin* (PHA-L). *J. Neurosci.* **8**, 4576–4588.

Cucchiaro, J. B., Uhlrich, D. J., and Sherman, S. M. (1991). Electron-microscopic analysis of synaptic input from the perigeniculate nucleus to the A-laminae of the lateral geniculate nucleus in cats. *J. Comp. Neurol.* **310**, 316–336.

Cucchiaro, J. B., Uhlrich, D. J., and Sherman, S. M. (1993). Ultrastructure of synapses from the pretectum in the A-laminae of the cat's lateral geniculate nucleus. *J. Comp. Neurol.* **334**, 618–630.

Cudeiro, J., Rivadulla, C., Rodriguez, R., Martinez-Conde, S., Martinez, L., Grieve, K. L., and Acuña, C. (1996). Further observations on the role of nitric oxide in the feline lateral geniculate nucleus. *Eur. J. Neurosci.* **8**, 144–152.

Cudeiro, J., and Sillito, A. M. (1996). Spatial frequency tuning of orientation-discontinuity-sensitive corticofugal feedback to the cat lateral geniculate nucleus. *J. Physiol. (Lond.)* **490**, 481–492.

Cusick, C. G., Steindler, D. A., and Kaas, J. H. (1985). Corticocortical and collateral thalamocortical connections of postcentral somatosensory cortical areas in squirrel monkeys: a double-labeling study with radiolabeled wheatgerm agglutinin and wheatgerm agglutinin conjugated to horseradish peroxidase. *Somatosensory Res.* **3**, 1–31.

Dagnino, N., Favale, E., Loeb, C., and Manfredi, M. (1965). Thalamic transmission changes during the rapid eye movements of deep sleep. *Arch. Intern. Physiol. Biochim.* **73**, 858–861.

Dagnino, N., Favale, E., Loeb, C., Manfredi, M., and Seitun, A. (1966). Pontine triggering of phasic changes in sensory transmission during deep sleep. *Arch. Intern. Physiol. Biochim.* **74**, 889–894.

Dagnino, N., Favale, E., Manfredi, M., Seitun, A., and Tartaglione, A. (1971). Tonic changes in excitability of thalamocortical neurons during the sleep-waking cycle. *Brain Res.* **29**, 354–357.

Daniel, P. M., and Whitteridge, D. (1961). The representation of the visual field on the cerebral cortex in monkeys. *J. Physiol. (Lond.)* **159**, 203–221.

Darian-Smith, C., and Darian-Smith, I. (1993). Thalamic projections to areas 3a, 3b, and 4 in the sensorimotor cortex of the mature and infant macaque monkey. *J. Comp. Neurol.* **335**, 173–199.

Darian-Smith, C., Tan, A., and Edwards, S. (1999). Comparing thalamocortical and corticothalamic microstructure and spatial reciprocity in the macaque ventral posterolateral nucleus (VPLc) and medial pulvinar. *J. Comp. Neurol.* **410**, 211–234.

De Biasi, S., Arcelli, P., and Spreafico, R. (1994). Parvalbumin immunoreactivity in the thalamus of guinea pig: Light and electron microscopic correlation with gamma-aminobutyric acid immunoreactivity. *J. Comp. Neurol.* **348**, 556–569.

de Curtis, M., Spreafico, R., and Avanzini, G. (1989). Excitatory amino acids mediate responses elicited in vitro by stimulation of cortical afferents to reticularis thalami neurons of the rat. *Neuroscience* **33**, 275–283.

de Lima, A. D., and Singer, W. (1987). The brainstem projection to the lateral geniculate nucleus in the cat: identification of cholinergic and monoaminergic elements. *J. Comp. Neurol.* **259**, 92–121.

Derrington, A. M., and Lennie, P. (1984). Spatial and temporal contrast sensitivities of neurones in lateral geniculate nucleus of macaque. *J. Physiol. (Lond.)* **357**, 219–240.

Deschênes, M., Bourassa, J., and Pinault, D. (1994). Corticothalamic projections from layer V cells in rat are collaterals of long-range corticofugal axons. *Brain Res.* **664**, 215–219.

Deschênes, M., Madariaga-Domich, A., and Steriade, M. (1985). Dendrodendritic synapses in the cat reticularis thalami nucleus: a structural basis for thalamic spindle synchronization. *Brain Res.* **334**, 165–168.

Deschênes, M., Paradis, M., Roy, J. P., and Steriade, M. (1984). Electrophysiology of neurons of lateral thalamic nuclei in cat: resting properties and burst discharges. *J. Neurophysiol.* **51**, 1196–1219.

Deschênes, M., Veinante, P., and Zhongwei, Z. (1998). The organization of corticothalamic projections: reciprocity versus parity. *Brain Res. Rev.* **28**, 286–308.

Destexhe, A., Contreras, D., and Steriade, M. (1998a). Mechanisms underlying the synchronizing action of corticothalamic feedback through inhibition of thalamic relay cells. *J. Neurophysiol.* **79**, 999–1016.

Destexhe, A., Contreras, D., Steriade, M., Sejnowski, T. J., and Huguenard, J. R. (1996). In vivo, in vitro, and computational analysis of dendritic calcium currents in thalamic reticular neurons. *J. Neurosci.* **16**, 169–185.

Destexhe, A., Neubig, M., Ulrich, D., and Huguenard, J. (1998b). Dendritic low-threshold calcium currents in thalamic relay cells. *J. Neurosci.* **18**, 3574–3588.

Diamond, I. T., Conley, M., Itoh, K., and Fitzpatrick, D. (1985). Laminar organization of geniculocortical projections in *Galago senegalensis* and *Aotus trivirgatus*. *J. Comp. Neurol.* **242**, 584–610.

Diamond, I. T., Fitzpatrick, D., and Schmechel, D. (1993). Calcium binding proteins distinguish large and small cells of the ventral posterior and lateral geniculate nuclei of the prosimian galago and the tree shrew (*Tupaia belangeri*). *Proc. Natl. Acad. Sci. USA* **90**, 1425–1429.

Diamond, M. E., Armstrong-James, M., Budway, M. J., and Ebner, F. F. (1992). Somatic sensory responses in the rostral sector of the posterior group (POm) and in the ventral posterior medial nucleus (VPM) of the rat thalamus: dependence on the barrel field cortex. *J. Comp. Neurol.* **319**, 66–84.

Dick, A., Kaske, A., and Creutzfeldt, O. D. (1991). Topographical and topological organization of the thalamocortical projection to the striate and prestriate cortex in the marmoset (*Callithrix jacchus*). *Exp. Brain Res.* **84**, 233–253.

Ding, Y., and Casagrande, V. A. (1997). The distribution and morphology of LGN K pathway axons within the layers and CO blobs of owl monkey V1. *Vis. Neurosci.* **14**, 691–704.

Ding, Y., and Casagrande, V. A. (1998). Synaptic and neurochemical characterization of parallel pathways to the cytochrome oxidase blobs of primate visual cortex. *J. Comp. Neurol.* **391**, 429–443.

Dowling, J. E. (1991). Retina. *In* "Encyclopedia of Human Biology," Vol. 6, pp. 615–631. Academic Press, San Diego, CA.

Duffy, K. R., Murphy, K. M., and Jones, D. G. (1998). Analysis of the postnatal growth of visual cortex. *Vis. Neurosci.* **15**, 831–839.

Dunlap, K., Luebke, J. I., and Turner, T. J. (1995). Exocytotic Ca^{2+} channels in mammalian central neurons. *Trends Neurosci.* **18**, 89–98.

Earle, K. L., and Mitrofanis, J. (1996). Genesis and fate of the perireticular thalamic nucleus during early development. *J. Comp. Neurol.* **367**, 246–263.

Eaton, S. A., and Salt, T. E. (1996). Role of N-methyl-D-aspartate and metabotropic glutamate receptors in corticothalamic excitatory postsynaptic potentials *in vivo. Neuroscience* **73**, 1–5.

Eccles, J. C., Ito, M., and Szentágothai, J. (1967). "The Cerebellum as a Neuronal Machine." Springer, New York, NY.

Elekessy, E. I., Campion, J. E., and Henry, G. H. (1973). Differences between the visual fields of Siamese and common cats. *Vis. Res.* **13**, 2533–2543.

Erişir, A., Van Horn, S. C., Bickford, M. E., and Sherman, S. M. (1997). Immunocytochemistry and distribution of parabrachial terminals in the lateral geniculate nucleus of the cat: A comparison with corticogeniculate terminals. *J. Comp. Neurol.* **377**, 535–549

Erişir, A., Van Horn, S. C., and Sherman, S. M. (1998). Distribution of synapses in the lateral geniculate nucleus of the cat: Differences between laminae A and A1 and between relay cells and interneurons. *J. Comp. Neurol.* **390**, 247–255.

Famiglietti, E. V., and Peters, A. (1972). The synaptic glomerulus and the intrinsic neuron in the dorsal lateral geniculate nucleus of the cat. *J. Comp. Neurol.* **144**, 285–334.

Favale, E., Loeb, C., and Manfredi, M. (1964). Cortical responses evoked by stimulation of the optic pathways during natural sleep and during arousal. *Arch. Intern. Physiol. Biochim.* **72**, 221–228.

Feig, S. L., and Guillery, R. W. (2000). Corticothalamic axons contact blood vessels as well as nerve cells in the thalamus. *Eur. J. Neurosci.* **12**, 2195–2198.

Feig, S. L., and Harting, J. K. (1994). Ultrastructural studies of the primate lateral geniculate nucleus: morphology and spinal relationships of axon terminals arising from the retina, visual cortex (area 17), superior colliculus, parabigeminal nucleus, and pretectum of *Galago crassicaudatus. J. Comp. Neurol.* **343**, 17–34.

Feig, S., and Harting, J. K. (1998). Corticocortical communication via the thalamus: ultrastructural studies of corticothalamic projections from area 17 to the lateral posterior nucleus of the cat and inferior pulvinar nucleus of the owl monkey. *J. Comp. Neurol.* **395**, 281–295.

Felleman, D. J., and Van Essen, D. C. (1991). Distributed hierarchical processing in the primate cerebral cortex. *Cerebral Cortex* **1**, 1–47.

Ferster, D. (1987). Origin of orientation-selective EPSPs in simple cells of cat visual cortex. *J. Neurosci.* **7**, 1780–1791.

Ferster, D. (1994). Linearity of synaptic interactions in the assembly of receptive fields in cat visual cortex. *Curr. Opin. Neurobiol.* **4**, 563–568.

Ferster, D., Chung, S., and Wheat, H. (1996). Orientation selectivity of thalamic input to simple cells of cat visual cortex. *Nature* **380**, 249–252.

Ferster, D., and LeVay, S. (1978). The axonal arborizations of lateral geniculate neurons in the striate cortex of the cat. *J. Comp. Neurol.* **182**, 923–944.

Fetz, E. E., Toyama, K., and Smith, W. (1991). Synaptic interactions between cortical neurons. *In* "Cerebral Cortex" Vol. 9, "Normal and Altered States of Function" (A. Peters and E. G. Jones, eds.), pp. 1–47. Plenum, New York, NY.

Fischer, W. H., Schmidt, M., and Hoffmann, K. P. (1998). Saccade-induced activity of dorsal lateral geniculate nucleus X- and Y-cells during pharmacological inactivation of the cat pretectum. *Vis. Neurosci.* **15,** 197–210.

Fitzpatrick, D., Diamond, I. T., and Raczkowski, D. (1989). Cholinergic and monoaminergic innervation of the cat's thalamus: comparison of the lateral geniculate nucleus with other principal sensory nuclei. *J. Comp. Neurol.* **288,** 647–675.

Fitzpatrick, D., Penny, G. R., and Schmechel, D. E. (1984). Glutamic acid decarboxylase-immunoreactive neurons and terminals in the lateral geniculate nucleus of the cat. *J. Neurosci.* **4,** 1809–1829.

Fitzpatrick, D., and Raczkowski, D. (1990). Innervation patterns of single physiologically identified geniculocortical axons in the striate cortex of the tree shrew. *Proc. Natl. Acad. Sci. USA* **87,** 449–453.

Fitzpatrick, D., Usrey, W. M., Schofield, B. R., and Einstein, G. (1994). The sublaminar organization of corticogeniculate neurons in layer 6 of macaque striate cortex. *Vis. Neurosci.* **11,** 307–315.

Fitzpatrick, D. C., Olsen, J. F., and Suga, N. (1998). Connections among functional areas in the mustached bat auditory cortex. *J. Comp. Neurol.* **391,** 366–396.

Florence, S. L., and Casagrande, V. A. (1987). Organization of individual afferent axons in layer IV of striate cortex in a primate. *J. Neurosci.* **7,** 3850–3868.

Francois, C., Percheron, G., Parent, A., Sadikot, A. F., Fenelon, G., and Yelnik, J. (1991). Topography of the projection from the central complex of the thalamus to the sensorimotor striatal territory in monkeys. *J. Comp. Neurol.* **305,** 17–34.

Freund, T. F., Martin, K. A. C., Somogyi, P., and Whitteridge, D. (1985). Innervation of cat visual areas 17 and 18 by physiologically identified X- and Y-type thalamic afferents. II. Identification of postsynaptic targets by GABA immunocytochemistry and Golgi impregnation. *J. Comp. Neurol.* **242,** 275–291.

Friedlander, M. J., Lin, C.-S., Stanford, L. R., and Sherman, S. M. (1981). Morphology of functionally identified neurons in lateral geniculate nucleus of the cat. *J. Neurophysiol.* **46,** 80–129.

Fries, W. (1981). The projection from the lateral geniculate nucleus to the prestriate cortex of the macaque monkey. *Proc. Roy. Soc. Lond.* B **213,** 73–86.

Frost, D. O., and Caviness, V. S. Jr. (1980). Radial organization of thalamic projections to the neocortex in the mouse. *J. Comp. Neurol.* **194,** 369–393.

Funke, K., and Eysel, U. T. (1995). Possible enhancement of GABAergic inputs to cat dorsal lateral geniculate relay cells by serotonin. *NeuroReport* **6,** 474–476.

Gandia, J. A., De Las Heras, S., Garcia, M., and Gimenez-Amaya, J. M. (1993). Afferent projections to the reticular thalamic nucleus from the globus pallidus and the substantia nigra in the rat. *Brain Res. Bull.* **32,** 351–358.

Gardner, E. P., and Fuchs, A. F. (1975). Single-unit responses to natural vestibular stimuli and eye movements in deep cerebellar nuclei of the alert rhesus monkey. *J. Neurophysiol.* **38,** 627–649.

Garey, L. J., and Powell, T. P. S. (1967). The projection of the lateral geniculate nucleus upon the cortex in the cat. *Proc. Roy. Soc. Lond.* B **179,** 41–63.

Geisert, E. E. (1980). Cortical projections of the lateral geniculate nucleus in the cat. *J. Comp. Neurol.* **190,** 793–812.

Geisert, E. E., Langsetmo, A., and Spear, P. D. (1981). Influence of the corticogeniculate pathway on response properties of cat lateral geniculate neurons. *Brain Res.* **208,** 409–415.

Gereau, R. W., and Conn, P. J. (1994). A cyclic AMP-dependent form of associative synaptic plasticity induced by coactivation of b-adrenergic receptors and metabotropic glutamate receptors in rat hippocampus. *J. Neurosci.* **14,** 3310–3318.

Ghelarducci, B., Pisa, M., and Pompeiano, O. (1970). Transformation of somatic afferent volleys across the prethalamic and thalamic components of the lemniscal system during the rapid eye movements of sleep. *Electroencephalogr. Clin. Neurophysiol.* **29,** 348–357.

Ghosh, A., and Shatz, C. J. (1994). Segregation of geniculocortical afferents during the critical period: a role for subplate neurons. *J. Neurosci.* **14,** 3862–3880.

Gilbert, C. D., and Kelly, J. P. (1975). The projections of cells in different layers of the cat's visual cortex. *J. Comp. Neurol.* **163,** 81–105.

Giménez-Amaya, J. M., McFarland, N. R., de las Heras, S., and Haber, S. N. (1995). Organization of thalamic projections to the ventral striatum in the primate. *J. Comp. Neurol.* **354,** 127–149.

Glickstein, M. (1988). The discovery of the visual cortex. *Sci. Am.* **259,** 118–127.

Godwin, D. W., Van Horn, S. C., Erişir, A., Sesma, M., Romano, C., and Sherman, S. M. (1996a). Ultrastructural localization suggests that retinal and cortical inputs access different metabotropic glutamate receptors in the lateral geniculate nucleus. *J. Neurosci.* **16,** 8181–8192.

Godwin, D. W., Vaughan, J. W., and Sherman, S. M. (1996b). Metabotropic glutamate receptors switch visual response mode of lateral geniculate nucleus cells from burst to tonic. *J. Neurophysiol.* **76,** 1800–1816.

Goldman-Rakic, P. S., and Porrino, L. J. (1985). The primate mediodorsal (MD) nucleus and its projection to the frontal lobe. *J. Comp. Neurol.* **242,** 535–560.

Gonzalo-Ruiz, A., Lieberman, A. R., and Sanz-Anquela, J. M. (1995). Organization of serotoninergic projections from the raphé nuclei to the anterior thalamic nuclei in the rat: A combined retrograde tracing and 5-HT immunohistochemical study. *J. Chem. Neuroanat.* **8,** 103–115.

Goodchild, A. K., and Martin, P. R. (1998). The distribution of calcium-binding proteins in the lateral geniculate nucleus and visual cortex of a New World monkey, the marmoset, *Callithrix jacchus. Vis. Neurosci.* **15,** 625–642.

Gray, E. G. (1959). Axo-somatic and Axo-dendritic synapses of the cerebral cortex. An electron microscope study. *J. Anat. (Lond.)* **93,** 420–433.

Gray, E. G. (1969). Electron microscopy of excitatory and inhibitory synapses: a brief review. *Progr. Brain Res.* **31,** 141–155.

Green, D. M., and Swets, J. A. (1966). "Signal Detection Theory and Psychophysics." Wiley, New York, NY.

Groenewegen, H. J. (1988). Organization of the afferent connections of the mediodorsal thalamic nucleus in the rat, related to the mediodorsal-prefrontal topography. *Neuroscience* **24,** 379–443.

Groenewegen, H. J., Berendse, H. W., Wolters, J. G., and Lohman, A. H. (1990). The anatomical relationship of the prefrontal cortex with the striato-pallidal system, the thalamus and the amygdala: evidence for a parallel organization. *Progr. Brain Res.* **85,** 95–118.

Gross, C. G. (1998). "Brain, Vision, Memory: Tales in the History of Neuroscience." MIT Press, Cambridge, MA.

Gudden, B von. (1874). Ueber die Kreuzung der Fasern im Chiasma nervorum opticorum. *Albrecht. von. Graefes. Arch. Ophthalmol.* **20 II,** 249–268.

Guido, W., Lu, S.-M., and Sherman, S. M. (1992). Relative contributions of burst and tonic responses to the receptive field properties of lateral geniculate neurons in the cat. *J. Neurophysiol.* **68,** 2199–2211.

Guido, W., Lu, S.-M., Vaughan, J. W., Godwin, D. W., and Sherman, S. M. (1995). Receiver operating characteristic (ROC) analysis of neurons in the cat's lateral geniculate nucleus during tonic and burst response mode. *Vis. Neurosci.* **12,** 723–741.

Guido, W., and Weyand, T. (1995). Burst responses in thalamic relay cells of the awake behaving cat. *J. Neurophysiol.* **74,** 1782–1786.

Guillery, R. W. (1966). A study of Golgi preparations from the dorsal lateral geniculate nucleus of the adult cat. *J. Comp. Neurol.* **128**, 21–50.

Guillery, R. W. (1967a). Patterns of fiber degeneration in the dorsal lateral geniculate nucleus of the cat following lesions in the visual cortex. *J. Comp. Neurol.* **140**, 197–222.

Guillery, R. W. (1967b). A light and electron microscopical study of neurofibrils and neurofilaments at neuroneuronal junctions in the dorsal lateral geniculate nucleus of the cat. *Am. J. Anat.* **120**, 583–604.

Guillery, R. W. (1969a). The organization of synaptic interconnections in the laminae of the dorsal lateral geniculate nucleus of the cat. *Z. Zellforsch.* **96**, 1–38.

Guillery, R. W. (1969b). A quantitative study of synaptic interconnections in the dorsal lateral geniculate nucleus of the cat. *Z. Zellforsch.* **96**, 39–48.

Guillery, R. W. (1995). Anatomical evidence concerning the role of the thalamus in corticocortical communication: a brief review. *J. Anat. (Lond.)* **187**, 583–592.

Guillery, R. W. (1996). Why do albinos and other hypopigmented mutants lack normal binocular vision, and what else is abnormal in their central visual pathways. *Eye* **10**, 217–221.

Guillery, R. W., and Casagrande, V. A. (1977). Studies of the modifiability of the visual pathways in Midwestern Siamese cats. *J. Comp. Neurol.* **174**, 15–46.

Guillery, R. W., and Colonnier, M. (1970). Synaptic patterns in the dorsal lateral geniculate nucleus of the monkey. *Zeitschrift Zellforsch.* **103**, 90–108.

Guillery, R. W., Feig, S. L., and Lozsádi, D. A. (1998). Paying attention to the thalamic reticular nucleus. *Trends Neurosci.* **21**, 28–32.

Guillery, R. W., Hickey, T. L., Kaas, J. H., Felleman, D. J., Debruyn, E. J., and Sparks, D. L. (1984). Abnormal central visual pathways in the brain of an albino green monkey (*Cercopithecus aethiops*). *J. Comp. Neurol.* **226**, 165–183.

Guillery, R. W., and Kaas, J. H. (1971). A study of normal and congenitally abnormal retinogeniculate projections in cats. *J. Comp. Neurol.* **143**, 73–100.

Guillery, R. W., Mason, C. A., and Taylor, J. S. H. (1995). Developmental determinants at the mammalian optic chiasm. *J. Neurosci.* **15**, 4727–4737.

Guiterrez, C., Cox, C. L., and Sherman, S. M. (1999). Dynamics of low threshold spike activation in relay neurons of the cat lateral geniculate nucleus. Submitted for publication.

Haberly, L. B., and Bower, J. M. (1989). Olfactory cortex: model circuit for study of associative memory? *Trends Neurosci.* **12**, 258–264.

Hallanger, A. E., Levey, A. I., Lee, H. J., Rye, D. B., and Wainer, B. H. (1987). The origins of cholinergic and other subcortical afferents to the thalamus in the rat. *J. Comp. Neurol.* **262**, 105–124.

Hallanger, A. E., Price, S. D., Lee, H. J., Steininger, T. L., and Wainer, B. H. (1990). Ultrastructure of cholinergic synaptic terminals in the thalamic anteroventral, ventroposterior, and dorsal lateral geniculate nuclei of the rat. *J. Comp. Neurol.* **299**, 482–492.

Hallanger, A. E., and Wainer, B. H. (1988). Ultrastructure of ChAT-immunoreactive synaptic terminals in the thalamic reticular nucleus of the rat. *J. Comp. Neurol.* **278**, 486–497.

Hamos, J. E., Van Horn, S. C., Raczkowski, D., Uhlrich, D. J., and Sherman, S. M. (1985). Synaptic connectivity of a local circuit neurone in lateral geniculate nucleus of the cat. *Nature* **317**, 618–621.

Hamos, J. E., Van Horn, S. C., Raczkowski, D., and Sherman, S. M. (1987). Synaptic circuits involving an individual retinogeniculate axon in the cat. *J. Comp. Neurol.* **259**, 165–192.

Harding, B. N. (1973). An ultrastructural study of the termination of afferent fibres within the ventrolateral and centre median nuclei of the monkey thalamus. *Brain Res.* **54**, 341–346.

Harting, J. K., Diamond, I. T., and Hall, W. C. (1973). Anterograde degeneration study of the cortical projections of the lateral geniculate and pulvinar nuclei in the tree shrew (*Tupaia glis*). *J. Comp. Neurol.* **150,** 393–440.

Harting, J. K., Hashikawa, T., and Van Lieshout, D. (1986). Laminar distribution of tectal, parabigeminal and pretectal inputs to the primate dorsal lateral geniculate nucleus: connectional studies in *Galago crassicaudatus*. *Brain Res.* **366,** 358–363.

Harting, J. K., Van Lieshout, D. P., Hashikawa, T., and Weber, J. T. (1991). The parabigeminogeniculate projection: connectional studies in eight mammals. *J. Comp. Neurol.* **305,** 559–581.

Hashikawa, T., Rausell, E., Molinari, M., and Jones, E. G. (1991). Parvalbumin- and calbindin-containing neurons in the monkey medial geniculate complex: differential distribution and cortical layer specific projections. *Brain Res.* **544,** 335–341.

Hassler, R. (1964). Spezifische und unspezifische Systeme des menschlichen Zwischenhirns. *Prog. Brain Res.* **5** (Lectures on the Diencephalon, edited by W. Bargmann and J. P. Schadé), 1–32.

Havton, L. A., and Ohara, P. T. (1994). Cell body and dendritic tree size of intracellularly labeled thalamocortical projection neurons in the ventrobasal complex of cat. *Brain Res.* **651,** 76–84.

Head, H. (1905). The afferent nervous system from a new aspect. *Brain* **28,** 99–116.

Hendry, S. H., and Calkins, D. J. (1998). Neuronal chemistry and functional organization in the primate visual system. *Trends Neurosci.* **21,** 344–349.

Hendry, S. H. C., and Reid, R. C. (2000). The koniocellular pathway in primate vision. *Annu. Rev. Neurosci.* **23,** 127–163.

Hendry, S. H. C., and Yoshioka, T. (1994). A neurochemically distinct third channel in the macaque dorsal lateral geniculate nucleus. *Science* **264,** 575–577.

Henneman, E., and Mendell, L. M. (1981). Functional organization of motoneuron pool and its inputs. *In* "Handbook of Physiology. Section 1: The Nervous System" Vol. II, "Motor Control," Part 1 (U. B. Brooks, ed.), pp. 423–507. American Physiological Society, Bethesda, MD.

Henschen, S. E. (1893). On the visual path and centre. *Brain* **16,** 170–180.

Hernández-Cruz, A., and Pape, H.-C. (1989). Identification of two calcium currents in acutely dissociated neurons from the rat lateral geniculate nucleus. *J. Neurophysiol.* **61,** 1270–1283.

Herrick, C. J. (1948). "The Brain of the Tiger Salamander." The University of Chicago Press, Chicago, IL.

Hodgkin, A. L., and Huxley, A. F. (1952). Currents carried by sodium and potassium ions through the membrane of the giant axon of *Loligo*. *J. Physiol. (Lond.)* **116,** 449–472.

Hoffman, D. A., and Johnston, D. (1998). Downregulation of transient K^+ channels in dendrites of hippocampal CA1 pyramidal neurons by activation of PKA and PKC. *J. Neurosci.* **18,** 3521–3528.

Hohl-Abrahao, J. C., and Creutzfeldt, O. D. (1991). Topographical mapping of the thalamocortical projections in rodents and comparison with that in primates. *Exp. Brain Res.* **87,** 283–294.

Holmes, W. R., and Woody, C. D. (1989). Effects of uniform and non-uniform synaptic 'activation-distributions' on the cable properties of modeled cortical pyramidal neurons. *Brain Res.* **505,** 12–22.

Hoogland, P. V., Wouterlood, F. G., Welker, E., and Van der Loos, H. (1991). Ultrastructure of giant and small thalamic terminals of cortical origin: a study of the projections from the barrel cortex in mice using *Phaseolus vulgaris leuco-agglutinin* (PHA-L). *Exp. Brain Res.* **87,** 159–172.

Houser, C. R., Vaughn, J. E., Barber, R. P., and Roberts, E. (1980). GABA neurons are the major cell type of the nucleus reticularis thalamic. *Brain Res.* **200,** 341–354.

Hubel, D. H., and Wiesel, T. N. (1971). Aberrant visual projections in the Siamese cat. *J. Physiol. (Lond.)* **218,** 33–62.

Hubel, D. H., and Wiesel, T. N. (1972). Laminar and columnar distribution of geniculo-cortical fibers in the macaque monkey. *J. Comp. Neurol.* **146,** 421–450.

Hubel, D. H., and Wiesel, T. N. (1977).

Functional architecture of macaque monkey visual cortex. *Proc. Roy. Soc. Lond. B* **198**, 1–59.

Huber, K. M., Sawtell, N. B., and Bear, M. F. (1998). Effects of the metabotropic glutamate receptor antagonist MCPG on phosphoinositide turnover and synaptic plasticity in visual cortex. *J. Neurosci.* **18**, 1–9.

Hughes, S. W., Cope, D. W., Tóth, T. I., Williams, S. R., and Crunelli, V. (1999). All thalamocortical neurones possess a T-type Ca^{2+} 'window' current that enables the expression of bistability-mediated activities. *J. Physiol. (Lond.)* **517**, 805–815.

Huguenard, J. R., and McCormick, D. A. (1992). Simulation of the currents involved in rhythmic oscillations in thalamic relay neurons. *J. Neurophysiol.* **68**, 1373–1383.

Huguenard, J. R., and Prince, D. A. (1992). A novel T-type current underlies prolonged Ca^{2+}-dependent burst firing in GABAergic neurons of rat thalamic reticular nucleus. *J. Neurosci.* **12**, 3804–3817.

Humphrey, A. L., Sur, M., Uhlrich, D. J., and Sherman, S. M. (1985a). Projection patterns of individual X- and Y-cell axons from the lateral geniculate nucleus to cortical area 17 in the cat. *J. Comp. Neurol.* **233**, 159–189.

Humphrey, A. L., Sur, M., Uhlrich, D. J., and Sherman, S. M. (1985b). Termination patterns of individual X- and Y-cell axons in the visual cortex of the cat: projections to area 18, to the 17/18 border region, and to both areas 17 and 18. *J. Comp. Neurol.* **233**, 190–212.

Hutchins, B., and Updyke, B. V. (1989). Retinotopic organization within the lateral posterior complex of the cat. *J. Comp. Neurol.* **285**, 350–398.

Ichida, J. M., Rosa, M. G. P., and Casagrande, V. A. (2000). Does the visual system of the flying fox resemble that of primates? The distribution of calcium-binding proteins in the primary visual pathway of *Pteropus poliocephalus*. *J. Comp. Neurol.* **417**, 73–87.

Ide, L. S. (1982). The fine structure of the perigeniculate nucleus in the cat. *J. Comp. Neurol.* **210**, 317–334.

Ilinsky, I. A., and Kultas-Ilinsky, K. (1990). Fine structure of the magnocellular subdivision of the ventral anterior thalamic nucleus (VAmc) of *Macaca mulatta*: I. Cell types and synaptology. *J. Comp. Neurol.* **294**, 455–478.

Ilinsky, I. A., Yi, H., and Kultas-Ilinsky, K. (1997). Mode of termination of pallidal afferents to the thalamus: a light and electron microscopic study with anterograde tracers and immunocytochemistry in *Macaca mulatta*. *J. Comp. Neurol.* **386**, 601–612.

Irvin, G. E., Casagrande, V. A., and Norton, T. T. (1993). Center/surround relationships of magnocellular, parvocellular, and koniocellular relay cells in primate lateral geniculate nucleus. *Vis. Neurosci.* **10**, 363–373.

Irvin, G. E., Norton, T. T., Sesma, M. A., and Casagrande, V. A. (1986). W-like response properties of interlaminar zone cells in the lateral geniculate nucleus of a primate (*Galago crassicaudatus*). *Brain Res.* **362**, 254–270.

Jack, J. J. B., Noble, D., and Tsien, R. W. (1975). "Electric Current Flow in Excitable Cells." Oxford University Press, Oxford, UK.

Jackson, H. J. (1873). On the anatomical and physiological localization of movements in the brain. *Lancet* **1**, 84–85, 162–164, 232–234.

Jahnsen, H., and Llinás, R. (1984a). Ionic basis for the electroresponsiveness and oscillatory properties of guinea-pig thalamic neurones *in vitro*. *J. Physiol. (Lond.)* **349**, 227–247.

Jahnsen, H., and Llinás, R. (1984b). Electrophysiological properties of guinea-pig thalamic neurones: an *in vitro* study. *J. Physiol. (Lond.)* **349**, 205–226.

Jasper, H. H. (1960). Unspecific thalamocortical relations. *In* "Handbook of Physiology, Section 1, Neurophysiology" (J. Field, H. W. Magoun, V. E. Hall, eds.), Vol. 2, pp. 1307–1321. American Physiological Society, Bethesda, MD.

Johnson, J. K., and Casagrande, V. A. (1995). Distribution of calcium-binding proteins within the parallel visual pathways of a primate (*Galago crassi-*

caudatus). *J. Comp. Neurol.* **356,** 238–260.

Johnston, D., Magee, J. C., Colbert, C. M., and Christie, B. R. (1996). Active properties of neuronal dendrites. *Annu. Rev. Neurosci.* **19,** 165–186.

Jones, E. G. (1983). Distribution patterns of individual medial lemniscal axons in the ventrobasal complex of the monkey thalamus. *J. Comp. Neurol.* **215,** 1–16.

Jones, E. G. (1985). "The Thalamus." Plenum Press, New York, NY.

Jones, E. G. (1998). Viewpoint: the core and matrix of thalamic organization. *Neuroscience* **85,** 331–345.

Jones, E. G., and Leavitt, R. Y. (1974). Retrograde axonal transport and the demonstration of non-specific projections to the cerebral cortex and striatum from thalamic intralaminar nuclei in the rat, cat and monkey. *J. Comp. Neurol.* **154,** 349–377.

Jones, E. G., and Powell, T. P. (1969). Electron microscopy of synaptic glomeruli in the thalamic relay nuclei of the cat. *Proc. Roy. Soc. Lond. B* **172,** 153–171.

Jones, E. G., and Rockel, A. J. (1971). The synaptic organization in the medial geniculate body of afferent fibres ascending from the inferior colliculus. *Z. Zellforsch.* **113,** 44–66.

Kaas, J. H. (1995). The evolution of isocortex. *Brain Behav. Evol.* **46,** 187–196.

Kaas, J. H., and Guillery, R. W. (1973). The transfer of abnormal visual field representations from the dorsal lateral geniculate nucleus to the visual cortex in Siamese cats. *Brain Res.* **59,** 61–95.

Kaas, J. H., Guillery, R. W., and Allman, J. M. (1972). Some principles of organization in the dorsal lateral geniculate nucleus. *Brain Behav. Evol.* **6,** 253–299.

Kaas, J. H., Hackett, T. A., and Tramo, M. J. (1999). Auditory processing in primate cerebral cortex. *Curr. Opin. Neurobiol.* **9,** 164–170.

Kalil, R. E., and Chase, R. (1970). Corticofugal influence on activity of lateral geniculate neurons in the cat. *J. Neurophysiol.* **33,** 459–474.

Karten, H. J., Hodos, W., Nauta, W. J., and Revzin, A. M. (1973). Neural connections of the "visual wulst" of the avian telencephalon. Experimental studies in the pigeon (*Columba livia*) and owl (*Speotyto cunicularia*). *J. Comp. Neurol.* **150,** 253–278.

Katz, L. C. (1987). Local circuitry of identified projection neurons in cat visual cortex brain slices. *J. Neurosci.* **7,** 1223–1249.

Kaufman, E. F., and Rosenquist, A. C. (1985). Efferent projections of the thalamic intralaminar nuclei in the cat. *Brain Res.* **335,** 257–279.

Kawano, J. (1998). Cortical projections of the parvocellular laminae C of the dorsal lateral geniculate nucleus in the cat: An anterograde wheat germ agglutinin conjugated to horseradish peroxidase study. *J. Comp. Neurol.* **392,** 439–457.

Kayama, Y., Shimada, S., Hishikawa, Y., and Ogawa, T. (1989). Effects of stimulating the dorsal raphe nucleus of the rat on neuronal activity in the dorsal lateral geniculate nucleus. *Brain Res.* **489,** 1–11.

Kennedy, H., and Bullier, J. (1985). A double-labeling investigation of the afferent connectivity to cortical areas V1 and V2 of the macaque monkey. *J. Neurosci.* **5,** 2815–2830.

Kharazia, V. N., and Weinberg, R. J. (1994). Glutamate in thalamic fibers terminating in layer IV of primary sensory cortex. *J. Neurosci.* **14,** 6021–6032.

Kim, H. G., and Connors, B. W. (1993). Apical dendrites of the neocortex: correlation between sodium- and calcium-dependent spiking and pyramidal cell morphology. *J. Neurosci.* **13,** 5301–5311.

Kim, U., and McCormick, D. A. (1998). The functional influence of burst and tonic firing mode on synaptic interactions in the thalamus. *J. Neurosci.* **18,** 9500–9516.

Koch, C., Poggio, T., and Torre, V. (1982). Retinal ganglion cells: a functional interpretation of dendritic morphology. *Phil. Trans. Roy. Soc. Lond. B* **298,** 227–264.

Koerner, F., and Teuber, H. L. (1973). Visual field defects after missile injuries to the geniculo-striate pathway in man. *Exp. Brain Res.* **18,** 88–113.

Kohonen, T., Lehtiö Rovamo, J., Hyvärinen Bry, K., and Vaino, L. (1977). A principle of neural associative memory. *Neuroscience* **2,** 1065–1076.

Kölliker, A. (1896). "Handbuch der Gewebelehre des Menschen. Nervensystemen des Menschen und der Thiere," Vol. 2, 6th ed. W. Engelmann, Leipzig.

Krubitzer, L. (1998). What can monotremes tell us about brain evolution? *Phil. Trans. Roy. Soc. Lond. B* **353,** 1127–1146.

Krubitzer, L. A., and Kaas, J. H. (1992). The somatosensory thalamus of monkeys: cortical connections and a redefinition of nuclei in marmosets. *J. Comp. Neurol.* **319,** 123–140.

Krug, K., Smith, A. L., and Thompson, I. D. (1998). The development of topography in the hamster geniculocortical projection. *J. Neurosci.* **18,** 5766–5776.

Krupa, D. J., Ghazanfar, A. A., and Nicolelis, M. A. (1999). Immediate thalamic sensory plasticity depends on corticothalamic feedback. *Proc. Natl. Acad. Sci. USA* **96,** 8200–8205.

Kudo, M., and Niimi, K. (1980). Ascending projections of the inferior colliculus in the cat: an autoradiographic study. *J. Comp. Neurol.* **191,** 545–556.

Kuhn, T. (1963). "The Structure of Scientific Revolutions." The University of Chicago Press, Chicago, IL.

Kultas-Ilinsky, K., and Ilinsky, I. A. (1991). Fine structure of the ventral lateral nucleus (VL) of the *Macaca mulatta* thalamus: cell types and synaptology. *J. Comp. Neurol.* **314,** 319–349.

Kultas-Ilinsky, K., Reising, L., Yi, H., and Ilinsky, I. A. (1997). Pallidal afferent territory of the *Macaca mulatta* thalamus: neuronal and synaptic organization of the VAdc. *J. Comp. Neurol.* **386,** 573–600.

Kuroda, M., and Price, J. L. (1991). Ultrastructure and synaptic organization of axon terminals from brainstem structures to the mediodorsal thalamic nucleus of the rat. *J. Comp. Neurol.* **313,** 539–552.

Kuroda, M., Murakami, K., Kishi, K., and Price, J. L. (1992). Distribution of the piriform cortical terminals to cells in the central segment of the mediodorsal thalamic nucleus of the rat. *Brain Res.* **595,** 159–163.

Kuroda, M., Murakami, K., Oda, S., Shinkai, M., and Kishi, K. (1993). Direct synaptic connections between thalamorcortical axon terminals from the mediodorsal thalamic nucleus (MD) and corticothalamic neurons to MD in the prefrontal cortex. *Brain Res.* **612,** 339–344.

Lachica, E. A., and Casagrande, V. A. (1988). Development of primate retinogeniculate axon arbors. *Vis. Neurosci.* **1,** 103–123.

Landò, L., and Zucker, R. S. (1994). Ca^{2+} cooperativity in neurosecretion measured using photolabile Ca^{2+} chelators. *J. Neurophysiol.* **72,** 825–830.

Lane, R. H., Kaas, J. H., and Allman, J. M. (1974). Visuotopic organization of the superior colliculus in normal and Siamese cats. *Brain Res.* **70,** 413–430.

Larkum, M. E., Zhu, J. J., and Sakmann, B. (1999). A new cellular mechanism for coupling inputs arriving at different cortical layers. *Nature* **398,** 338–341.

Latawiec, D., Martin, K. A. C., and Meskenaite, V. (2000). Termination of the geniculocortical projection in the striate cortex of macaque monkey: A quantitative immunoelectron microscopic study. *J. Comp. Neurol.* **419,** 306–319.

LeDoux, J. E., Ruggiero, D. A., and Reis, D. J. (1985). Projections to the subcortical forebrain from anatomically defined regions of the medial geniculate body in the rat. *J. Comp. Neurol.* **242,** 182–213.

Le Gros Clark, W. E. (1932). The structure and connections of the thalamus. *Brain* **55,** 406–470.

Lee, K., and McCormick, D. A. (1998). Histamine inhibits perigeniculate neurons through activation of a Cl^- conductance. *Soc. Neurosci. Abstracts* **22,** 1477.

Lennie, P. (1980). Perceptual signs of parallel pathways. *Phil. Trans. Roy. Soc. Lond. B* **290,** 23–37.

Lenz, F. A., Garonzik, I. M., Zirh, T. A., and Dougherty, P. M. (1998). Neuronal activity in the region of the thalamic principal sensory nucleus (ventralis caudalis) in patients with pain following amputations. *Neuroscience* **86,** 1065–1081.

Le Vay, S., and Ferster, D. (1977). Relay cell classes in the lateral geniculate nucleus of the cat and the effects of visual deprivation. *J. Comp. Neurol.* **172,** 563–584.

Le Vay, S., and Ferster, D. (1979). Proportion of interneurons in the cat's lateral geniculate nucleus. *Brain Res.* **164,** 304–308.

Le Vay, S., and McConnell, S. K. (1982). ON and OFF layers in the lateral geniculate nucleus of the mink. *Nature* **300,** 350–351.

Leventhal, A. G., and Creel, D. J. (1985). Retinal projections and functional architecture of cortical areas 17 and 18 in the tyrosinase-negative albino cat. *J. Neurosci.* **5,** 795–807.

Levitt, J. B., Takashi, Y., and Lund, J. S. (1995). Connections between the pulvinar complex and cytochrome oxidase-defined compartments in visual area V2 of macaque monkey. *Exp. Brain Res.* **104,** 419–430.

Lieberman, A. R. (1973). Neurons with presynaptic perikarya and presynaptic dendrites in the rat lateral geniculate nucleus. *Brain Res.* **59,** 35–59.

Lieberman, A. R., and Spacek, J. (1997). Filamentous contacts: the ultrastructure and three-dimensional organization of specialized non-synaptic interneuronal appositions in thalamic relay nuclei. *Cell Tissue Res.* **288,** 43–57.

Ling, C., Schneider, G. E., Northmore, D., and Jhaveri, S. (1997). Afferents from the colliculus, cortex, and retina have distinct terminal morphologies in the lateral posterior thalamic nucleus. *J. Comp. Neurol.* **388,** 467–483.

Lisman, J. E. (1997). Bursts as a unit of neural information: making unreliable synapses reliable. *Trends Neurosci.* **20,** 38–43.

Liu, X. B., Warren, R. A., and Jones, E. G. (1995). Synaptic distribution of afferents from reticular nucleus in ventroposterior nucleus of cat thalamus. *J. Comp. Neurol.* **352,** 187–202.

Livingstone, M. S., and Hubel, D. H. (1981). Effects of sleep and arousal on the processing of visual information in the cat. *Nature* **291,** 554–561.

Livingstone, M. S., and Hubel, D. H. (1984). Anatomy and physiology of a color system in the primate visual cortex. *J. Neurosci.* **4,** 309–356.

Llinás, R. (1988). The intrinsic electrophysiological properties of mammalian neurons: insights into central nervous system. *Science* **242,** 1654–1664.

Llinás, R. R., and Pare, D. (1991). Of dreaming and wakefulness. *Neuroscience* **44,** 521–535.

Lo, F.-S., and Sherman, S. M. (1994). Feedback inhibition in the cat's lateral geniculate nucleus. *Exp. Brain Res.* **100,** 365–368.

Lorente de Nó, R. (1938). Cerebral cortex: Architecture, intracortical connections, motor projections. *In* "Physiology of the Nervous System" (J. Fulton, ed.), pp. 291–325. Oxford University Press, Oxford, UK.

Lozsádi, D. A. (1995). Organization of connections between the thalamic reticular and the anterior thalamic nuclei in the rat. *J. Comp. Neurol.* **358,** 233–246.

Lozsádi, D. A., Gonzalez-Soriano, J., and Guillery, R. W. (1996). The course and termination of corticothalamic fibres arising in the visual cortex of the rat. *Eur. J. Neurosci.* **8,** 2416–2427.

Lu, S.-M., Guido, W., and Sherman, S. M. (1992). Effects of membrane voltage on receptive field properties of lateral geniculate neurons in the cat: contributions of the low threshold Ca^{2+} conductance. *J. Neurophysiol.* **68,** 2185–2198.

Lu, S.-M., Guido, W., and Sherman, S. M. (1993). The brainstem parabrachial region controls mode of response to visual stimulation of neurons in the cat's lateral geniculate nucleus. *Vis. Neurosci.* **10,** 631–642.

Lübke, J. (1993). Morphology of neurons in the thalamic reticular nucleus (TRN) of mammals as revealed by intracellular injections into fixed brain slices. *J. Comp. Neurol.* **329,** 458–471.

Lujan, R., Nusser, Z., Roberts, J. D., Shigemoto, R., and Somogyi, P. (1996). Perisynaptic location of metabotropic glutamate receptors mGluR1 and mGluR5 on dendrites and dendritic spines in the rat hippocampus. *Eur. J. Neurosci.* **8,** 1488–1500.

Luppino, G., Matelli, M., Carey, R. G., Fitzpatrick, D., and Diamond, I. T. (1988). New view of the organization of the pulvinar nucleus in *Tupaia* as revealed by tectopulvinar and pulvinarcortical projections. *J. Comp. Neurol.* **273,** 67–86.

Lysakowski, A., Standage, G. P., and Benevento, L. A. (1988). An investigation of collateral projections of the dorsal lateral geniculate nucleus and other subcortical structures to cortical areas V1 and V4 in the macaque monkey: a double label retrograde tracer study. *Exp. Brain Res.* **69,** 651–661.

Ma, W., Peschanski, M., and Ralston, H. J. 3d. (1987a). The differential synaptic organization of the spinal and lemniscal projections to the ventrobasal complex of the rat thalamus. Evidence for convergence of the two systems upon single thalamic neurons. *Neuroscience* **22,** 925–934.

Ma, W., Peschanski, M., and Ralston, H. J. 3d. (1987b). Fine structure of the spinothalamic projections to the central lateral nucleus of the rat thalamus. *Brain Res.* **414,** 187–191.

Macchi, G. (1983). Old and new anatomofunctional criteria in the subdivision of the thalamic nuclei. *In* "Somatosensory Integration in the Thalamus" (G. Macchi, A. Rustioni, and R. Spreafico, eds.), pp. 3–16. Elsevier, Amsterdam, The Netherlands.

Macchi, G. (1993). The intralaminar system revisited. *In* "Thalamic Networks for Relay and Modulation" (D. Minciacchi, M. Molinari, G. Macchi, and E. G. Jones, eds.), pp. 175–184. Pergamon Press, Oxford, UK.

Macchi, G., and Bentivoglio, M. (1982). The organization of the efferent projections of the thalamic intralaminar nuclei: past, present and future of the anatomical approach. *Ital. J. Neurol. Sci.* **3,** 83–96.

Macchi, G., Bentivoglio, M., Molinari, M., and Minciacchi, D. (1984). The thalamo-caudate versus thalamo-cortical projections as studied in the cat with fluorescent retrograde double labeling. *Exp. Brain Res.* **54,** 225–239.

Macmillan, N. A., and Creelman, C. D. (1991). "Detection Theory: A User's Guide." Cambridge University Press, Cambridge, UK.

Magee, J., Hoffman, D., Colbert, C., and Johnston, D. (1998). Electrical and calcium signaling in dendrites of hippocampal pyramidal neurons. *Annu. Rev. Physiol.* **60,** 327–346.

Majorossy, K., and Kiss, A. (1976). Specific patterns of neuron arrangement and of synaptic articulation in the medial geniculate body. *Exp. Brain Res.* **26,** 1–17.

Malpeli, J. G., Lee, D., and Baker, F. H. (1996). Laminar and retinotopic organization of the macaque lateral geniculate nucleus: magnocellular and parvocellular magnification functions *J. Comp. Neurol.* **375,** 363–377.

Manning, K. A., Wilson, J. R., and Uhlrich, D. J. (1996). Histamine-immunoreactive neurons and their innervation of visual regions in the cortex, tectum, and thalamus in the primate *Macaca mulatta. J. Comp. Neurol.* **373,** 271–282.

Marks, G. A., Farber, J., and Roffwarg, H. P. (1981). Phasic influences during REM sleep upon dorsal lateral geniculate nucleus unit activity in the rat. *Brain Res.* **222,** 388–394.

Martin, K. A. C. (1985). Neuronal circuits in cat striate cortex. *In* "Cerebral Cortex," Vol. 2 (E. G. Jones and A. Peters, eds.), pp. 241–284. Plenum, New York, NY.

Mason, R. (1978). Functional organization in the cat's pulvinar complex. *Exp. Brain Res.* **31,** 51–66.

Mastronarde, D. N. (1987). Two classes of single-input X-cells in cat lateral geniculate nucleus. I. Receptive field properties and classification of cells. *J. Neurophysiol.* **57,** 357–380.

Mathers, L. H. (1971). Tectal projection to the posterior thalamus of the squirrel monkey. *Brain Res.* **35,** 295–298.

Mathers, L. H. (1972). The synaptic organization of the cortical projection to the pulvinar of the squirrel monkey. *J. Comp. Neurol.* **146,** 43–60.

Matthews, G. (1996). Neurotransmitter release. *Annu. Rev. Neurosci.* **19,** 219–233.

Matthews, M. R. (1964). Further observations on transneuronal degeneration in the lateral geniculate nucleus of the macaque monkey. *J. Anat. (Lond.)* **98,** 255–263.

Mayer, M. L., and Westbrook, G. L. (1987). The physiology of excitatory amino acids in the vertebrate central nervous system. *Progr. Neurobiol.* **28,** 197–276.

McCarley, R. W., Benoit, O., and Barrionuevo, G. (1983). Lateral geniculate nucleus unitary discharge in sleep and waking: state- and rate-specific aspects. *J. Neurophysiol.* **50,** 798–818.

McClurkin, J. W., and Marrocco, R. T. (1984). Visual cortical input alters spatial tuning in monkey lateral geniculate nucleus cells. *J. Physiol. (Lond.)* **348,** 135–152.

McClurkin, J. W., Optican, L. M., and Richmond, B. J. (1994). Cortical feedback increases visual information transmitted by monkey parvocellular lateral geniculate nucleus neurons. *Vis. Neurosci.* **11,** 601–617.

McConnell, S. K., and LeVay, S. (1984). Segregation of on- and off-center afferents in mink visual cortex. *Proc. Natl. Acad. Sci. USA* **81,** 1590–1593.

McCormick, D. A. (1989). Cholinergic and noradrenergic modulation of thalamocortical processing. *Trends Neurosci.* **12,** 215–221.

McCormick, D. A. (1990). Membrane properties and neurotransmitter actions. *In* "The Synaptic Organization of the Brain" 3rd ed. (G. M. Shepherd, ed.), pp. 32–66. Oxford University Press, New York, NY.

McCormick, D. A. (1991a). Cellular mechanisms underlying cholinergic and noradrenergic modulation of neuronal firing in the cat and guinea pig dorsal lateral geniculate nucleus. *Neuroscience* **12,** 278–289.

McCormick, D. A. (1991b). Functional properties of a slowly inactivating potassium current in guinea pig dorsal lateral geniculate relay neurons. *J. Neurophysiol.* **66,** 1176–1189.

McCormick, D. A. (1992). Neurotransmitter actions in the thalamus and cerebral cortex and their role in neuromodulation of thalamocortical activity. *Progr. Neurobiol.* **39,** 337–388.

McCormick, D. A., and Bal, T. (1997). Sleep and arousal: Thalamocortical mechanisms. *Annu. Rev. Neurosci.* **20,** 185–215.

McCormick, D. A., and Feeser, H. R. (1990). Functional implications of burst firing and single spike activity in lateral geniculate relay neurons. *Neuroscience* **39,** 103–113.

McCormick, D. A., and Huguenard, J. R. (1992). A model of the electrophysiological properties of thalamocortical relay neurons. *J. Neurophysiol.* **68,** 1384–1400.

McCormick, D. A., and Pape, H.-C. (1988). Acetycholine inhibits identified interneurons in the cat lateral geniculate nucleus. *Nature* **334,** 246–248.

McCormick, D. A., and Pape, H.-C. (1990a). Noradrenergic and serotonergic modulation of a hyperpolarization-activated cation current in thalamic relay neurones. *J. Physiol. (Lond.)* **431,** 319–342.

McCormick, D. A., and Pape, H.-C. (1990b). Properties of a hyperpolarization-activated cation current and its role in rhythmic oscillation in thalamic relay neurones. *J. Physiol. (Lond.)* **431,** 291–318.

McCormick, D. A., and Prince, D. A. (1986). Acetylcholine induces burst firing in thalamic reticular neurones by activating a potassium conductance. *Nature* **319,** 402–405.

McCormick, D. A., and Prince, D. A. (1988). Noradrenergic modulation of firing pattern in guinea-pig and cat thalamic neurons, *in vitro*. *J. Neurophysiol.* **59,** 978–996.

McCormick, D. A., and Von Krosigk, M. (1992). Corticothalamic activation modulates thalamic firing through glutamate "metabotropic" receptors. *Proc. Natl. Acad. Sci. USA* **89,** 2774–2778.

McCormick, D. A., and Wang, Z. (1991). Serotonin and noradrenaline excite

GABAergic neurones in the guinea-pig and cat nucleus reticularis thalami. *J. Physiol. (Lond.)* **442**, 235–255.

McCormick, D. A., and Williamson, A. (1991). Modulation of neuronal firing mode in cat and guinea pig LGNd by histamine: possible cellular mechanisms of histaminergic control of arousal. *J. Neurosci.* **11**, 3188–3199.

Meeren, H. K., Van Luijtelaar, E. L., and Coenen, A. M. (1998). Cortical and thalamic visual evoked potentials during sleep-wake states and spike-wave discharges in the rat. *Electroencephalogr. Clin. Neurophysiol.* **108**, 306–319.

Melchner, L. V., Pallas, S. L., and Sur, M. (2000). Visual behaviour mediated by retinal projections directed to the auditory pathway. *Nature* **404**, 871–876.

Merabet, L., Desautels, A., Minville, K., and Casanova, C. (1998). Motion integration in a thalamic visual nucleus. *Nature* **396**, 265–268.

Middleton, F. A., and Strick, P. L. (1997). Cerebellar output channels: motor and cognitive components. *Progr. Brain Res.* **114**, 553–566.

Minciacchi, D., Granato, A., and Santarelli Macchi, G. (1993). Different weights of subcortical-cortical projections upon primary sensory areas: the thalamic anterior intralaminar system. *In* "Thalamic Networks for Relay and Modulation" (D. Minciacchi, M. Molinari, G. Macchi, and E. G. Jones, eds.), pp. 209–226. Pergamon Press, Oxford, UK.

Minkowski, M. (1914). Experimentelle Untersuchungen über die Beziehungen der Grosshirnrinde und der Netzhaut zu den primären optischen Zentren, besonder zum Corpus geniculatum externum. *Arbeiten Hirnanat. Inst. Zürich* **7**, 259–362.

Mitrofanis, J. (1994). Development of the thalamic reticular nucleus in ferrets with special reference to the perigeniculate and perireticular cell groups. *Eur. J. Neurosci.* **6**, 253–263.

Mitrofanis, J., and Guillery, R. W. (1993). New views of the thalamic reticular nucleus in the adult and the developing brain. *Trends Neurosci.* **16**, 240–245.

Molinari, M., Dell'Anna, M. E., Rausell, E., Leggio, M. G., Hashikawa, T., and Jones, E. G. (1995). Auditory thalamocortical pathways defined in monkeys by calcium-binding protein immunoreactivity. *J. Comp. Neurol.* **362**, 171–194.

Molinari, M., Leggio, M. G., Dell'Anna, M. E., Giannetti, S., and Macchi, G. (1993). Structural evidence in favour of a relay function for the anterior intralaminar nuclei. *In* "Thalamic Networks for Relay and Modulation" (D. Minciacchi, M. Molinari, G. Macchi, and E. G. Jones, eds.), pp. 197–208. Pergamon Press, Oxford, UK.

Molotchnikoff, S., Casanova, C., and Cerat, A. (1988). The consequences of the superior colliculus output on lateral geniculate and pulvinar responses. *Progr. Brain Res.* **75**, 67–74.

Mombaerts, P., Wang, F., Dulac, C., Chao, S. K., Nemes, A., Mendelsohn, M., Edmondson, J., and Axel, R. (1996). Visualizing an olfactory sensory map. *Cell* **87**, 675–686.

Monakow, C. von. (1889). Experimentelle und pathologische-anatomische Untersuchungen über die optischen Centren und Bahnen. *Arch. Psychiat. Nervenkr.* **20**, 714–787.

Montero, V. M. (1986). The interneuronal nature of GABAergic neurons in the lateral geniculate nucleus of the rhesus monkey: a combined HRP and GABA-immunocytochemical study. *Exp. Brain Res.* **64**, 615–622.

Montero, V. M. (1987). Ultrastructural identification of synaptic terminals from the axon of type 3 interneurons in the cat lateral geniculate nucleus. *J. Comp. Neurol.* **264**, 268–283.

Montero, V. M. (1989). The GABA-immunoreactive neurons in the interlaminar regions of the cat lateral geniculate nucleus: light and electron microscopic observations. *Exp. Brain Res.* **75**, 497–512.

Montero, V. M. (1991). A quantitative study of synaptic contacts on interneurons and relay cells of the cat lateral geniculate nucleus. *Exp. Brain Res.* **86**, 257–270.

Montero, V. M., and Guillery, R. W. (1978). Abnormalities of the corticogeniculate pathway in Siamese cats. *J. Comp. Neurol.* **179**, 1–12.

Montero, V. M., Guillery, R. W., and Woolsey, C. N. (1977). Retinotopic organization within the thalamic reticular nucleus demonstrated by a double label autoradiographic technique. *Brain Res.* **138**, 407–421.

Montero, V. M., and Scott, G. L. (1981). Synaptic terminals in the dorsal lateral geniculate nucleus from neurons of the thalamic reticular nucleus: a light and electron microscope autoradiographic study. *Neuroscience* **6**, 2561–2577.

Montero, V. M., and Zempel, J. (1985). Evidence for two types of GABA-containing interneurons in the A-laminae of the cat lateral geniculate nucleus: a double-label HRP and GABA-immunocytochemical study. *Exp. Brain Res.* **60**, 603–609.

Morest, D. K. (1964). The neuronal architecture of the medial geniculate body of the cat. *J. Anat. (Lond.)* **98**, 611–630.

Morest, D. K. (1975). Synaptic relationships of Golgi type II cells in the medial geniculate body of the cat. *J. Comp. Neurol.* **162**, 157–193.

Morrison, J. H., and Foote, S. L. (1986). Noradrenergic and serotoninergic innervation of cortical, thalamic, and tectal visual structures in Old and New World monkeys. *J. Comp. Neurol.* **243**, 117–138.

Mott, D. D., and Lewis, D. V. (1994). The pharmacology and function of central $GABA_B$ receptors. *Int. Rev. Neurobiol.* **36**, 97–223.

Mukherjee, P., and Kaplan, E. (1995). Dynamics of neurons in the cat lateral geniculate nucleus: *in vivo* electrophysiology and computational modeling. *J. Neurophysiol.* **74**, 1222–1243.

Murphy, P. C., Duckett, S. G., and Sillito, A. M. (2000). Comparison of the laminar distribution of input from areas 17 and 18 of the visual cortex to the lateral geniculate nucleus of the cat. *J. Neurosci.* **20**, 845–853.

Murphy, P. C., and Sillito, A. M. (1996). Functional morphology of the feedback pathway from area 17 of the cat visual cortex to the lateral geniculate nucleus. *J. Neurosci.* **16**, 1180–1192.

Murphy, P. C., Uhlrich, D. J., Tamamaki, N., and Sherman, S. M. (1994). Brainstem modulation of the response properties of cells in the cat's perigeniculate nucleus. *Vis. Neurosci.* **11**, 781–791.

Mushiake, H., and Strick, P. L. (1995). Pallidal neuron activity during sequential arm movements. *J. Neurophysiol.* **74**, 2754–2758.

Nakanishi, S., Nakajima, Y., Masu, M., Ueda, Y., Nakahara, K., Watanabe, D., Yamaguchi, S., Kawabata, S., and Okada, M. (1998). Glutamate receptors: brain function and signal transduction. *Brain Res. Rev.* **26**, 230–235.

Négyessy, L., Hámori, J., and Bentivoglio, M. (1998). Contralateral cortical projection to the mediodorsal thalamic nuclei: origin and synaptic organization in the rat. *Neuroscience* **84**, 741–753.

Nelson, J. I., Salin, P. A., Munk, M. H. J., Arzi, M., and Bullier, J. (1992). Spatial and temporal coherence in corticocortical connections: A cross-correlation study in areas 17 and 18 in the cat. *Vis. Neurosci.* **9**, 21–37.

Nelson, S. B., and LeVay, S. (1985). Topographic organization of the optic radiation of the cat. *J. Comp. Neurol.* **240**, 322–330.

Nicolelis, M. A., Baccala, L. A., Lin, R. C., and Chapin, J. K. (1995). Sensorimotor encoding by synchronous neural ensemble activity at multiple levels of the somatosensory system. *Science* **268**, 1353–1358.

Nicoll, R. A., Malenka, R. C., and Kauer, J. A. (1990). Functional comparison of neurotransmitter receptor subtypes in mammalian central nervous system. *Physiol. Rev.* **70**, 513–565.

Niimi, K., Ono, K., and Kusunose, M. (1984). Projections of the medial geniculate nucleus to layer 1 of the auditory cortex in the cat traced with horseradish peroxidase. *Neurosci. Lett.* **45**, 223–228.

Niimi, K., Yamazaki, Y., Matsuoka, H., Kusunose, M., Imataki, T., and Ono, K. (1985). Thalamic projections to the posterior suprasylvian gyrus and the ventrally adjacent cortex in the cat traced with horseradish peroxidase. *J. für Hirnforschung.* **26,** 497–508.

Nissl, F. (1913). Die Grosshirnanteile des Kanninchens. *Arch Psychiat. Nervenkr.* **52,** 867–953.

Norton, T. T., Casagrande, V. A., Irvin, G. E., Sesma, M. A., and Petry, H. M. (1988). Contrast-sensitivity functions of W-, X-, and Y-like relay cells in the lateral geniculate nucleus of bush baby, *Galago crassicaudatus. J. Neurophysiol.* **59,** 1639–1656.

Ogren, M. P., and Hendrickson, A. E. (1979). The morphology and distribution of striate cortex terminals in the inferior and lateral subdivisions of the *Macaca* monkey pulvinar. *J. Comp. Neurol.* **188,** 179–199.

Ohara, P. T. (1988). Synaptic organization of the thalamic reticular nucleus. *J. Electron Microscopy Technique* **10,** 283–292.

Ohara, P. T., Chazal, G., and Ralston, H. J. 3d. (1989). Ultrastructural analysis of GABA-immunoreactive elements in the monkey thalamic ventrobasal complex. *J. Comp. Neurol.* **283,** 541–558.

Ohara, P. T., and Havton, L. A. (1994). Preserved features of thalamocortical projection neuron dendritic architecture in the somatosensory thalamus of the rat, cat and macaque. *Brain Res.* **648,** 259–264.

Ohara, P. T., and Lieberman, A. R. (1985). The thalamic reticular nucleus of the adult rat: experimental anatomical studies. *J. Neurocytol.* **14,** 365–411.

Ohara, P. T., Sefton, A. J., and Lieberman, A. R. (1980). Mode of termination of afferents from the thalamic reticular nucleus in the dorsal lateral geniculate nucleus of the rat. *Brain Res.* **197,** 503–506.

Ojima, H. (1994). Terminal morphology and distribution of corticothalamic fibers originating from layers 5 and 6 of cat primary auditory cortex. *Cerebral Cortex* **4,** 646–663.

Ojima, H., Murakami, K., and Kishi, K. (1996). Dual termination modes of corticothalamic fibers originating from pyramids of layers 5 and 6 in cat visual cortical area 17. *Neurosci. Lett.* **208,** 57–60.

Olausson, B., Shyu, B. C., and Rydenhag, B. (1989). Projection from the thalamic intralaminar nuclei on the isocortex of the rat: a surface potential study. *Exp. Brain Res.* **75,** 543–554.

Olavarria, J., and Montero, V. M. (1984). Relation of callosal and striate-extrastriate cortical connections in the rat: Morphological definition of extrastriate visual areas. *Exp. Brain Res.* **54,** 240–252.

Olavarria, J., and Montero, V. M. (1989). Organization of visual cortex in the mouse revealed by correlating callosal and striate-extrastriate connections. *Vis. Neurosci.* **3,** 59–69.

Ozawa, S., Kamiya, H., and Tsuzuki, K. (1998). Glutamate receptors in the mammalian central nervous system *Progr. Neurobiol.* **54,** 581–618.

Pallas, S. L., Hahm, J., and Sur, M. (1994). Morphology of retinal axons induced to arborize in a novel target, the medial geniculate nucleus. I. Comparison with arbors in normal targets. *J. Comp. Neurol.* **349,** 343–362.

Pallas, S. L., and Sur, M. (1994). Morphology of retinal axon arbors induced to arborize in a novel target, the medial geniculate nucleus. II. Comparison with axons from the inferior colliculus. *J. Comp. Neurol.* **349,** 363–376.

Pandya, D. N., and Rosene, D. L. (1993). Laminar termination patterns of thalamic, callosal, and association afferents in the primary auditory area of the rhesus monkey. *Exp. Neurol.* **119,** 220–234.

Pape, H.-C., Budde, T., Mager, R., and Kisvárday, Z. E. (1994). Prevention of Ca^{2+}-mediated action potentials in GABAergic local circuit neurones of rat thalamus by a transient K^+ current. *J. Physiol. (Lond.)* **478,** 403–422.

Pape, H.-C., and Mager, R. (1992). Nitric oxide controls oscillatory activity in thalamocortical neurons. *Neuron* **9,** 441–448.

Pape, H.-C., and McCormick, D. A. (1995). Electrophysiological and pharmacological properties of interneurons in the cat dorsal lateral geniculate nucleus. *Neuroscience* **68,** 1105–1125.

Paré, D., Hazrati, L. N., Parent, A., and Steriade, M. (1990). Substantia nigra pars reticulata projects to the reticular thalamic nucleus of the cat: a morphological and electrophysiological study. *Brain Res.* **535,** 139–146.

Parra, P., Gulyas, A. I., and Miles, R. (1998). How many subtypes of inhibitory cells in the hippocampus? *Neuron* **20,** 983–993.

Pearson, J. C., and Haines, D. E. (1980). Somatosensory thalamus of a prosimian primate *(Galago senegalensis).* II. An HRP and Golgi study of the ventral posterolateral nucleus (VPL). *J. Comp. Neurol.* **190,** 559–580.

Penney, J. B. Jr., and Young, A. B. (1981). GABA as the pallidothalamic neurotransmitter: implications for basal ganglia function. *Brain Res.* **207,** 195–199.

Penny, G. R., Fitzpatrick, D., Schmechel, D. E., and Diamond, I. T. (1983). Glutamic acid decarboxylase-immunoreactive neurons and horseradish peroxidase-labeled projection neurons in the ventral posterior nucleus of the cat and Galago senegalensis. *J. Neurosci.* **3,** 1868–1887.

Penny, G. R., Itoh, K., and Diamond, I. T. (1982). Cells of different sizes in the ventral nuclei project to different layers of the somatic cortex in the cat. *Brain Res.* **242,** 55–65.

Perkel, D. H., Gerstein, G. L., and Moore, G. P. (1967). Neuronal spike trains and stochastic point processes. II. Simultaneous spike trains. *Biophys. J.* **7,** 419–440.

Peruzzi, D., Bartlett, E., Smith, P. H., and Oliver, D. L. (1997). A monosynaptic GABAergic input from the inferior colliculus to the medial geniculate body in rat. *J. Neurosci.* **17,** 3766–3777.

Peschanski, M., Lee, C. L., and Ralston, H. J. 3d. (1984). The structural organization of the ventrobasal complex of the rat as revealed by the analysis of physiologically characterized neurons injected intracellularly with horseradish peroxidase. *Brain Res.* **297,** 63–74.

Peschanski, M., Ralston, H. J., and Roudier, F. (1983). Reticularis thalami afferents to the ventrobasal complex of the rat thalamus: an electron microscope study. *Brain Res.* **270,** 325–329.

Peters, A., and Palay, S. L. (1966). The morphology of laminae A and A1 of the dorsal nucleus of the lateral geniculate body of the cat. *J. Anat. (Lond.)* **100,** 451–486.

Peters, A., Palay, S. L., and Webster, H. de F. (1991). "The Fine Structure of the Nervous System." Oxford University Press, New York, NY.

Pfrieger, F. W., and Barres, B. A. (1996). New views on synapse-glia interactions. *Curr. Opin. Neurobiol.* **6,** 615–621.

Pin, J. P., and Bockaert, J. (1995). Get receptive to metabotropic glutamate receptors. *Curr. Opin. Neurobiol.* **5,** 342–349.

Pin, J. P., and Duvoisin, R. (1995). The metabotropic glutamate receptors: structure and functions. *Neuropharmacology* **34,** 1–26.

Pinault, D., Bourassa, J., and Deschênes, M. (1995a). The axonal arborization of single thalamic reticular neurons in the somatosensory thalamus of the rat. *Eur. J. Neurosci.* **7,** 31–40.

Pinault, D., Bourassa, J., and Deschênes, M. (1995b). Thalamic reticular input to the rat visual thalamus: a single fiber study using biocytin as an anterograde tracer. *Brain Res.* **670,** 147–152.

Pinault, D., and Deschênes, M. (1998a). Projection and innervation patterns of individual thalamic reticular axons in the thalamus of the adult rat: a three-dimensional, graphic, and morphometric analysis. *J. Comp. Neurol.* **391,** 180–203.

Pinault, D., and Deschênes, M. (1998b). Anatomical evidence for a mechanism of lateral inhibition in the rat thalamus. *Eur. J. Neurosci.* **10,** 3462–3469.

Pinault, D., Smith, Y., and Deschênes, M. (1997). Dendrodendritic and axoaxonic synapses in the thalamic reticular nucleus of the adult rat. *J. Neurosci.* **17,** 3215–3233.

Polyak, S. (1957). "The Vertebrate Visual System." The University of Chicago Press, Chicago, IL.

Powell, T. P. S., and Cowan, W. M. (1954). The connexions of the midline and intralaminar nuclei of the thalamus of the rat. *J. Anat. Lond.* **88,** 307–319.

Powell, T. P. S., and Cowan, W. M. (1956). A study of thalamo-striate relations in the monkey. *Brain* **79,** 364–390.

Powers, R. K., Sawczuk, A., Musick, J. R., and Binder, M. D. (1999). Multiple mechanisms of spike-frequency adaptation in motoneurones. *J. Physiol. (Paris)* **93,** 101–114.

Price, J. L. (1986). Subcortical projections from the amygdaloid complex. *Adv. Exp. Med. Biol.* **203,** 19–33.

Raczkowski, D., and Fitzpatrick, D. (1989). Organization of cholinergic synapses in the cat's dorsal lateral geniculate and perigeniculate nuclei. *J. Comp. Neurol.* **288,** 676–690.

Radhakrishnan, V., Tsoukatos, J., Davis, K. D., Tasker, R. R., Lozano, A. M., and Dostrovsky, J. O. (1999). A comparison of the burst activity of lateral thalamic neurons in chronic pain and non-pain patients. *Pain* **80,** 567–575.

Rall, W. (1977). Core conductor theory and cable properties of neurons. In "Handbook of Physiology: The Nervous System I" (E. Kandel, S. Geiger, eds.), pp. 39–97. American Physiological Society, Bethesda, MD.

Ralston, H. J. III. (1969). The synaptic organization of lemniscal projections to the ventrobasal thalamus of the cat. *Brain Res.* **14,** 99–116.

Ralston, H. J. III. (1971). Evidence for presynaptic dendrites and a proposal for their mechanism of action. *Nature* **230,** 585–587.

Ralston, H. J. III, and Herman, M. M. (1969). The fine structure of neurons and synapses in ventrobasal thalamus of the cat. *Brain Res.* **14,** 77–97.

Ramcharan, E. J., Gnadt, J. W., and Sherman, S. M. (2000). Burst and tonic firing in thalamic cells of unanesthetized, behaving monkeys. *Vis. Neurosci.* **17,** 55–62.

Rapisardi, S. C., and Miles, T. P. (1984). Synaptology of retinal terminals in the dorsal lateral geniculate nucleus of the cat. *J. Comp. Neurol.* **223,** 515–534.

Rausell, E., Bae, C. S., Vinuela, A., Huntley, G. W., and Jones, E. G. (1992). Calbindin and parvalbumin cells in monkey VPL thalamic nucleus: distribution, laminar cortical projections, and relations to spinothalamic terminations. *J. Neurosci.* **12,** 4088–4111.

Rausell, E., and Jones, E. G. (1991). Chemically distinct compartments of the thalamic VPM nucleus in monkeys relay principal and spinal trigeminal pathways to different layers of the somatosensory cortex. *J. Neurosci.* **11,** 226–237.

Ray, J. P., and Price, J. L. (1992). The organization of the thalamocortical connections of the mediodorsal thalamic nucleus in the rat, related to the ventral forebrain-prefrontal cortex topography. *J. Comp. Neurol.* **323,** 167–197.

Reale, R. A., and Imig, T. J. (1980). Tonotopic organization in auditory cortex of the cat. *J. Comp. Neurol.* **192,** 265–291.

Recasens, M., and Vignes, M. (1995). Excitatory amino acid metabotropic receptor subtypes and calcium regulation. *Ann. NY Acad. Sci.* **757,** 418–429.

Reid, R. C., and Alonso, J. M. (1995). Specificity of monosynaptic connections from thalamus to visual cortex. *Nature* **378,** 281–284.

Reid, R. C., and Alonso, J. M. (1996). The processing and encoding of information in the visual cortex. *Curr. Opin. Neurobiol.* **6,** 475–480.

Reid, S. N. M., Romano, C., Hughes, T., and Daw, N. W. (1997). Developmental and sensory-dependent changes of phosphoinositide-linked metabotropic glutamate receptors. *J. Comp. Neurol.* **389,** 577–583.

Reinagel, P., Godwin, D. W., Sherman, S. M., and Koch, C. (1999). Encoding of visual information by LGN bursts. *J. Neurophysiol.* **81,** 2558–2569.

Reuter, H. (1996). Diversity and function of presynaptic calcium channels in the brain. *Curr. Opin. Neurobiol.* **6,** 331–337.

Richard, D., Gioanni, Y., Kitsikis, A., and Buser, P. (1975). A study of geniculate unit activity during cryogenic blockade of the primary visual cortex in the cat. *Exp. Brain Res.* **22,** 235–242.

Rinvik, E., and Grofova, I. (1974a). Cerebellar projections to the nuclei ventralis lateralis and ventralis anterior thalami. Experimental electron microscopical and light microscopical studies in the cat. *Anat. Embryol.* **146,** 95–111.

Rinvik, E., and Grofova, I. (1974b). Light and electron microscopical studies of the normal nuclei ventralis lateralis and ventralis anterior thalami in the cat. *Anat. Embryol.* **146,** 57–93.

Rittenhouse, C. D., Shouval, H. Z., Paradiso, M. A., and Bear, M. F. (1999). Monocular deprivation induces homosynaptic long-term depression in visual cortex. *Nature* **397,** 347–350.

Rivadulla, C., Rodriguez, R., Martinez-Conde, S., Acuña, C., and Cudeiro, J. (1996). The influence of nitric oxide on perigeniculate GABAergic cell activity in the anaesthetized cat. *Eur. J. Neurosci.* **8,** 2459–2466.

Robson, J. A. (1983). The morphology of corticofugal axons to the dorsal lateral geniculate nucleus in the cat. *J. Comp. Neurol.* **216,** 89–103.

Robson, J. A., and Hall, W. C. (1977). The organization of the pulvinar in the grey squirrel (*Sciurus carolinensis*). II Synaptic organization and comparisons with the lateral geniculate nucleus. *J. Comp. Neurol.* **173,** 389–416.

Rockel, A. J., Hiorns, R. W., and Powell, T. P. S. (1980). The basic uniformity in structure of the neocortex. *Brain* **103,** 221–144.

Rockland, K. S. (1996). Two types of corticopulvinar termination: round (type 2) and elongate (type 1). *J. Comp. Neurol.* **368,** 57–87.

Rockland, K. S. (1998). Convergence and branching patterns of round, type 2 corticopulvinar axons. *J. Comp. Neurol.* **390,** 515–536.

Rockland, K. S., Andresen, J., Cowie, R. J., and Robinson, D. L. (1999). Single axon analysis of pulvinocortical connections to several visual areas in the macaque. *J. Comp. Neurol.* **406,** 221–250.

Rodieck, R. W. (1998). "The First Steps in Seeing." Sinauer, Sunderland, MA.

Rodieck, R. W., and Brening, R. K. (1983). Retinal ganglion cells: properties, types, genera, pathways and trans-species comparisons. *Brain Behav. Evol.* **23,** 121–164.

Rogawski, M. A. (1985). The A-current: How ubiquitous a feature of excitable cells is it? *Trends Neurosci.* **8,** 214–219.

Rose, J. E., and Woolsey, C. N. (1948). Structure and relations of limbic cortex and anterior thalamic nuclei in rabbit and cat. *J. Comp. Neurol.* **89,** 79–347.

Rose, J. E., and Woolsey, C. N. (1949). The relations of thalamic connections, cellular structure and evocable electrical activity in the auditory region of the cat. *J. Comp. Neurol.* **91,** 441–466.

Rose, J. E., and Woolsey, C. N. (1958). Cortical connections and functional organization of the thalamic auditory system of the cat. *In* "Biological and Biochemical Bases of Behavior" (H. F. Harlow and C. N. Woolsey, eds.), pp. 127–150. University of Wisconsin Press, Madison, WI.

Rouiller, E. M., Tanne, J., Moret, V., Kermadi, I., Boussaoud, D., and Welker, E. (1998). Dual morphology and topography of the corticothalamic terminals originating from the primary, supplementary motor, and dorsal premotor cortical areas in macaque monkeys. *J. Comp. Neurol.* **396,** 169–185.

Royce, G. J. (1983). Cells of origin of corticothalamic projections upon the centromedian and parafascicular nuclei in the cat. *Brain Res.* **258,** 11–21.

Royce, G. J., Bromley, S., Gracco, C., and Beckstead, R. M. (1989). Thalamocortical connections of the rostral intralaminar nuclei: an autoradiographic analysis in the cat. *J. Comp. Neurol.* **288,** 555–582.

Royce, G. J., and Mourney, R. J. (1985). Efferent connections of the centremedian and parafascicular thalamic nuclei: and autoradiographic investigation in the cat. *J. Comp. Neurol.* **235,** 277–300.

Russchen, F. T., Amaral, D. G., and Price, J. L. (1987). The afferent input to the magnocellular division of the mediodorsal thalamic nucleus in the monkey, *Macaca fascicularis*. *J. Comp. Neurol.* **256**, 175–210.

Salin, P. A., and Bullier, J. (1995). Corticocortical connections in the visual system: structure and function. *Physiol. Rev.* **75**, 107–154.

Salt, T. E., and Eaton, S. A. (1996). Functions of ionotropic and metabotropic glutamate receptors in sensory transmission in the mammalian thalamus. *Progr. Neurobiol.* **48**, 55–72.

Sanchez-Vives, M. V., Bal, T., Kim, U., Von Krosigk, M., and McCormick, D. A. (1996). Are the interlaminar zones of the ferret dorsal lateral geniculate nucleus actually part of the perigeniculate nucleus? *J. Neurosci.* **16**, 5923–5941.

Sanchez-Vives, M. V., and McCormick, D. A. (1997). Functional properties of perigeniculate inhibition of dorsal lateral geniculate nucleus thalamocortical neurons in vitro. *J. Neurosci.* **17**, 8880–8893.

Sanderson, K. J. (1971). The projection of the visual field to the lateral geniculate and medial interlaminar nuclei in the cat. *J. Comp. Neurol.* **143**, 101–108.

Sanderson, K. J., Bishop, P. O., and Darian-Smith, I. (1971). The properties of the binocular receptive fields of lateral geniculate neurons. *Exp. Brain Res.* **13**, 178–207.

Schall, J. D., Ault, S. J., Vitek, D. J., and Leventhal, A. G. (1988). Experimental induction of an abnormal ipsilateral visual field representation in the geniculocortical pathway of normally pigmented cats. *J. Neurosci.* **8**, 2039–2048.

Schambra, U. B., Lauder, J. M., and Silver, J. (1992). "Atlas of the Prenatal Mouse Brain." Academic Press, San Diego, CA.

Scharfman, H. E., Lu, S.-M., Guido, W., Adams, P. R., and Sherman, S. M. (1990). N-methyl-D-aspartate (NMDA) receptors contribute to excitatory postsynaptic potentials of cat lateral geniculate neurons recorded in thalamic slices. *Proc. Natl. Acad. Sci. USA* **87**, 4548–4552.

Scheibel, M. E., and Scheibel, A. B. (1966). The organization of the nucleus reticularis thalami: a Golgi study. *Brain Res.* **1**, 43–62.

Schiller, J., Schiller, Y., Stuart, G., and Sakmann, B. (1997). Calcium action potentials restricted to distal apical dendrites of rat neocortical pyramidal neurons. *J. Physiol. (Lond.)* **505**, 605–616.

Schiller, P. H., and Malpeli, J. G. (1978). Functional specificity of lateral geniculate nucleus laminae of the rhesus monkey. *J. Neurophysiol.* **41**, 788–797.

Schmielau, F., and Singer, W. (1977). The role of visual cortex for binocular interactions in the cat lateral geniculate nucleus. *Brain Res.* **120**, 354–361.

Schwartz, M. L., Dekker, J. J., and Goldman-Rakic, P. S. (1991). Dual mode of corticothalamic synaptic termination in the mediodorsal nucleus of the rhesus monkey. *J. Comp. Neurol.* **309**, 289–304.

Sefton, A. J., and Burke, W. (1966). Mechanism of recurrent inhibition in the lateral geniculate nucleus of the rat. *Nature* **211**, 1276–1278.

Shapley, R., and Lennie, P. (1985). Spatial frequency analysis in the visual system. *Annu. Rev. Neurosci.* **8**, 547–583.

Shatz, C. J. (1994). Role for spontaneous neural activity in the patterning of connections between retina and LGN during visual system development. *Int. J. Dev. Neurosci.* **12**, 531–546.

Shatz, C. J. (1996). Emergence of order in visual system development. *Proc. Natl. Acad. Sci. USA* **93**, 602–608.

Shatz, C. J., and LeVay, S. (1979). Siamese cat: altered connections of visual cortex. *Science* **204**, 328–330.

Sherman, S. M. (1985). Functional organization of the W-, X-, and Y-cell pathways in the cat: a review and hypothesis. *In* "Progress in Psychobiology and Physiological Psychology" Vol. 11 (J. M. Sprague, A. N. Epstein, eds.), pp. 233–314. Academic Press, Orlando, FL.

Sherman, S. M. (1996). Dual response modes in lateral geniculate neurons: mechanisms and functions. *Vis. Neurosci.* **13**, 205–213.

Sherman, S. M., and Friedlander, M. J.

(1988). Identification of X versus Y properties for interneurons in the A-laminae of the cat's lateral geniculate nucleus. *Exp. Brain Res.* **73,** 384–392.

Sherman, S. M., and Guillery, R. W. (1996). The functional organization of thalamocortical relays. *J. Neurophysiol.* **76,** 1367–1395.

Sherman, S. M., and Guillery, R. W. (1998). On the actions that one nerve cell can have on another: Distinguishing "drivers" from "modulators." *Proc. Natl. Acad. Sci. USA* **95,** 7121–7126.

Sherman, S. M., and Koch, C. (1986). The control of retinogeniculate transmission in the mammalian lateral geniculate nucleus. *Exp. Brain Res.* **63,** 1–20.

Sherman, S. M., and Koch, C. (1998). Thalamus. In "The Synaptic Organization of the Brain, Fourth Edition" (G. M. Shepherd, ed.), pp. 289–328. Oxford University Press, New York, NY.

Sherman, S. M., and Spear, P. D. (1982). Organization of visual pathways in normal and visually deprived cats. *Physiol. Rev.* **62,** 738–855.

Sidibe, M., Bevan, M. D., Bolam, J. P., and Smith, Y. (1997). Efferent connections of the internal globus pallidus in the squirrel monkey: I. Topography and synaptic organization of the pallidothalamic projection. *J. Comp. Neurol.* **382,** 323–347.

Sillito, A. M., Jones, H. E., Gerstein, G. L., and West, D. C. (1994). Feature-linked synchronization of thalamic relay cell firing induced by feedback from the visual cortex. *Nature* **369,** 479–482.

Silver, R. A., Cull-Candy, S. G., and Takahashi, T. (1996). Non-NMDA glutamate receptor occupancy and open probability at a rat cerebellar synapse with single and multiple release sites. *J. Physiol. (Lond.)* **494,** 231–250.

Singer, W., and Gray, C. M. (1995). Visual feature integration and the temporal correlation hypothesis. *Annu. Rev. Neurosci.* **18,** 555–586.

Smith, G. D., Cox, C. L., Sherman, S. M., and Rinzel, J. (2000). Fourier analysis of sinusoidally-driven thalamocortical relay neurons and a minimal integrate-and-fire-or-burst model. *J. Neurophysiol.* **83,** 588–610.

Smith, G. D., Cox, C. L., Sherman, S. M., and Rinzel, J. (1999). A firing-rate model of spike-frequency adaptation in sinusoidally driven thalamocortical relay cells. Submitted for publication.

So, Y.-T., and Shapley, R. (1981). Spatial tuning of cells in and around lateral geniculate nucleus of the cat: X and Y relay cells and perigeniculate interneurons. *J. Neurophysiol.* **45,** 107–120.

Somogyi, G., Hajdu, F., and Tömböl, T. (1978). Ultrastructure of the anterior ventral and anterior medial nuclei of the cat thalamus. *Exp. Brain Res.* **31,** 417–431.

Spillane, J. D. (1981). "The Doctrine of the Nerves." Oxford University Press, New York, NY.

Spreafico, R. (1997). GABAergic neurons in mammalian thalamus: A marker of thalamic complexity? *Brain Res. Bull.* **42,** 27–37.

Spreafico, R., Battaglia, G., and Frassoni, C. (1991). The reticular thalamic nucleus (RTN) of the rat: cytoarchitectural, Golgi, immunocytochemical, and horseradish peroxidase study. *J. Comp. Neurol.* **304,** 478–490.

Spreafico, R., de Curtis, M., Frassoni, C., and Avanzini, G. (1988). Electrophysiological characteristics of morphologically identified reticular thalamic neurons from rat slices. *Neuroscience* **27,** 629–638.

Spreafico, R., Hayes, N. L., and Rustioni, A. (1981). Thalamic projections to the primary and secondary somatosensory cortices in cat: single and double retrograde tracer studies. *J. Comp. Neurol.* **203,** 67–90.

Stackman, R. W., and Taube, J. S. (1998). Firing properties of rat lateral mammillary single units: head direction, head pitch, and angular head velocity. *J. Neurosci.* **18,** 9020–9037.

Stanford, L. R., Friedlander, M. J., and Sherman, S. M. (1983). Morphological and physiological properties of geniculate W-cells of the cat: a comparison with X- and Y-cells. *J. Neurophysiol.* **50,** 582–608.

Stent, G. S. (1978). "Paradoxes of Progress." W. H. Freeman, San Francisco, CA.

Steriade, M., and Contreras, D. (1995). Relations between cortical and thalamic cellular events during transition from sleep patterns to paroxysmal activity. *J. Neurosci.* **15**, 623–642.

Steriade, M., and Deschênes, M. (1984). The thalamus as a neuronal oscillator. *Brain Res. Rev.* **8**, 1–63.

Steriade, M., Deschênes, M., Domich, L., and Mulle, C. (1985). Abolition of spindle oscillations in thalamic neurons disconnected from nucleus reticularis thalami. *J. Neurophysiol.* **54**, 1473–1497.

Steriade, M., Jones, E. G., and Llinás, R. (1990). "Thalamic Oscillations and Signalling." Wiley, New York, NY.

Steriade, M., and Llinás, R. (1988). The functional states of the thalamus and the associated neuronal interplay. *Physiol. Rev.* **68**, 649–742.

Steriade, M., and McCarley, R. W. (1990). "Brainstem Control of Wakefulness and Sleep." Plenum Press, New York, NY.

Steriade, M., McCormick, D. A., and Sejnowski, T. J. (1993). Thalamocortical oscillations in the sleeping and aroused brain. *Science* **262**, 679–685.

Storm, J. F. (1987). Action potential repolarization and a fast after-hyperpolarization in rat hippocampal pyramidal cells. *J. Physiol. (Lond.)* **385**, 733–759.

Storm, J. F. (1990). Potassium currents in hippocampal pyramidal cells. *Progr. Brain Res.* **83**, 161–187.

Stratford, K. J., Tarczy-Hornoch, K., Martin, K. A. C., Bannister, N. J., and Jack, J. J. B. (1996). Excitatory synaptic inputs to spiny stellate cells in cat visual cortex. *Nature* **382**, 258–261.

Strick, P. L. (1988). Anatomical organization of multiple motor areas in the frontal lobe: implications for recovery of function. *Adv. Neurol.* **47**, 293–312.

Stryker, M. P., and Zahs, K. R. (1983). On and off sublaminae in the lateral geniculate nucleus of the ferret. *J. Neurosci.* **3**, 1943–1951.

Stuart, G., Schiller, J., and Sakmann, B. (1997a). Action potential initiation and propagation in rat neocortical pyramidal neurons. *J. Physiol. (Lond.)* **505**, 617–632.

Stuart, G., Spruston, N., Sakmann, B., and Hausser, M. (1997b). Action potential initiation and backpropagation in neurons of the mammalian CNS. *Trends Neurosci.* **20**, 125–131.

Sur, M., Esguerra, M., Garraghty, P. E., Kritzer, M. F., and Sherman, S. M. (1987). Morphology of physiologically identified retinogeniculate X- and Y-axons in the cat. *J. Neurophysiol.* **58**, 1–32.

Sur, M., Garraghty, P. E., and Roe, A. W. (1988). Experimentally induced visual projections into auditory thalamus and cortex. *Science* **242**, 1437–1441.

Sur, M., and Sherman, S. M. (1982). Linear and nonlinear W-cells in C-laminae of the cat's lateral geniculate nucleus. *J. Neurophysiol.* **47**, 869–884.

Suzuki, H., and Kato, H. (1966). Binocular interaction at cat's lateral geniculate body. *J. Neurophysiol.* **29**, 909–920.

Symonds, L. L., and Kaas, J. H. (1978). Connections of striate cortex in the prosimian, *Galago senegalensis*. *J. Comp. Neurol.* **181**, 477–512.

Szentágothai, J. (1963). The structure of the synapse in the lateral geniculate nucleus. *Acta. Anat.* **55**, 166–185.

Taube, J. S. (1995). Head direction cells recorded in the anterior thalamic nuclei of freely moving rats. *J. Neurosci.* **15**, 70–86.

Thach, W. T., Goodkin, H. P., and Keating, J. G. (1992). The cerebellum and the adaptive coordination of movement. *Annu. Rev. Neurosci.* **15**, 403–442.

Thomson, A. M. (2000). Molecular frequency filters at central synapses. *Progr. Neurobiol.*, in press.

Thomson, A. M., and Deuchars, J. (1994). Temporal and spatial properties of local circuits in neocortex. *Trends Neurosci.* **17**, 119–126.

Thomson, A. M., and Deuchars, J. (1997). Synaptic interactions in neocortical local circuits: Dual intracellular recordings *in vitro*. *Cerebral Cortex* **7**, 510–522.

Tömböl, T. (1967). Short neurons and their synaptic relations in the specific thalamic nuclei. *Brain Res.* **3**, 307–326.

Tömböl, T. (1969). Two types of short axon (Golgi 2nd) interneurons in the specific thalamic nuclei. *Morphol. Acad. Scien. Hungaricae.* **17**, 285–297.

Torrealba, F., Guillery, R. W., Eysel, U., Polley, E. H., and Mason, C. A. (1982). Studies of retinal representations within the cat's optic tract. *J. Comp. Neurol.* **211**, 377–396.

Tóth, T. I., Hughes, S. W., and Crunelli, V. (1998). Analysis and biophysical interpretation of bistable behaviour in thalamocortical neurons. *Neuroscience* **87**, 519–523.

Towns, L. C., Tigges, J., and Tigges, M. (1990). Termination of thalamic intralaminar nuclei afferents in visual cortex of squirrel monkey. *Vis. Neurosci.* **5**, 151–154.

Trevelyan, A. J., and Thompson, I. D. (1992). Altered topography in the geniculocortical projection of the golden hamster following neonatal enucleation. *Eur. J. Neurosci.* **4**, 1104–1111.

Tsumoto, T., Creutzfeldt, O. D., and Legendy, C. R. (1978). Functional organization of the cortifugal system from visual cortex to lateral geniculate nucleus in the cat. *Exp. Brain Res.* **32**, 345–364.

Tsumoto, T., and Suda, K. (1980). Three groups of cortico-geniculate neurons and their distribution in binocular and monocular segments of cat striate cortex. *J. Comp. Neurol.* **193**, 223–236.

Tusa, R. J., and Palmer, L. A. (1980). Retinotopic organization of areas 20 and 21 in the cat. *J. Comp. Neurol.* **193**, 147–164.

Tusa, R. J., Palmer, L. A., and Rosenquist, A. C. (1978). The retinotopic organization of area 17 (striate cortex) in the cat. *J. Comp. Neurol.* **177**, 213–235.

Tusa, R. J., Rosenquist, A. C., and Palmer, L. A. (1979). Retinotopic organization of areas 18 and 19 in the cat. *J. Comp. Neurol.* **185**, 657–678.

Uhlrich, D. J., Cucchiaro, J. B., Humphrey, A. L., and Sherman, S. M. (1991). Morphology and axonal projection patterns of individual neurons in the cat perigeniculate nucleus. *J. Neurophysiol.* **65**, 1528–1541.

Uhlrich, D. J., Cucchiaro, J. B., and Sherman, S. M. (1988). The projection of individual axons from the parabrachial region of the brainstem to the dorsal lateral geniculate nucleus in the cat. *J. Neurosci.* **8**, 4565–4575.

Uhlrich, D. J., Manning, K. A., and Pienkowski, T. P. (1993). The histaminergic innervation of the lateral geniculate complex in the cat. *Vis. Neurosci.* **10**, 225–235.

Ullan, J. (1985). Cortical topography of thalamic intralaminar nuclei. *Brain Res.* **328**, 333–340.

Updyke, B. V. (1975). The patterns of projection of cortical areas 17, 18, and 19 onto the laminae of the dorsal lateral geniculate nucleus in the cat. *J. Comp. Neurol.* **163**, 377–396.

Updyke, B. V. (1977). Topographic organization of the projections from cortical areas 17, 18 and 19 onto the thalamus, pretectum and superior colliculus in the cat. *J. Comp. Neurol.* **173**, 81–122.

Updyke, B. V. (1979). A Golgi study of the class V cell in the visual thalamus of the cat. *J. Comp. Neurol.* **186**, 603–619.

Updyke, B. V. (1981). Projections from visual areas of the middle suprasylvian sulcus onto the lateral posterior complex and adjacent thalamic nuclei in cat. *J. Comp. Neurol.* **201**, 477–506.

Usrey, W. M., Alonso, J. M., and Reid, R. C. (2000). Synaptic interactions between thalamic inputs to simple cells in cat visual cortex. *J. Neurosci.*, in press.

Usrey, W. M., Reppas, J. B., and Reid, R. C. (1998). Paired-spike interactions and synaptic efficacy of retinal inputs to the thalamus. *Nature* **395**, 384–387.

Valdivia, O. (1971). Methods of fixation and the morphology of synaptic vesicles. *J. Comp. Neurol.* **142**, 257–273.

Van Essen, D. C., Anderson, C. H., and Felleman, D. J. (1992). Information processing in the primate visual system: an integrated systems perspective. *Science* **225**, 419–423.

Van Essen, D. C., Felleman, D. J., DeYoe,

E. A., Olavarria, J., and Knierim, J. (1990). Modular and hierarchical organization of extrastriate visual cortex in the macaque monkey. *Cold Spring Harbor Symp. Quant. Biol.* **55,** 679–696.

Van Horn, S. C., Erişir, A., and Sherman, S. M. (2000). The relative distribution of synapses in the A-laminae of the lateral geniculate nucleus of the cat. *J. Comp. Neurol.* **416,** 509–520.

Vidnyánszky, Z., Borostyánkoi, Z., Görcs, T. J., and Hámori, J. (1996a). Light and electron microscopic analysis of synaptic input from cortical area 17 to the lateral posterior nucleus in cats. *Exp. Brain Res.* **109,** 63–70.

Vidnyánszky, Z., Görcs, T. J., Negyessy, L., Borostyánkio, Z., Knopfel, T., and Hámori, J. (1996b). Immunocytochemical visualization of the mGluR1a metabotropic glutamate receptor at synapses of corticothalamic terminals originating from area 17 of the rat. *Eur. J. Neurosci.* **8,** 1061–1071.

Vidnyánszky, Z., and Hámori, J. (1994). Quantitative electron microscopic analysis of synaptic input from cortical areas 17 and 18 to the dorsal lateral geniculate nucleus in cats. *J. Comp. Neurol.* **349,** 259–268.

von Krosigk, M., Bal, T., and McCormick, D. A. (1993). Cellular mechanisms of a synchronized oscillation in the thalamus. *Science* **261,** 361–364.

Wagner, S., Castel, M., Gainer, H., and Yarom, Y. (1997). GABA in the mammalian suprachiasmatic nucleus and its role in diurnal rhythmicity. *Nature* **387,** 598–603.

Walker, A. E. (1938). "The Primate Thalamus." The University of Chicago Press, Chicago, IL.

Wall, J. T., Symonds, L. L., and Kaas, J. H. (1982). Cortical and subcortical projections of the middle temporal area (MT) and adjacent cortex in galagos. *J. Comp. Neurol.* **211,** 193–214.

Walls, G. L. (1953). "The Lateral Geniculate Nucleus and Visual Histophysiology." Univ. Calif. Publ. Physiol. No. 1, 1–100.

Walshe, F. M. R. (1948). "Critical Studies in Neurology." Livingstone, Edinburgh, UK.

Weber, A. J., Kalil, R. E., and Behan, M. (1989). Synaptic connections between corticogeniculate axons and interneurons in the dorsal lateral geniculate nucleus of the cat. *J. Comp. Neurol.* **289,** 156–164.

Welker, W. (1987). Comparative study of cerebellar somatosensory representations. The importance of micromapping and natural stimulation. *In* "Cerebellum and Neuronal Plasticity" (M. Glickstein, C. Yeo, and J. Stein, eds.), Nato ASI Series, Vol. 148, pp. 109–118. Plenum Press, New York, NY.

Wiesel, T. N., and Hubel, D. H. (1963). Effects of visual deprivation on morphology and physiology of cells in the cat's lateral geniculate body. *J. Neurophysiol.* **26,** 978–993.

Williamson, A. M., Ohara, P. T., and Ralston, H. J. 3d. (1993). Electron microscopic evidence that cortical terminals make direct contact onto cells of the thalamic reticular nucleus in the monkey. *Brain Res.* **631,** 175–179.

Williamson, A. M., Ohara, P. T., Ralston, D. D., Milroy, A. M., and Ralston, H. J. III. (1994). Analysis of gamma-aminobutyric acidergic synaptic contacts in the thalamic reticular nucleus of the monkey. *J. Comp. Neurol.* **349,** 182–192.

Wilson, J. R. (1986). Synaptic connections of relay and local circuit neurons in the monkey's dorsal lateral geniculate nucleus. *Neurosci. Lett.* **66,** 79–84.

Wilson, J. R., Forestner, D. M., and Cramer, R. P. (1996). Quantitative analyses of synaptic contacts of interneurons in the dorsal lateral geniculate nucleus of the squirrel monkey. *Vis. Neurosci.* **13,** 1129–1142.

Wilson, J. R., Friedlander, M. J., and Sherman, S. M. (1984). Fine structural morphology of identified X- and Y-cells in the cat's lateral geniculate nucleus. *Proc. Roy. Soc. Lond.* B **221,** 411–436.

Wilson, J. R., Manning, K. A., Forestner, D. M., Counts, S. E., and Uhlrich, D. J. (1999). Comparison of cholinergic and histaminergic axons in the lateral geniculate nucleus of the macaque monkey. *Anat. Rec.* **255,** 295–305.

Wilson, P. D., Rowe, M. H., and Stone, J. (1976). Properties of relay cells in cat's lateral geniculate nucleus: a comparison of W-cells with X- and Y-cells. *J. Neurophysiol.* **39**, 1193–1209.

Winckler, B., and Mellman, I. (1999). Neuronal polarity: Controlling the sorting and diffusion of membrane components. *Neuron* **23**, 637–640.

Winer, J. A. (1985). The medial geniculate body of the cat. *Adv. Anat. Embryol. Cell Biol.* **86**, 1–97.

Winer, J. A., and Larue, D. T. (1988). Anatomy of glutamic acid decarboxylase immunoreactive neurons and axons of the rat medial geniculate body. *J. Comp. Neurol.* **278**, 47–68.

Winer, J. A., and Morest, D. K. (1983). The neuronal architecture of the dorsal division of the medial geniculate body of the cat. A study with the rapid Golgi method. *J. Comp. Neurol.* **221**, 1–30.

Winer, J. A., and Morest, D. K. (1984). Axons of the dorsal division of the medial geniculate body of the cat: a study with the rapid Golgi method. *J. Comp. Neurol.* **224**, 344–370.

Wong, R. O., Chernjavsky, A., Smith, S. J., and Shatz, C. J. (1995). Early functional neural networks in the developing retina. *Nature* **374**, 716–718.

Wong-Riley, M. (1979). Changes in the visual system of monocularly sutured or enucleated cats demonstrable with cytochrome oxidase histochemistry. *Brain Res.* **171**, 11–28.

Wu, L. G., Borst, J. G., and Sakmann, B. (1998). R-type Ca^{2+} currents evoke transmitter release at a rat central synapse. *Proc. Natl. Acad. Sci. USA* **95**, 4720–4725.

Yen, C.-T., Conley, M., and Jones, E. G. (1985a). Morphological and functional types of neurons in cat ventral posterior thalamic nucleus. *J. Neurosci.* **5**, 1316–1338.

Yen, C.-T., Conley, M., Hendry, S. H. C., and Jones, E. G. (1985b). The morphology of physiologically identified GABAergic neurons in the somatic sensory part of the thalamic reticular nucleus in the cat. *J. Neurosci.* **5**, 2254–2268.

Yeterian, E. H., and Pandya, D. N. (1997). Corticothalamic connections of extrastriate visual areas in rhesus monkeys. *J. Comp. Neurol.* **378**, 562–585.

Yukie, M., and Iwai, E. (1981). Direct projection from the dorsal lateral geniculate nucleus to the prestriate cortex in macaque monkeys. *J. Comp. Neurol.* **201**, 81–97.

Zahs, K. R., and Stryker, M. P. (1988). Segregation of ON and OFF afferents to ferret visual cortex. *J. Neurophysiol.* **59**, 1410–1429.

Zeki, S. M., and Shipp, S. (1988). The functional logic of cortical connections. *Nature* **335**, 311–317.

Zeki, S. M. (1993). "A Vision of the Brain." Blackwell Scientific Publications, Oxford, UK.

Zeki, S. M. (1969). Representation of central visual fields in prestriate cortex of monkey. *Brain Res.* **14**, 271–291.

Zhan, X. J., Cox, C. L., Rinzel, J., and Sherman, S. M. (1999). Current clamp and modeling studies of low threshold calcium spikes in cells of the cat's lateral geniculate nucleus. *J. Neurophysiol.* **81**, 2360–2373.

Zhan, X. J., Cox, C. L., and Sherman, S. M. (2000). Dendritic depolarization efficiently attenuates low threshold calcium spikes in thalamic relay cells. *J. Neurosci.* **20**, 3909–3914.

Zhou, Q., Godwin, D. W., O'Malley, D. M., and Adams, P. R. (1997). Visualization of calcium influx through channels that shape the burst and tonic firing modes of thalamic relay cells. *J. Neurophysiol.* **77**, 2816–2825.

Zhu, J. J., Lytton, W. W., Xue, J. T., and Uhlrich, D. J. (1999). An intrinsic oscillation in interneurons of the rat lateral geniculate nucleus. *J. Neurophysiol.* **81**, 702–711.

Index

A

Acetylcholine, 144–145, 157
Action potential, 109–111, 117, 136–137
   conductances, 118–122
   refractory period, 120
Activation curves, 126–127, 138
A current, (see I_A)
Afferents,
   ascending drivers, 3, 147–151, 231
   ascending modulators, 3, 156–157
   axon terminals, 66, 68, 86f
   brain stem, 75–78, 86, 156–158, 244
   cerebellar, 10, 60
   cholinergic, 76, 156–157, 162–164, 188
   classification, 60
   convergence, 151, 251–253, 256, 259–261
   corticothalamic,
     distinctions, 68–69
     functions, 155–156
     laminar origins, 36f, 70
     topographical organization, 13–14, 72–74, 207–208
   drivers, (see Drivers)
   GABAergic, 100–101, 151–154, 164–165, 263–264
   glutamatergic, 147–151, 154–156, 160–162
   histaminergic, 76, 96, 157–158, 166, 252–253
   lemniscal, 62, 68
   light microscopy, 68–72
   mamillothalamic, 87, 207,
   modulators, (see Modulators)
   noradrenergic, 76, 157, 165
   pallidal, 12, 65, 100–102, 165
   parabrachial, 76, 93f
   reticulothalamic, 77–82, 94, 99–100
   retinal, 36–37, 60, 64f, 66f, 255
   serotonergic, 76, 157–158, 166, 189–190
   tectogeniculate, 60
   type I, 66, 67f
   type II, 64f, 65, 66f
Afterhypolarization, 119–120, 136, 139
Albinos, 213–220
Amygdala, thalamic connections, 20, 232
Anterior thalamic nuclei, 3, 87, 223, 239
Arrhythmic bursting (see Relay cells, arrhythmic bursting)
Astrocytes, 97–99
Attention, 6, 28, 105, 178, 193–194, 258
Awake state, 139–141, 170–172, 194, 261–263
Axon hillock, 110–112, 115–118, 121–122, 139–140
Axons,
   ascending afferents, 62–68, 100, 229, 235
   basket cell axons, 63
   corticofugal, 70, 236, 239
   corticothalamic, 68–72
     distribution, 72, 104, 106, 208, 225–227
     function, 155–156

Axons (*continued*)
  lemniscal, 62
  retinal, 36–37, 60, 64f, 66f, 72, 98
  terminals,
    E type, 70
    F1, 83
    F2, 83
    RL, 83, 86–91
    RS, 83, 91–93, 104
    R type, 70, 83
    type I, 65–68
    type II, 68
  thalamocortical,
    cortical distribution, 39–41, 42f, 46
    functions, 61, 179, 183
  W cell, 252
  X cell, 160
  Y cell, 160

B
Basal forebrain, 102, 164
Basket cells, 63
Binocular interaction, 104,
Binocular matching, 207, 208
Boston connections, 215–218, 220
Brain stem,
  afferents, 75–76, 244
    cholinergic, 156–157, 162–164, 188
    modulatory inputs, 156–157
    noradrenergic, 157, 165
    serotonergic, 157–158, 166, 189–190
Burst mode,
  arrhythmic, 132–133, 171–172, 194
  cell types,
    interneurons, 137
    relay cells, 122–133, 150–157, 169–195
    reticular cells, 137–139
  conductances, 122–133, 137–139
  control, 149–151, 153–154, 155–158, 188–192
  detectability, 175, 178–179, 187–188, 192–194
  linearity, 123, 129, 172–175, 178–181
  rhythmic, 127, 131–133, 169–172, 194, 261–262
  sleep, 6, 131–133, 153–154, 170–171, 261–262
  spontaneous activity, 112, 171–172, 175, 264
  visual responses, 171–179
  wakefulness, 6, 154, 171, 261
Bushy cells, 30–31

C
Cable properties, 109–117
Calbindin, 45
Calcium-binding proteins, 45
Calcium channels,
  distribution, 139
  I_T,
    interneurons, 137
    relay cells, 122–133, 135–136
    reticular cells, 137–139
    transmitter release, 135–136, 181
  types,
    L, 135
    N, 135
    T, 123
  voltage dependency, 122–127, 135
Callosal connections, 209
Caudate nucleus, 5
Center median nucleus, 4
Cerebellum, 3
  climbing fibers, 63
  mossy fibers, 98
  parallel fibers, 63
  pathways, 60
  Purkinje cells, 63, 85f
Cl⁻ channels, 118, 152–153, 166
Colliculus, 100, 215 (see also Tectum)
  inferior, 76
  superior, 76
Conductance, (see Membrane conductance)
Contacts,
  filamentous, 88f
  synaptic, (see Synaptic contacts)
Convergence,
  afferents, 151–153, 212, 250–253
  drivers, 250–253, 260–261
  maps, 225
  modulators, 260–261
Cortex,
  auditory, 22, 61, 200
  cells, 186–187, 193
  corticofugal axons, 67–68
  evolution, 3
  extrastriate, 40, 212
  functional areas, 209–210
  higher areas, 208
  laminar arrangements, 43
  layers,
    layer 1, 41
    layers 3 & 4, 41, 258–259, 264
    layer 5, 41, 70, 236, 258, 265
    layer 6, 14, 41–42, 68, 154–156, 236, 249, 252, 256
  lesions, 13, 20, 205

Cortex (*continued*)
  olfactory, 2
  prefrontal, 40, 220, 232
  somatosensory, 3, 61
  striate, 205–206, 242, 250, 258
  subplate, 211
  thalamocortical afferents, 62–67, 182–188
  visual,
    area 17, 14, 104, 205
    area 18, 14, 104
    disrupted representation, 216
    extrastriate, 40, 212
    geniculocortical pathways, 44–45
    maps, 198–200
    receptive fields, 11–13, 27, 71
    reversed connections, 216
    visual field losses, 203
Corticocortical pathways,
  classification, 265
  functions, 210
  hierarchy, 61, 74
  interconnections, 8
  thalamic input, 240–242
Corticothalamic pathways,
  basic properties, 154–155
  cell response mode, 155–156
  description, 69–70
  distinctions, 235, 237–245
  distribution, 43–44, 208
  function, 227
  laminar origin, 235–237
  mapping, 13–14, 73f
  terminations, 6
  topographical organization, 72–76
Cross-correlograms,
  definition, 254
  driver, 254–257
  modulator, 254–257
Current clamp, 127–128
Current flow, 110

D
Degeneration,
  retrograde, 1, 20, 21f, 205
  transneuronal, 20
Dendrites,
  active, 114
  analysis, 24–25
  appendages, 33, 48f, 88, 163
  arbors, 24
  back propagation, 122, 141, 186
  cable properties, 109–117
  cell bodies and, 47–49
  contacts, 115
  current flow, 115
  current injection, 112
  interneuron, 96, 114–117
  ion channels, 121–122
  presynaptic, 90, 115, 136–137, 141, 163
  relay cell, 95f, 96
  reticular cells, 48, 55f, 221, 224
  Sholl circles, 24, 26
  terminals, 48, 152, 158
  voltage attenuation, 113
Depolarization,
  synaptic, 147–149, 154–155, 156–158
  voltage sensitive channels, and, 118–141
Development, 199, 207
Diencephalon, 8–10
Diffuse projections, 12–13, 189–190
Divergence, 104, 212, 225
Dorsal column nuclei, 98, 239
Dorsal thalamus,
  afferents, 10–11, 61–76
  description, 8–10
  nuclei, 5 (see also named nuclei and Thalamic nuclei)
  topographical maps, 12–15
Dreaming, 244
Drivers, 11, 220
  categorization, 60, 265–266
  characteristics, 251–257, 260–261
  convergence, 27, 250–255, 260–261
  corticortical pathways, 210
  cross-correlograms, 253–257
  definition, 60, 250–251
  function in sleep, 261–263
  functions, 147–149, 238–239, 250–251, 260–261, 268
  GABAergic inputs as, 101, 263–264
  geniculate input, 259–260
  glutamatergic, 270
  identification, 232, 260–261, 271
  inputs, number, 257–259
  lateral geniculate nucleus, 250–253
  modulators, differences, 60, 251–253, 260–261
  multiple inputs, 16
  receptive fields, 250–251
  receptors, 251–253
  relay cell response, 149–151
Dynamic polarization, law of, 50

E
Electron microscopy, 82–87
Epicritic pathways, 39

Epilepsy, 7
Epithalamus, 9–10
EPSPs,
  conductances, 149, 154–157
  kinetics, 146
  receptors, 144–149, 156–158
  temporal summation, 150–151, 182, 254
E-type axons, 70
Evolution, 3
Extraglomerular synapses, 103

F
Facilitation, paired-pulse, 179–185
Feedback, 81, 127, 186–187, 192–193
Filamentous junctions, 88f
First order circuits, 11, 61, 229–234
  connections, 7–8, 233f, 270
  cortical afferents, 67–68, 233f
  criteria, 5, 231–232
  driver afferents, 99, 231, 239
  examples, 4f, 229–231
  RL terminals, 87–90
  role, 8
Fourier techniques, description, 174
F-type profiles, 81, 94

G
GABA, 12
  Cl- channel, 152
  extradiencephalic afferents, 100–101, 263–264
  F profiles (F1 and F2), 83f, 87f, 96, 115–117, 141, 160–168
  interneurons, 22–23, 47, 96, 151–154
  receptors,
    $GABA_A$, 152–154
    $GABA_B$, 152–154, 167, 189
  reticular cells, 151–154
Ganglion cells,
  retinal, 24, 37
  dorsal root, 50
Geniculocortical pathways,
  axon patterns, 41, 212
  cortical terminations,
    layers, 41–44, 86–188
    numbers of synapses, 258
    synaptic properties, 179–185
  divergence, 212
  geniculate layers, 42–43
  mapping, 203–206
  relay cells, 44–45
  response modes, 122–123, 131–140, 149–194, 260–263
Gitterkern, 16, 224
Glia, 22, 83, 97–98
Globus pallidus, 65, 100, 231, 239
Glomeruli, 97–99
  astrocytes, 84f, 97–98
  axon types, 83, 84f
  characterization, 83–85
  connectivity patterns, 102–103
  driver afferents, 83, 87, 99
  modulator afferents, 92–93
  synapses, 98
  terminal types,
    cholinergic, 92–92
    F1, 84f, 93–94
    F2, 84f, 93–94
    RL, 84f, 87
    RS, 84f, 91–93
  triads, 89f, 97–100, 160–163, 251–252
  X cells, 103, 160
  Y cells, 160
Glutamate,
  action, postsynaptic, 147–151, 154–156, 160–162
  inputs, relation to
    cortical, 154–156
    retinal, 147–151
  receptors,
    ionotropic,
      AMPA, 144, 148–149
      kainate, 144, 148–149
      NMDA, 144, 146–149
    metabotropic, 144–146, 154–156
Glycine receptors, 149
Golgi preparations,
  axonal arbors, 24, 65
  dendritic arbors, 24–25
  grape-like appendages, 33
  interneurons, 47–49, 51
  relay cell types, 31
  reticular cells, 54, 55f

H
Habenular nuclei, 5, 10
Higher order circuits, 11, 61, 208
  connections, 7–8, 233, 270
  driver afferents, 231, 239
  RL terminals, 87–91
  role, 5–6, 8, 237–239
Hippocampus, 2–3, 104
Histamine,
  action, 133

Histamine (*continued*)
   release, 96
   source, 157–158, 189
Hyperpolarization, 119
   synaptic, 151–154
   voltage sensitive channels, and, 118–141
Hypothalamus, 9–10, 76

I
I_A, 133–135, 137, 140
I_h, 131–133, 138–140, 153–154, 158, 170
$I_{K\text{-leak}}$, 118, 153–166
Inactivation curves, 123, 125–126
Inhibition,
   acetylcholine, 163
   conductances and,
      Cl⁻ conductance, 152–154, 166
      K⁺ conductance, 152–153, 162–163
   feedback, 81
   feedforward, 74, 160
   GABA, 152–154
   hyperpolarization and, 151–154
   lateral, 74
   serotonin, 158
   shunting, 152
   sources,
      brain stem, 100–101
      interneurons, 99, 116
      reticular cells, 75f, 99–100, 140, 263–264
Interneurons, 10, 19, 47–54, 99–100
   action potentials, 136–137
   afferents to, 99–100
   axonless, 52, 117
   axons, 51–52, 96
   cable properties, 114–116
   cell bodies, 47–49
   classification, 52–54
   degeneration, 22
   dendrites, 47–49
   dendritic terminals, 47–49, 96, 137, 152, 158
   description, 19
   distribution, 270
   identification, 22–23
   innervation, 164
   interlaminar, 160
   local afferents, 59
   multiplexing, 117, 140
   relay cell, connections, 82
   synaptic inputs,
      basic properties, 158–159

   cholinergic, 162–164
   GABAergic, 164–165
   glutamatergic, 160–161
   histaminergic, 166
   noradrenergic, 165
   serotonergic, 166
   triads, 115, 160–163, 251–252
Interspike intervals, 179–180
Intralaminar nuclei,
   cortical projections, 39–41
   first order connections, 234–235
   functions, 19, 255
   higher order connections, 234–235
   striatal projections, 10, 22
Ion channels,
   action potential, 120–121
   Ca^{2+},
      L channel, 135
      N channel, 135
      T channel, 123, 126–129, 139
   cation, 131
   Cl⁻, 118, 152–154
   conductance, 110
   distribution effects, 111
   function, 6
   K⁺,
      Ca^{2+} dependent, 122, 130–131
      "leak", 118, 152–166
      voltage-dependent, 118–122, 133–135
   location, 121–122
   low-threshold conductance, 122–131
   Na⁺, 118–122
   response modes, 123–124
   states, 120
      activation, 120–121
      deactivation, 120–121
      de-inactivation, 120–121, 123, 129–135, 150
      inactivation, 120–121, 123, 126–140, 150–156
   voltage dependent, 118–140
   voltage independent, 118
Ionotropic receptors, (see Receptors, ionotropic)
IPSPs,
   conductances, 152–156
   kinetics, 146
   receptors,
      ionotropic, 152–154
      metabotropic, 152–154
   response modes, 167, 189
I_T, 123–135, 137–140, 150–151, 166–170

K

Koniocellular pathways,
   Ca-binding proteins, 46
   cortical terminations, 43, 64f
   geniculate layers, 36f, 76–77

L

Lamination,
   cortical, 39–43, 236, 243
   lateral geniculate, 33, 35–43, 208–209
   functions, 208
   structure, 235–237
Lateral posterior nucleus, 27
Lemniscus, medial, 60
Local collaterals, 53
Local sign,
   axonal pathways, 13, 81
   brain stem afferents, 13
   corticothalamic pathways, 199–200, 209
   definition, 12
   olfaction, 200, 203
   reticular connections, 6, 7f, 221–226
   taste, 200
   thalamocortical pathways, 199–200
   modulators, 208
Locus ceruleus, 76

M

Magnocellular pathways, 21f, 64f
   Ca-binding proteins, 46
   cortical terminations, 41–43
   geniculate layers, 21f, 36f
Mamillary bodies, 60, 65, 239
Mamillothalamic tract, 87, 207
Mamillothalamocortical pathways, 12
Maps, 197–227
   abnormal, 213–220
   albinos, 213–220
   alignment, 206–209
   cortical, 198–200
   functional significance, 197–199
   multiple, 209–213
   reticular nuclei, 220–226, 246
   retinal, 205
   sensory, 198–199
   thalamocortical modifications, 220
Marchi method, 1
Mediodorsal nucleus, 4f, 65, 104
Membrane channels,
   Ca^{2+}, 123, 126–129, 135, 139
   cation, 131
   Cl^-, 118, 152–154
   K^+, 118–122, 130–135, 152–166
   Na^+, 118–122
Membrane conductance,
   interneurons, 114–116
   K "leak", 118, 152–166
   membrane resistance, 111–117
   relay cells, 118–136
   reticular cells, 137–140
   voltage dependent,
      action potential, 118–122, 136–137
      I_A, 133–135, 137
      I_{CAN}, 139
      I_h, 131–133, 153–154
      $I_{K[Ca]}$, 122, 130–131
      I_T, 123–135, 137–140, 150–151, 166–170
      persistent Na^+, 136
   voltage independent, 118
Membrane currents,
   action potential, 118–122, 136–137
   I_A, 133–135, 137
   I_h, 131–133, 153–154
   $I_{K\text{-leak}}$, 118, 152–166
   I_T, 123–135, 137–140, 150–151, 166–170
Metabotropic receptors, (see Receptors, metabotropic)
Microtubules, 51
Midline nuclei, 5, 19, 39–41
Midwestern connections, 215–216, 218
Mirror reversals, 14, 210–213
   topographic order, 226–227
   visual fields, 225
Mitochondria, 87, 92
Modulators, 11, 208, 220
   characteristics, 260–261
   convergence, 104, 225–226, 256, 260
   corticocortical pathways, 165, 210
   cross-correlograms, 256, 259–261
   definition, 11, 260–261
   divergence, 104
   drivers, differences, 251–253, 260–261
   functions, 27, 238–239, 260–261
   identification, 232, 271
   inputs, number, 257–259
   lateral geniculate nucleus, 250–253
   multiple inputs, 16
   receptors, 260–261
Monocular enucleation, 220
Muscarinic receptors,
   M1, 156, 163
   non-M1, 63
Mutations, 213–220
Myelinated axons, 121

Index

N
Neocortex, 15, 19–20, 146
Neuroglia, 22, 83, 97–98
Neuron,
  classifications, 24–26
  doctrine, 50
  dynamic polarization, 50
  interneuron, 47–53, 95, 99–100
  invertebrate, 50
  relay, 19
Neurotransmitters,
  activation, 144–145
  binding, 145–146
  postsynaptic effects, 146
  release,
    action potentials, 179–181
    Ca concentration, 181
  types,
    acetylcholine, 144–145, 157
    GABA, 144–145, 151–154, 163–167, 263–264
    glutamate, 147–151, 154–156, 160–162
    histamine, 96, 133
    nitric oxide, 76, 93f, 157
    noradrenalin, 133, 144–145, 157–158, 165–166
    serotonin, 133, 144–145, 157–158, 166
Nissl method, 1, 22
Nitric oxide, 76, 93f, 157
NMDA, 144, 146–149
Node of Ranvier, 121
Noradrenalin, 133, 144–145, 157–158, 165–166
Nucleus, (see Thalamic nuclei)

O
Occipital cortex, 203
Olfaction, 200, 203, 208
Olfactory cortex, 2, 200
Optic chiasm, 200
Optic nerve, 201

P
Pain, 3
Paired-pulse depression, 179–185
Paired-pulse facilitation, 179–185
Parabigeminal nucleus, 76
Parabrachial region, 76f
  inputs to thalamus, 76f, 93f
  modulatory effects, 156
  response mode control, 188–190
  transmitters, 76, 156
Parvalbumin, 45
Parvocellular pathways,
  Ca-binding proteins, 46
  cortical terminations, 41–43, 64f
  geniculate layers, 21f, 36f
Pathways,
  auditory, 22, 198, 200
  cerebellar, 10, 60, 87
  corticocortical, 8, 210, 242–243
  corticothalamic, 6, 11, 14, 62–63, 68–72, 103–106, 155–156, 208
  koniocellular, 64f
  lemniscal, 39, 98
  magnocellular, 64f
  motor, 209, 239
  off-center, 36f, 38
  olfactory, 2, 204, 208
  on-center, 36f, 38
  parvocellular, 64f
  second messenger, 144–147
  somatosensory, 44, 198, 207, 209
  spinothalamic, 39, 98
  taste, 10, 201, 208, 258
  thalamocortical, 39–43, 198
  visual, 198
Perigeniculate nucleus,
  axons, 79–80
  relation to thalamic reticular nucleus, 53
  retinal mapping, 79, 224
Perikaryal sizes, 29–35, 45–46
Perireticular nucleus, 211, 224–225
Pons, 71
Posterior nucleus, 5, 236
Postsynaptic potentials, (see EPSPs and IPSPs)
Pretectal nuclei, 100
Proteins, calcium binding, (see Calcium-binding proteins)
Protopathic pathways, 39
Pulvinar, 5, 27, 125–126, 240
Purkinje cells, 63, 85f
Pyramidal cells, 71

R
Radiate cells, 30–31, 35
Raphé nucleus, 76
Rapid eye movement,
  dreaming, 244
  initiation, 261
  thalamic activity, 263

Receiver operating characteristics, 176–177
Receptive fields, 11
 cortical, 104, 185, 191, 250
 corticothalamic drivers, 238
 cross-correlograms and, 253–257
 properties, 250–251
 relay cell, 27, 185–188, 250–254
Receptors,
 acetylcholine,
  muscarinic, 156, 163
  nicotinic, 156, 163
 GABA,
  $GABA_A$, 152–154
  $GABA_B$, 152–154, 167, 189
 glutamate,
  AMPA, 144, 148–149
  metabotropic, 144–146, 154–156
  NMDA, 144, 146–149
 ionotropic,
  distribution, 270
  metabotropic, comparison to, 145–147
  types, 144, 152, 156
 metabotropic,
  distribution, 270
  ionotropic, comparison to, 145–147
  types, 144–146, 152–154, 156
 noradrenalin, 157
 serotonin, 158
Refractory period,
 action potential, 120
 low-threshold Ca^{2+} spike, 130, 138
Relay cells,
 arrhythmic bursting, 132–133, 171–172, 194
 cable properties (see Cable properties)
 classification, 19–24
 cortical terminals, 38–43
 inputs,
  drivers, 60
   cortical layer 5, 70, 105, 147, 265
   subcortical, 147–151
  modulators, 60
   brain stem, 156–158
   cortical layer 6, 68, 103, 105, 154–156
 interneuronal afferents, 82, 95f
 low-threshold spike, 123, 127, 130, 135, 138–139
 membrane properties, (see Membrane conductance, relay cells)
 receptors, (see Receptors)
 response modes,
  burst, 131–133, 169–172, 194, 261–262
  tonic, 122–124, 129–140, 150–158
  rhythmic bursting, 127, 131–133, 169–172, 194, 261–262
  sleep, 131–133, 153–154, 170–171, 261–262
 synaptic inputs, 147–158, 265
 transmitter actions,
  acetylcholine, 156–158
  GABA, 151–154
  glutamate, 147–151, 154–156
  histamine, 157–158
  nitric oxide, 157
  noradrenaline, 157–158
  serotonin, 157–158
  wakefulness, 154, 171, 261
Resistance,
 input, 112, 121, 152, 155
 membrane, 111–117
Response modes, (see Burst mode; Tonic mode)
Reticular cells,
 afferents, 81, 101
 cable properties, 114–115
 connections, 13, 71
 innervation, 82, 164
 response modes,
  burst, 137–139
  tonic, 137–139
 synaptic inputs, 159f
Reticulogeniculate axons, 80f, 207
Reticulothalamic axons, 77–82
 F-type profiles, 94
 inhibitory inputs, 223
 local sign, 81
Retina, 24, 37, 259–260
Retinal ganglion cells, (see Ganglion cells)
Retinogeniculate axons,
 circuitry, 161–162
 cross-correlograms, 255
 glomeruli, 98
 mutants, 214
 relay cell inputs, 154–155
 terminal arbor, 66
 triads, 89f
Retrograde degeneration,
 cortical lesions, 1, 20–21, 205
 lateral geniculate nucleus, 21, 205
 methods, 22f
Reversal of maps, 225
Rhythmic bursting (see Relay cells, rhythmic bursting)
R-type axons, 70

Index

S

Sag current, 131
Saltatory conduction, 121
Scotomas, 203, 205
Serotonin, 76, 158, 166
Sholl circles, 24, 26f, 33
Siamese cats, 214–218
Signal detection theory, 175
Sleep, 12, 56
   burst firing, 132–133
   dreaming, 244
   rapid eye movement (REM), 261
   slow wave, 244, 261–263
Spinothalamic pathways, 39
Squid giant axon, 118–119
Stellate cells of cortex, 31
Striate cortex, (see Visual cortex)
Striatum, 10, 22
   intralaminar nuclei, 10, 22, 41
   midline nuclei, 41
   thalamic connection, 8
Substantia nigra, 65, 100, 231
Superior colliculus, 215
Sustaining projection, 22
Synaptic contacts,
   asymmetrical, 87
   axodendritic, 95f
   quantitative relationships, 94–96
   symmetrical, 87
   triadic, 97–99
Synaptic properties, (see also Neurotransmitters; Receptors)
   drivers, 147–151, 265
   modulators, 151–158
   paired-pulse effects,
     depression, 179–185
     facilitation, 179–185
   postsynaptic potentials,
     EPSPs, 144–149, 150–151, 156–158
     IPSPs, 152–156
Synaptic terminals, 86f
   F1 terminals, 84f, 93–94
   F2 terminals, 84f, 93–94
   RL terminals, 84f, 87
   RS terminals, 84f, 86, 91–93, 104
     brain stem afferents, 86
     cortical afferents, 86
Synchrony, 169–170

T

Tactile stimuli, 3
T current, (see I_T)
Telencephalon, 9f, 10
Temporal summation, 150, 254
Thalamic nuclei,
   anterior, 87, 223
     anterodorsal, 4f, 239
     anteromedial, 4f
     anteroventral, 4f
   center median, 4f, 65, 100
   habenular, 4f, 10
   intralaminar, 4f, 10, 22, 39–40
   lateral dorsal, 4f
   lateral geniculate, 3, 4f, 49, 65
   lateral posterior, 4f, 240
   medial geniculate, 4f, 40, 65, 87, 91
   mediodorsal, 4f, 65, 104, 223, 232
   midline, 4f, 39, 41
   perigeniculate, 53, 80
   perireticular, 211, 224–225
   posterior, 4f
   pulvinar, 4f, 40, 240
   reticular, 4f, 20, 54–56, 77, 101, 211–213, 220–225, 245
   ventral anterior, 3, 4f, 65, 100
   ventral lateral, 3, 4f, 65, 87, 97
   ventral posterior, 3, 40, 87
     ventral posterior inferior, 4f
     ventral posterior lateral, 4f
     ventral posterior medial, 4f
Thalamocortical connections,
   abnormal patterns, 215–218
   axons, 39–41
   Boston patterns, 217f, 218
   cortical layers, 39–43, 269
   distribution, 43–44
   experimental modifications, 220
   function, 227, 237–239
   first order circuits, 67, 229, 230f, 233f, 244
   higher order circuits, 229, 230f, 233f, 244
   integrative function, 8, 26, 102–103
   laminar distributions, 39–43
   mapping, 13–14, 209–210
   Midwestern pattern, 217f, 218
   non-specific, 39–41
   reversals, 220, 225
   specific, 39–41
Thalamus,
   dorsal, 10–15
     afferents, 10–11
     description, 8–10
     nuclei, 5 (see also individual named nuclei)
     topographical maps, 12–15, 97, 198

Thalamus (*continued*)
  ventral,
    description, 8–10, 15–16
    location, 15
Threshold, 110
Tonic mode,
  cell types,
    interneurons, 137
    relay cells, 122–124, 129–140, 150–158
    reticular cells, 137–139
  control, 188–192
  detectability, 175, 178–179, 187–188, 192–194
  linearity, 123, 129, 172–175, 178–181
  sleep, 261
  spontaneous activity, 175
  visual responses, 171–179
  wakefulness, 154, 171, 261
Topographical maps, 12–15, 198–199
Topographical organization, 72
Transient channels, 123–128
Transmembrane currents, 131 (see also Membrane currents)
Transmitters, (see Neurotransmitters)
Triads, 97–99
  components, 92
  function, 161–162
  glomeruli, 97–99
  inputs, 160–161
  receptors, 160, 162

V
Ventral anterior nucleus, 5, 65, 125–126
Ventral posterior nucleus,
  corticothalamic inputs, 155–156
  interneurons, 47f, 49
  relay cells, 31,
  reticulothalamic axons, 77, 78f, 94
  topographical organization, 207
Ventral thalamus, 15–16
  description, 8–10
  location, 15
Vesicles, synaptic, 87
  flattened, 83
  pleomorphic, 83
  round, 83
Vision,
  behavioral response, 219
  binocular, 201
Visual cortex,
  afferents, 60–61
  extrastriate, 212
  geniculocortical pathways, 44–45
  laminar structure, 43
  reversed connections, 216
  thalamocortical pathways, 39–41, 211

Visual fields,
  binocular, 202
  losses, 203, 205–206
  maps, 203, 225
  mismatched, 219
  reversal, 218
Visual pathway,
  maps,
    abnormal, 213–220
    albinos, 213–220
    reversed connections, 216

W
Wakefulness, 154, 171, 261
W cell pathways,
  axons, 42–43, 46, 68
  cell bodies, 37, 46
  geniculocortical pathways, 36
  laminar distributions, 76–77
    cortical, 42f
    geniculate, 36f

X
X cell pathways,
  axons, 42–43
  cell bodies, 25f, 26f, 37
  characterization, 24–26, 34f
  dendritic appendages, 33, 48
  geniculocortical pathways, 42f
  glomerular synapses, 103
  glomerular triad, 98
  interneuronal connections, 271
  laminar distributions, 37
    cortical, 42f
    geniculate, 36f
  synaptic connections, 103

Y
Y cell pathways,
  axons, 42–43
  cell bodies, 25f, 26f, 37
  characterization, 24–26, 34f
  extraglomerular synapses, 103
  geniculocortical pathways, 44–45
  interneuronal connection, 271
  laminar distributions,
    geniculate, 36f
    cortical, 42f
  synaptic connections, 95, 103, 271

Z
Zona incerta, 100

Lightning Source UK Ltd.
Milton Keynes UK
UKOW06n0615130115

244402UK00016B/232/P